Archaeoastronomy in the 1990s

ARCHAEOASTRONOMY IN THE 1990s

Papers derived from the third 'Oxford' International Symposium on Archaeoastronomy, St. Andrews, U.K, September 1990.

Edited by
CLIVE L.N. RUGGLES

School of Archaeological Studies
University of Leicester
United Kingdom

GROUP D PUBLICATIONS LTD.
LOUGHBOROUGH, UK

Published by Group D Publications Ltd.
81, Park Road, Loughborough, Leicestershire LE11 2HD
United Kingdom

Copyright © 1993 by Group D Publications Ltd

All rights reserved. No part of this book may be reproduced in any form, by photostat, microfilm, retrieval system, or any other means, without the prior permission of the publisher.

First published 1993

Printed and bound in the United Kingdom by Audio Visual Services, Loughborough University of Technology.

British Library Cataloguing–in–Publication data

A catalogue record for this book is available
from the British Library

ISBN 1-874152-01-2

The Mandarin fonts used to print this work are available from Linguists' Software Inc., PO Box 580, Edmonds, Washington 98020, USA, tel. (206)-775-1130.

IDRISI™ is a trademark of Clark State University, 950 Main Street, Worcester, Massachusetts, USA.

GRASS™ is a trademark of the US Army Construction Engineering Research Laboratory, PO Box 9005, Champaign, Illinois, USA.

Copies of this book can be ordered from the publisher.

Contents

	Preface	ix
	Dedication	xi
	List of contributors	xii
1	**Introduction:** Archaeoastronomy—the way ahead. *Clive Ruggles*	1

I: Thematic contributions

2	Archaeoastronomy in the Americas since Oxford 2. *Anthony F. Aveni*	15
3	Space, time and the calendar in the traditional cultures of America. *Stephen C. McCluskey*	33
4	Some social correlates of directional symbolism. *Stanisław Iwaniszewski*	45

II: New horizons

5	Moon Man and Sea Woman: the cosmology of the Central Inuit. *Susan M. Pearce*	59
6	Time-reckoning in Iceland before literacy. *Thorsteinn Vilhjálmsson*	69

7   The geometry of pastoral stone octagons: the Basque *sarobe*.
    Roslyn M. Frank and Jon D. Patrick   77

8   The moon and Indo-European calendar structure. *Emily Lyle*   92

9   Some remarks on the moon cult of Teutonic tribes.
    Emília Pásztor   98

10 Astronomical knowledge in Bulgarian lands during the
    Neolithic and Early Bronze Age. *Tsvetanka Radoslavova*   107

11 Four approaches to the Borana calendar. *Clive Ruggles*   117

12 Astronomy in the ancient written sources of the Far East.
    Ildikó Ecsedy and Katalin Barlai   123

13 Orientations of religious and ceremonial structures in
    Polynesia. *William Liller*   128

14 Aboriginal sky-mapping? Possible astronomical interpretation of Australian Aboriginal ethnographic and
    archaeological material. *Hugh Cairns*   136

III: New techniques, methods and approaches

15 Basic research in astronomy and its applications to
    archaeoastronomy. *Bradley E. Schaefer*   155

16 A method for determining limits on the accuracy of
    naked-eye locations of astronomical events. *Rolf M.
    Sinclair and Anna Sofaer*   178

17 An integrated approach to the investigation of astronomical
    evidence in the prehistoric record: the North Mull project.
    Clive Ruggles and Roger Martlew   185

18 The astronomy and geometry of Irish passage grave
    cemeteries: a systematic approach. *Jon D. Patrick*   198

IV: Continuing research: new results

19 Sun and sun serpents: continuing observations in south-eastern
    Utah. *Carol W. Ambruster and Ray A. Williamson*   219

Contents

20	The origin and meaning of Navajo star ceilings. *Von Del Chamberlain and Polly Schaafsma*	227
21	Organisation of large settlements of the northern Anasazi. *John McKim Malville and James Walton*	242
22	Summer solstice: a Chumash basket case. *Edwin C. Krupp*	251
23	Counting and sky-watching at Boca de Potrerillos, Nuevo León, Mexico: clues to an ancient tradition. *Wm. Breen Murray*	264
24	Venus orientations in ancient Mesoamerican architecture. *Ivan Šprajc*	270
25	Mesoamerican geometry combined with astronomy and calendar: the way to realise orientation. *Franz Tichy*	278
26	Mesoamerican cross-circle designs revisited. *Stanisław Iwaniszewski*	288
27	Were the Incas able to predict lunar eclipses? *Mariusz S. Ziółkowski and Arnold Lebeuf*	298
28	Callanish: maximising the symbolic and dramatic potential of the landscape at the southern extreme moon. *Margaret Curtis and Ronald Curtis*	309
29	The Bush Barrow gold lozenge: a solar and lunar calendar for Stonehenge? *Archibald S. Thom*	317
30	New evidence concerning possible astronomical orientations of 'Tombe di Giganti'. *Edoardo Proverbio*	324

V: Education and dissemination

31	An image database for learning archaeoastronomy. *Clive Ruggles*	335
	Appendix: Abstracts of papers published in the companion Oxford 3 volume 'Astronomies and cultures'	343
	Index	350

The conference logo depicts a version of the glyphic element often referred to as the 'St. Andrews Cross', frequently found in celestial bands and forming part of the Mayan sky glyph *ca(a)n*. A similar design dates back to Olmec times.

Preface

The third 'Oxford' international conference on archaeoastronomy was held in St. Andrews, Scotland, in September 1990. The 'Oxford' conferences, held at approximately four-yearly intervals, have aimed to provide a truly interdisciplinary focus on the study of astronomical practice in its cultural context, as well as a forum for discussing issues of theory and method. 'Oxford 1', held in 1981, was notable for bringing together, for the first time in substantial numbers, archaeoastronomers from the Old and New Worlds, as well as bringing face to face two very different methodological approaches, reflected in the contents of the two volumes arising from the Oxford 1 conference (Heggie 1982; Aveni 1982). Over the decade since Oxford 1, the scope and variety of cultural investigations involving astronomy has increased widely. 'Oxford 2' held in Mérida, Mexico in 1986 (Aveni 1989), and now 'Oxford 3', have demonstrated the substantial progress that has been made in building bridges between the diversity of academic disciplines which have an interest in, and must inevitably be involved in, the study of astronomy in culture.

A further set of issues was raised at Oxford 3 involving the relationship of archaeoastronomy to the mainstream of its parent, or constituent, disciplines. It is clear that if archaeoastronomy is not to remain an 'island discipline', then communication between archaeoastronomy and these mainstream disciplines is of critical importance. It was with a view to ensuring that archaeoastronomers do not merely communicate amongst themselves that two separate volumes of papers were planned to arise out of the contributions at Oxford 3. This, the first, is a volume of conference papers intended primarily for an 'internal' readership within the archaeoastronomical community. The second contains ten longer review papers, specially selected from contributions at Oxford 3, and is aimed at a wider readership.

'Archaeoastronomy in the 1990s' contains 31 papers covering a wide variety of current research topics in archaeoastronomy. Further details are given in the introduction which follows. The companion Oxford 3 volume, 'Astronomies and Cultures', aims to give an overview both of the scope of archaeoastronomy and of the broad range of problems being addressed within it. It is published by the University Press of Colorado, PO Box 849, Niwot, Colorado 80544, USA. Abstracts of the papers contained there are given in the Appendix of the current volume.

The editor wishes to express his gratitude to the many people who made Oxford 3 possible, most notably Aubrey Burl, Michael Hoskin and Nicholas Saunders for their work within the National Organising Committee, and Tristan Platt for his work as local organiser. Special and warm thanks are due to the late Douglas Gifford, whose enthusiasm and foresight made the conference possible at St. Andrews. The conference is grateful to the British Academy and the British Council for financial support.

References

Aveni, A. F., ed. (1982). *Archaeoastronomy in the New World*. Cambridge: Cambridge University Press.

Aveni, A. F., ed. (1989). *World archaeoastronomy*. Cambridge: Cambridge University Press.

Heggie, D. C., ed. (1982). *Archaeoastronomy in the Old World*. Cambridge: Cambridge University Press.

Dedication

The interval between Oxford 2 and Oxford 3 has witnessed the passing of Horst Hartung, a pioneer in the interdisciplinary field of archaeoastronomy. From his beginnings as an engineer (Diploma in Engineering, Stuttgart, 1948; Doctor of Engineering, Stuttgart, 1971) he became Professor of Pre-Columbian Architecture and Urbanism and a member of the Faculty of Architecture in the University of Guadalajara. What is little known to his academic colleagues is that Horst also was a practising modern architect of considerable skill and reputation.

An early guiding inspiration to all archaeoastronomers was induced by a note from one of Hartung's papers that read: 'Astronomers might be able to find out whether the orientations and directions that are noted are related to any astronomical events' (*Centro de Investigaciones Históricas y Estéticas, Universidad Central* boletín 11 (1969), 127–37; my translation from the Spanish). Horst had the wisdom to follow up his own suggestion and as a result he produced more than three dozen papers as well as a co-authored monograph and book.

Beginning with an early interest in site planning and settlement patterns, Hartung recognised in his earliest publications, dating from the 1960s, that it would be possible to frame purposive hypotheses to explain the often unusual situation and orientation of certain ensembles of Pre-Columbian buildings. Among the interesting factors he brought into play were the sky and landscape surrounding the ceremonial centre. This led to an interest in particular specialised buildings that may have been designed deliberately for astronomical observing. Some of Hartung's earliest work was concerned with the Zapotecan sites and with Teotihuacan, where he attempted to lay out systematically perspective views that might have made sense in terms of other documentary evidence on cosmovision and the encoding of its message in architecture. It is with great respect and deepest thanks to our colleague Horst Hartung for all he has contributed to archaeoastronomy that we dedicate this volume of writings from the third Oxford Conference on Archaeoastronomy.

Anthony F. Aveni, November 8, 1991

List of contributors

CAROL W. AMBRUSTER, Department of Astronomy and Astrophysics, Villanova University, Villanova, PA 19085, USA.

ANTHONY F. AVENI, Russell B. Colgate Professor of Astronomy and Anthropology, Colgate University, Hamilton, NY 13346, USA.

KATALIN BARLAI, Konkoly Observatory, PO Box 67, H–1525 Budapest XII, Hungary.

HUGH CAIRNS, 23 Wallaroy Road, Double Bay, NSW 2028, Australia.

VON DEL CHAMBERLAIN, Hansen Planetarium, 15 South State Street, Salt Lake City, UT 84111, USA.

MARGARET AND RONALD CURTIS, 'Olcote', New Park, Callanish, Isle of Lewis PA86 9DZ, UK.

ILDIKÓ ECSEDY, Oriental Research Centre, Országház u. 30, H–1014 Budapest, Hungary.

ROSLYN M. FRANK, Department of Spanish and Portuguese, University of Iowa, Iowa City, IA 52242, USA.

STANISŁAW IWANISZEWSKI, Państwowe Muzeum Archeologiczne, Skr. poczt. 69, 00-590 Warszawa, Poland.

EDWIN C. KRUPP, Griffith Observatory, Los Angeles, CA 90027, USA.

ARNOLD LEBEUF, Department of Historical Anthropology, Institute of Archaeology, University of Warsaw, Krakowskie Przed. 26/28, Warszawa 64, Poland.

List of contributors

WILLIAM LILLER, Instituto Isaac Newton, Ministerio de Educación de Chile, Santiago, Chile.

EMILY LYLE, School of Scottish Studies, University of Edinburgh, 27 George Square, Edinburgh EH8 9LD, UK.

STEPHEN C. MCCLUSKEY, Department of History, West Virginia University, Morgantown, WV 26506, USA.

J. MCKIM MALVILLE, Department of Astrophysical, Planetary and Atmospheric Sciences, Campus Box 391, University of Colorado, Boulder, CO 80309-0391, USA.

ROGER D. MARTLEW, Department of Adult Continuing Education, University of Leeds, Leeds LS2 9JT, UK.

WM. BREEN MURRAY, Universidad de Monterrey, Monterrey, N.L., Mexico.

EMÍLIA PÁSZTOR, Intercisa Múzeum, Lenin tér 10, H-2401 Dunaújváros, Hungary.

JON D. PATRICK, Department of Computing and Mathematics, Deakin University, Geelong, Victoria 3217, Australia.

SUSAN M. PEARCE, Department of Museum Studies, University of Leicester, 105 Princess Road East, Leicester LE1 7LG, UK.

EDOARDO PROVERBIO, Osservatorio Astronomico, Via Ospedale 72, 09100 Cagliari, Italy.

TSVETANKA RADOSLAVOVA, Department of Astronomy, Bulgarian Academy of Sciences, 72 Lenin Boulevard, 1784 Sofia, Bulgaria.

CLIVE L.N. RUGGLES, School of Archaeological Studies and Department of Mathematics and Computer Science, University of Leicester, Leicester LE1 7RH, UK.

BRADLEY E. SCHAEFER, NASA/Goddard Space Flight Center, Code 661, Greenbelt, MD 20771, USA.

POLLY SCHAAFSMA, Laboratory of Anthropology, Museum of New Mexico, Santa Fe, NM 87505, USA.

ROLF M. SINCLAIR, Physics Division, National Science Foundation, Washington, DC 20550, USA.

ANNA SOFAER, Solstice Project, PO Box 9619, Washington, DC 20016, USA.

IVAN ŠPRAJC, Subdirección de Registro Público de Monumentos y Zonas Arqueológicos, Instituto Nacional de Antropología y Historia, Mexico City, Mexico.

FRANZ TICHY, Spardorferstr. 51, D-8520 Erlangen, Germany.

ARCHIBALD S. THOM, The Hill, Dunlop, Kilmarnock, Ayrshire KA3 4DH, UK.

THORSTEINN VILHJÁLMSSON, Physics Department, University of Iceland, IS–107 Reykjavík, Iceland.

JAMES WALTON, University of Colorado Museum, Campus Box 218, University of Colorado, Boulder, CO 80309, USA.

RAY A. WILLIAMSON, Office of Technology Assessment, US Congress, Washington, DC 20510, USA.

MARIUSZ S. ZIÓŁKOWSKI, Department of Historical Anthropology, Institute of Archaeology, University of Warsaw, Krakowskie Przed. 26/28, Warszawa 64, Poland.

1

Introduction: Archaeoastronomy—the way ahead

CLIVE RUGGLES

Archaeoastronomy is undergoing rapid change. Indeed, the very term is inadequate, and arguably misleading, for describing what is practised today within the bounds of this 'interdiscipline', as Aveni designates it. In his introduction to the Oxford 2 volume, he develops a definition of archaeoastronomy as 'the study of the practice and use of astronomy among the ancient cultures of the world based upon all forms of evidence, written and unwritten' (Aveni 1989: 9). Its sister discipline, ethnoastronomy, now increasingly features alongside archaeoastronomy in the titles of meetings and publications, and the word 'ancient' can safely be removed from Aveni's definition to encapsulate a rich, wide, and coherent field of enquiry.

There is no doubt that a considerable amount of high-quality research is now being generated under the twin banners of archaeo- and ethnoastronomy. We have come of age in that we recognise the multifaceted nature of the evidence and appreciate the need to consider astronomical practice in its full cultural context. We realise that it is meaningless to study astronomy and calendrics in isolation from religion and society. Where astronomical alignments are found amongst monumental architecture, they are likely to be interpreted as symbolic representations of relationships existing within a culture's 'world view', not as the elements of ancient observatories. Valuable as his role was in getting archaeoastronomy on the road, the days of 'megalithic man' are now over.

Nonetheless, archaeoastronomy and ethnoastronomy are still in a state of great flux, striving to find their identity, to define their precise scope and goals, and to establish firm methodological foundations. A key issue, still unresolved, is whether they really merit an independent existence at all: while it is clear that archaeologists, ethnographers and others need to do archaeoastronomical work from time to time, it is less clear—certainly from the outside—that the combination of archaeoastronomy and ethnoastronomy necessarily needs to be recognised as a discipline in its own right. So, to paraphrase Aveni's question at Oxford 2: whether archaeoastronomy? If we cannot justify our existence within the wider structure of established academic disciplines, then we must question what we do and why we do it, and question the role and value of our specialist conferences and journals. If, on the other hand, we feel we can justify ourselves, then first and foremost we must go out and do so (see also Aveni, this volume). In addition, we must begin to address some serious questions concerning the academic future of the interdiscipline. Either way, a range of questions needs to be addressed with some urgency if the quality work currently classified under archaeo- and/or ethnoastronomy is to flourish and to achieve the academic recognition it deserves in the 1990s and beyond.

This short introductory paper seeks to address such questions in the light of developments during recent years, and to give one view of how archaeoastronomy should proceed in the 1990s.

Whether archaeoastronomy?

It may surprise many working within the field that the question of archaeoastronomy's continued existence is being asked at all in the 1990s. Certainly, if levels of activity alone were the deciding factor then the answer would be in no doubt. In the late 1980s American archaeoastronomy saw what Aveni (this volume) aptly describes as a period of 'hyperactivity'. Conferences and publications have abounded. In Europe, a myopic focus on the prehistoric north-west has given way to a set of topics of interest which range widely, both geographically and chronologically. Since 1988, at least one archaeoastronomical symposium has been held in Europe each year: at Tolbuchin, Bulgaria, in 1988 (Tolbuchin 1988), Venice in 1989 (Romano and Traversari 1991), and more recently the annual series on 'Current problems and future of archaeoastronomy', begun in 1990 at Warsaw (Iwaniszewski 1992). There has also been a substantial increase in awareness and activity world-wide since Oxford 2, to which the breadth of new culture areas touched upon at St. Andrews amply bears witness. Oxford 3 had over eighty participants from over twenty countries. We have certainly come a long way in the decade since the first faltering cross-fertilisation at Oxford 1 between, essentially, the Americanists on one hand and the 'megalithic astronomers' on the other.

Viewed from outside, however—for example from mainstream archaeology or anthropology—the picture is less euphoric. First, archaeo- and ethnoastronomy have consistently projected a bad 'public image' to the outside academic world, and to a considerable extent continue to do so. Archaeoastronomy originally arose as such in the climate of the 1960s and has carried a significant quantity of fringe baggage ever since, a burden that continues to colour attitudes amongst those outside the field to mainstream work within it. The fact, for example, that relatively little work in archaeoastronomy is presented at archaeological conferences or published in archaeological journals does nothing to aid communication (see also Aveni, this volume). Neither does the fact that the British journal *Archaeoastronomy* only exists, for historical reasons, as the supplement to *Journal for the History of Astronomy*, and hence is filed not under archaeology but under history of science, for most archaeologists and anthropologists an obscure and unvisited corner of the library.

The second problem is that it is generally unclear precisely what archaeo- and ethnoastronomy encompass and precisely what they are trying to achieve. A widespread perception of the two disciplines amongst their mainstream proponents appears simply to be the collection of data on astronomical practice in different cultural settings. From an anthropological or archaeological point of view the value of merely collecting data, as opposed to trying to derive some meaning from those data, is of very limited interest (see, for example, Kintigh 1992). In a great many cultural settings astronomical practice is far more likely to be closely related to a shamanic world-view than to be any form of precursor to twentieth-century astronomy in the western world. Thus, the fact that modern astronomers have played a significant part in archaeoastronomical studies has aroused considerable (and often justified) suspicions of ethnocentrism.

Finally, cultural needs and perceptions are all-pervading and social systems are notoriously open-ended; so why should astronomy merit special attention over and above many other factors such as economy, trade, politics, burial practice, diet, and so on? Few would dispute that astronomy should not be studied in isolation from the many other potential influences on a cultural system; the point at issue is whether it deserves to be given special emphasis. If it does, then the reason must be that the study of astronomical observations can perform a special role in the study of cultural systems as a whole.

Ruggles and Saunders (1993) argue this is indeed the case. The key, they suggest, is in viewing the sky as a cultural resource. It is a resource affording many uses to a culture, particularly because it is the only resource within the environment which is not susceptible to physical alteration by human beings. Furthermore, the recurrent phenomena within it are directly accessible to us: they form a part of the environment of another culture that we can accurately reconstruct (within determinable margins of error). Finally, various features within it are common, at various levels, to different cultures. These three properties permit wide-ranging structuralist studies of topics such as the correlation between astronomy and culture, cross-cultural parallels, and the ways in which different cultures interpret and manipulate the same immutable database for different political reasons and ends.

It is surely this line of argument above all others that justifies the significance, and *ipso facto* the existence, of archaeo- and ethnoastronomy as endeavours in their own right, with their own identity, and the general goal of exploring and exploiting the special nature of the sky as a cultural resource in order to make a distinctive contribution to cultural studies as a whole. However, recognising this is just the first step, and several further developments are now needed. One is to argue the case with our colleagues within the mainstream of the parent cultural disciplines. This involves developing the theoretical arguments outlined above, and developing case studies.[1] Another is to clarify the scope of the our own endeavours. Finally, and perhaps most importantly of all, we must develop suitable underlying theoretical and methodological principles. If we do not do this, we cannot expect archaeoastronomy to be taken seriously as an academic discipline.

Underlying principles

The very quality that gives archaeoastronomy its originality and vitality is also the source of the most fundamental problem in trying to establish theoretical and methodological principles: the mix of mainstream disciplines that are relevant and contribute to studies in this area. I have argued elsewhere (Ruggles 1988) that the root cause of many of the misunderstandings and arguments between archaeologists and numerate scientists about 'megalithic astronomy' in the early days of archaeoastronomy was a conflict of methodological approaches. Nowadays the problem is much wider, with the involvement of disciplines as diverse as astronomy, history of science, archaeology, anthropology, ethnohistory, ethnography, geography, architecture, art history and the history of religions, each with its own different epistemological perspectives and methodological principles.

---

[1] For existing case studies of the correlation between astronomy and culture see Iwaniszewski (1989; this volume). For an excellent cross-cultural comparison see Roe (1993).

In this context, it becomes clear that the 'green' versus 'brown' methodological divide identified by Aveni (1986; 1989) is really symptomatic of a more general problem: the lack of a rigorous methodology for *combining* evidence from the main constituent disciplines. For example, it is arguable that Mesoamerican archaeoastronomers have tended to depend too heavily upon ethnohistoric documents as opposed to other forms of evidence. In the absence of any methodological foundations it is impossible to judge. Ruggles and Saunders (1993) suggest that appropriate methodology for data integration might be founded upon the use of analogical inference in conjunction with a Bayesian statistical methodology.[2] This suggestion requires further investigation.

It is clear at this stage, though, that the need to formulate acceptable methodologies for the integration of diverse data itself gives the archaeo- and ethnoastronomy a further *raison d'être* and potential strength. This is because the interdisciplines may now be able to lead the way to new cross-disciplinary approaches within their parent disciplines. We are forced to face issues which established, compartmentalised disciplines are able (and even keen) to avoid, despite recognising the problem (Charlton 1981). By seriously tackling such issues we can turn necessity into a virtue. However, the price to be paid is that we must be the ones who attempt to forge new methodological frontiers.

Scope (and name)

The recognition that the separate existence of archaeo- and ethnoastronomy may have value on more than one level—the advance of methodology as well as the advance of knowledge of human culture—only serves to highlight the still unresolved question of scope. What is in the domain of archaeoastronomy or ethnoastronomy and what is not? Where are the interfaces between the interdisciplines and their mainstream counterparts?

The most immediate question relates to the boundary between ethno- and archaeoastronomy. This is often quoted as simply the distinction between present and past. In practice, however, the distinction is based upon the type of input data, and in the past there was also a distinction

---

[2] The Bayesian paradigm is to express our prior model of the world in terms of probability distributions of suitable parameters, and to develop a formalism for how this model should be modified in the light of new data. The main advantages of such an approach are (i) that data of many different kinds may be considered in the formulation of a prior model ('classical' statistics, in contrast, lacks any means of taking into account 'background' or 'corroborating' evidence relating to a hypothesis being tested) and (ii) that data may legitimately be reused, to examine how they affect different prior ideas.

in motivation. Archaeoastronomy emerged as a result of people investigating the possible astronomical significance of architectural alignments. Other types of data—ethnohistoric, folkloric, and so on—were subsequently found to be relevant in particular cases and added to the general armoury. However, an interest in astronomical practice remained the main motive force, at least for the scientifically-minded researcher. Most ethnoastronomical data, on the other hand, have been obtained incidentally: examples such as Urton's (1981) work at Misminay are the exception, not the rule.

On closer examination, the boundary between 'ethno-' and 'archaeo-' appears to become more and more contrived. Where, for example, do we place Urton's (1990) ethnographic work on contemporary Andean societies aimed at interpreting ethnohistorical and archaeological data on the Nazca lines? Where do we place Frank and Patrick's examination (this volume), through field study and historical research, of the stone octagon complexes associated with pastoralism in the Basque Country? Much archaeoastronomical work in the New World has been based on ethnohistorical rather than archaeological research. If there is a distinction to be drawn between types of input data, then many New World archaeoastronomers would surely more aptly be termed ethnohistoastronomers. We could also identify folkloricoastronomers, mythoastronomers, arthistoastronomers, and so on. Clearly this way lies lunacy.

The great potential strength of studies in cultural astronomy is the breadth of types of input data involved. It clearly serves a better scientific purpose to combine all these views of one resource in a single mode of enquiry. The only logical conclusion seems to be that what we might call the 'study of cultural astronomy' should be a single study area.

'Cultural astronomy' as a name also has the considerable advantage that it better communicates to our colleagues in the mainstream disciplines what it is we actually strive to do. The term 'archaeoastronomy' is confusing when compared with, say, archaeo- (or palaeo-) botany, which describes the application of botanical techniques to the study of the material record rather than the study of botanical practice amongst the society concerned. To many of our colleagues, it also continues to suggest alignment studies alone, as Aveni (this volume) avers.

The boundary between history of astronomy and archaeoastronomy also deserves mention at this point. It has arisen for historical reasons and seems artificial and inconsistent. Thus, for example, studies of Babylonian astronomy traditionally come within the realm of history of astronomy whereas those of Classic Maya astronomy are regarded as a topic within archaeoastronomy. On grounds of socio-economic and political correlates there can be little justification for this. The reason is, of course, that while the basic concern of history of astronomy is cultural astronomy, its scope is restricted to those cases that are considered in some sense to be precursors of twentieth century western 'scientific' astronomy. The self-evident solution, it seems, is to regard the history of astronomy as a part of the general study of cultural astronomy, and in this one heartily concurs with the view expressed by Aveni (1989), Gingerich (1989) and others at Oxford 2. However, we have to recognise that this is a complex issue about which there is likely to be a good deal of debate in the future.

An infrastructure for cultural astronomy

If we accept the value of, and need for, cultural astronomy as an interdiscipline, and assume for the moment that its theoretical foundations, scope and goals are sufficiently clear, then we must begin to address some serious questions concerning its academic future. The central question is how we set up an infrastructure by which cultural astronomy takes its place and is recognised and supported within the general framework of academic disciplines. Working in cross-disciplinary areas is not easy, as all those who have worked in archaeo-astronomy from an academic base in a mainstream discipline will know only too well, so how can this support be achieved?

One possibility that was raised at Oxford 3 is that of an international academic body to represent cultural astronomy. Two opposing views were expressed. One was that the vitality and spontaneity of cultural astronomy owes much to the unique and ever-changing mix of people involved, and the establishment of fixed bodies are likely to impose a fixed base of ideas and hence produce a relative stagnation and kill off cultural astronomy as we now know it. The opposing view is that, as accountability becomes a greater and greater concern within academic institutions, budgets are continually devolved and there are great pressures to become increasingly compartmentalised and specialised, some sort of institutional support, both academic and financial, is needed if work in cultural astronomy is to continue at all. This will be discussed further at Oxford 4.

Meanwhile, the 1992 European conference on 'Current problems and future of archaeoastronomy' voted to establish a European Society for Cultural Astronomy' (Société Européene pour l'Astronomie dans la Culture). There are also moves to establish special commissions within the International Astronomical Union and the International Union for the History of Science, although, significantly and perhaps regrettably, there is not yet any similar move in the domain of anthropology or archaeology.

In general terms, though, the most crucial issue for ensuring the academic future of cultural astronomy is that of education. This covers the question of educating our colleagues in what we do as well as educating the next generation. Clearly to some extent the former influences the latter, for we can only teach course options in cultural astronomy within mainstream disciplines such as anthropology, archaeology, astronomy or history of science if we can persuade our colleagues in those disciplines that it is valuable to do so.

On the subject of informing our colleagues, Aveni (this volume) asks: 'To whom, to what groups of scholars, does our work have value? Exactly what does it illuminate in their fields of concern?' The answer, he suggests, is to communicate our work not to each other but to our colleagues in the mainstream, through publication in disciplinary journals and attendance at disciplinary conferences. This communication is clearly crucial.

What of the undergraduate curriculum? Currently, courses in cultural astronomy are a rarity. The only example in Britain is one that has been run by the present author at Leicester since 1990 as a third-year optional module within the BA and BSc Single Honours Archaeology Degrees. It is strongly methodology-based, and the hope is that in two years, when Leicester switches to a modular degree structure along American lines, it will possible to open the course to a mixture of students from, say, archaeology and sociology together with astronomy and history of science, to examine their different approaches in a critical way in the light of some archaeoastronomical evidence.

One of the great attractions of teaching archaeoastronomy at this level is the ability to focus upon high-level questions and methodological issues that arise from the necessity to deal with highly interdisciplinary data. A disadvantage is the basic factual background that is needed in areas with which students in a particular discipline will not be familiar. If one is not careful, communicating these basic factual data can assume a disproportionate amount of time on a short course. In the case of archaeology students this means basic astronomy. (On the Leicester course planetarium demonstrations are used.)

Perhaps this points to a broader way in which cultural astronomy can be 'marketed' educationally—its vocational value being limited, of course, owing to its intense specialisation—namely as a vehicle for increasing cross-disciplinary awareness and critical perspectives. There is much talk at the moment about fostering and encouraging interdisciplinary research, and the fact that interdisciplinary work is as essential to the development of ideas as highly specialised work is now widely recognised. This suggests that interdisciplinary teaching might also receive increased support. However, the constraints of the discipline-based 'system' act against interdisciplinary developments at every level. On the other hand again, effective interdisciplinary teaching might be the first way to move this mountain.

A related question on the subject of education is whether we need to train specialised 'cultural astronomers'. Many of today's archaeoastronomers, the present author included, have arrived in their present positions by changing discipline and retraining at some point in their career, sometimes more than once. There is clearly a need for some people well versed in all aspects of the subject, theoretical and practical. It is to these people that we must look to push forward the theoretical and methodological frontiers in the years to come. Arguably, we need to persuade at least one University somewhere to support cultural astronomy as a mainstream activity in itself, with a concentration of research and perhaps a specialist taught Masters course.

The way ahead

Where, then, should cultural astronomy should be headed in the 1990s and into the next millennium? I would argue, then, that our priorities for the near future should be as follows.

- Develop methodologies for the joint assessment and integration of diverse types of data obtained through different disciplinary approaches. This will both help to build suitable methodological foundations within cultural astronomy and also, hopefully, contribute to theoretical developments in a much wider domain. In my view, the development of suitable methodological foundations for dealing with problems of data integration represents the single most important item on the agenda for cultural astronomy in the coming few years.

- Give priority to research demonstrating the potential of structuralist studies of correlations between astronomy and other cultural attributes, and to cross-cultural studies.

- Spend much more time and effort communicating our work to the mainstream disciplines, and receiving their critical feedback (cf. Aveni, this volume).

- Establish suitable academic bodies to support work in cultural astronomy.

- Make more effort to disseminate work in cultural astronomy through teaching and training.

Finally, while not proposing that every person needs to be familiar with all the details of the theoretical foundations and developments in order to practice archaeoastronomy, any more than the average field archaeologist needs to have more than a passing familiarity with developments in theoretical archaeology, I wonder if perhaps every aspiring 'cultural astronomer' should try if possible to foster a great breadth of contribution, say in folk astronomy as well as architecture; in a subpolar or tropical region as well as a temperate one, a protohistoric state culture as well as a prehistoric tribal one? Doing so might, for example, encourage 'green' archaeoastronomers to consider symbolism other than that encoded in alignments upon horizon phenomena—shadow phenomena, for example—or 'brown' archaeoastronomers to reconsider how much weight they are putting on ethnohistoric accounts in relation to the other types of evidence available to them.

Organisation of papers in this volume

In order to encourage a wider awareness of the breadth of the discipline, the organisers tried to encourage thematic contributions at Oxford 3, with the whole of the first day of the conference being devoted to thematic papers. Contributors were also encouraged to emphasise the comparative element in studies of the social role of astronomy. Some contributions focused upon evidence for astronomical practice, as manifested in the material record (architecture, artefacts, written record) or in the ethnographic record (recorded directly through fieldwork or indirectly through the interpretation of others' fieldwork or of historical accounts). Others concentrated more on the interpretation of astronomical practice in its social context: cosmologies and calendrical systems; the sky as a cultural and ideological resource; structuralist attempts to derive correlates between astronomy and the nature of society (e.g. state or non-state). Three major themes at the conference were astronomy and architecture, astronomy and calendrical systems, and astronomy in myth and ideology.

This diversity suggested a number of ways in which the papers in this volume could be arranged, following Aveni's decision not to organise the volume of papers from Oxford 2 on purely geographical grounds. This was not an easy task. The papers included here span many disciplines and types of data, and tackle many different integration problems; small groups of hunter-gatherers through to large state cultures; modern times back to the Neolithic; northern sub-polar regions down to the Antipodes; and a huge variety of approaches ranging from structuralist studies of native cosmologies through to mathematical studies of the layout of passage grave cemeteries, not to mention several papers which are thematic or concentrate on new techniques.

In the end, I have chosen to divide the book into five sections. The first contains three thematic papers that specifically address some of the issues raised in this introduction. The second contains papers which concern culture areas largely unfamiliar within the archaeoastronomical literature to date. The third contains two papers which introduce new techniques and methods from mainstream astronomy together with two papers that describe new approaches to the analysis of familiar data from prehistoric north-west Europe. The fourth section contains the latest research results using research methods and in topic areas already well familiar to archaeoastronomers. Finally, the fifth section contains a single paper addressing the issue of education and training in a specific context.

Together with its companion volume 'Astronomies and cultures', I hope that this collection of papers goes some way to recreating the vitality of Oxford 3 and gives an adequate impression of the scope and quality of the contributions presented there. Oxford 4, planned to be held in Bulgaria in 1993, may show us how far we have progressed along some of the paths suggested here.

Acknowledgements

I am most grateful to Anthony Aveni and Nicholas Saunders for their comments on earlier drafts of this manuscript.

References

Aveni, A.F. (1986). Archaeoastronomy: past, present, and future. *Sky and Telescope* **72**, 456–60.

Aveni, A.F. (1989). Introduction: whither archaeoastronomy? In *World archaeoastronomy*, ed. A.F. Aveni, pp. 3–12. Cambridge: Cambridge University Press.

Charlton, T.H. (1981). Archaeology, ethnohistory and ethnology: interpretive interfaces. In *Advances in archaeological method and theory, vol. 4*, ed. M.B. Schiffer, pp 129–174. London: Academic Press.

Gingerich, O. (1989). Reflections on the role of archaeoastronomy in the history of astronomy. In *World archaeoastronomy*, ed. A.F. Aveni, pp. 38–44. Cambridge: Cambridge University Press.

Iwaniszewski, S. (1989). Exploring some anthropological theoretical foundations for archaeoastronomy. In *World archaeoastronomy*, ed. A.F. Aveni, pp. 27–37. Cambridge: Cambridge University Press.

Iwaniszewski, S. (1992). *Readings in archaeoastronomy. Papers presented at the international conference 'Current problems and future of archaeoastronomy' held at the State Archaeological Museum in Warsaw, 15-16 November 1990*. Warsaw: State Archaeological Museum and Department of Historical Anthropology, Institute of Archaeology, Warsaw University.

Kintigh, K.W. (1992). Archaeoastronomy and archaeology. *Archaeoastronomy and Ethnoastronomy News* (Center for Archaeoastronomy) no. 5, 1 & 4.

Roe, P.G. (1993). The Pleiades in comparative perspective: the Waiwai *shirkoimo* and the Shipibo *huishmabo*. In *Astronomies and cultures*, eds. C.L.N. Ruggles and N.J. Saunders. Niwot CO: University Press of Colorado. In press.

Romano, G. and Traversari, G., eds. (1991). *Colloquio Internazionale Archeologia e Astronomia*. Rome: Giorgio Bretschneider Editore (Supplementi alla RdA, 9).

Ruggles, C.L.N. (1988). The stone alignments of Argyll and Mull: a perspective on the statistical approach in archaeoastronomy. In *Records in stone*, ed. C.L.N. Ruggles, pp. 232–50. Cambridge: Cambridge University Press.

Ruggles, C.L.N. and Saunders, N.J. (1993). The study of cultural astronomy. In *Astronomies and cultures*, eds. C.L.N. Ruggles and N.J. Saunders. Niwot CO: University Press of Colorado. In press.

Tolbuchin (1988). Първи национален симпозиум археоастрономия [First national symposium on archaeoastronomy]. *Интердисциплинарни изследвания [Interdisciplinary Studies]* **15**, 1–136.

Urton, G. (1981). *At the crossroads of the earth and sky: an Andean cosmology*. Austin TX: University of Texas Press.

Urton, G. (1990). Andean social organization and the maintenance of the Nazca lines. In *The lines of Nazca*, ed. A.F. Aveni, pp. 173–206. Philadelphia PA: American Philosophical Society.

I

THEMATIC CONTRIBUTIONS

2

Archaeoastronomy in the Americas since Oxford 2

ANTHONY F. AVENI

Overview

The primary developments on our two continents since 1986 have been characterised by three things.

- The realisation that the interpretation of the material we have been dealing with may be far more complex than we had previously thought. Thus, we began to move away from pure reportage of our discoveries as ends in themselves and towards addressing the meaning of what we found, to which disciplines it is important, and why.

- The development of a level of dialogue with one another (as an archaeoastronomical peer group), as well as with the traditional disciplines, that is demonstrably more sophisticated than it had been. This is attributable in part to the reassessment of the data of some investigators by others, a process well familiar to students of megalithic astronomy in the 1980s.

- The expansion of archaeoastronomical inquiry to new places and cultures in the Americas.

The several examples from the recent literature that I shall employ in an attempt to elaborate these developments are drawn from the most active areas of study: the search for a cultural context in the archaeoastronomy of the US South-West and the Mexica world; the proliferation of data on the pecked cross petroglyph in Mesoamerica; alignment studies in Cuzco, Nazca, and Copan; and the inquiry into the empirical basis for the astronomical portions of the Maya codices.

Finally, I shall discuss some of our prolonged shortcomings, most notably

- the need to continue to try to address the question of the relationship between archaeo- and ethnoastronomy; and

- the necessity of broadcasting our news to professional audiences that traditionally have not been as receptive as they might be (e.g. the anthropologists and archaeologists).

Introduction

When I edited the Oxford 2 proceedings (Aveni 1989), the first single volume on world archaeoastronomy as such, I tried to organise the presentations by subject area as follows:

I. review papers that deal with places and cultures found to offer information of interest to archaeoastronomy;

II. papers that draw on textual evidence; and

III. interdisciplinary works in which both cultural and textual sources are used to develop testable hypotheses that relate to the archaeological and architectural corpus.

Some reviewers have portrayed such a reorganisation as radical;[1] nevertheless I took this course to try to close a gap that had become evident in the Oxford 1 conference between British megalithic astronomy and archaeoastronomy in the Americas. This so-called 'green-brown' dichotomy (Aveni 1986) received its label from the colours of the covers of the pair of Oxford 1 proceedings volumes (Heggie 1982; Aveni 1982). It referred to the fact that New World studies were more frequently able to take advantage of written evidence bearing on the practice of astronomy while Old World studies stressed a more rigorous statistical approach.

Though the nature of the evidence amongst the various cultures archaeoastronomers deal with may vary, I feel that, while astronomers and anthropologists might not articulate precisely the same agendas, we all evidently share a collective goal, lest why would we reassemble again

---

[1] For reviews, see Leaver (1989), Saunders (1989) and Heggie (1990). Heggie, an astronomer, pronounced this organisation radical and enforced, while anthropologist Saunders, who called it a 'real attempt at a world perspective', stated that 'a coherent justification is taking shape as the universality and implications of human uses of the sky are established beyond doubt'.

and again? What we have in common is that we all desire, where possible, to make use of both the written and the unwritten record to comprehend the astronomical, cosmological and, consequently, ideological systems of societies past. I believe we also have assembled to critically reassess our collective practice and to explore its relevance to other endeavours such as ethnoastronomy or 'blue archaeoastronomy' (see Farrer and Williamson 1991).

Let me expand upon this point about different disciplines having different agendas by referring briefly to three reviews (see note 1) of the Oxford 2 volume, 60% of which contained research on the Americas.

Heggie, in his brief review in *The Observatory*, an astronomical publication, seems concerned that one ought not to encompass too much of the sociocultural context of astronomy. He regards Carrasco's description of the Aztec celestial god Tezcatlipoca (Carrasco 1989: 49–50) as irrelevant to archaeoastronomy. Heggie's agenda, like that of most astronomers and scientifically-trained people, seems to be concerned only with the facts, i.e. with collecting and describing the astronomical phenomena the ancients witnessed while giving particular emphasis to the observational methodology they employed. On the other hand he praises Floyd Lounsbury's paper, in the same volume, on the Maya calendar because the 'apposite' cultural considerations are held to a minimum. The focus, just as Heggie would like, is reduced almost to pure astronomy. Leaver, a chemist with an interest in astronomy writing in the *Journal of the British Astronomical Association*, also an astronomical publication, looks primarily for astronomical content as well. However, what matters more to Saunders, the archaeologist-anthropologist, is how archaeoastronomical work 'reflects the diversity of human experience and the relationship between natural phenomena and social premise'. His agenda is more expansive than reductive. He is concerned with the way in which human societies deal with astronomical material. How are natural phenomena liable to social manipulation? What cultural metaphors are manifested in different kinds of astronomical event? It is on this agenda—the cultural agenda—that the description of Tezcatlipoca becomes relevant, for precisely what aspects of the night sky this god does or does not represent are indicated in the discussion of what he looks like. Moreover, the repetitive flow of the Nahuatl dialogue captures the continuous cyclic nature of the Aztec way of knowing time. For this writer, archaeoastronomy belongs solidly in the cultural camp because it is concerned with the interaction between people and nature. If we have witnessed any significant development on the American side in the past five years, it is that investigators have begun to realise how richly complex and diverse the dialogue between human societies and the celestial vault really was in pre-contact times.

	North America	Meso-america	Other America	All America (%age of total)
Journal				
JHA Supplement, UK	7	5	1	37
Center for Archaeoastronomy, USA	6	3	3	71
Conference				
Oxford 2 (Mérida)	8	13	2	60
Oxford 3 (St Andrews)	10	10	8	32

TABLE 1. Distribution of papers on American archaeoastronomy across the primary archaeoastronomical literature, 1984–90.

Rather than bore the reader with a one-sentence summary of each paper on American archaeoastronomy that has been contributed to the literature in the past five years, I think it more meaningful to tailor this review to the kinds of questions the recent contributions to American archaeoastronomy have addressed, and to stress how these questions differ from those being asked from five to ten years ago. I also want to address the question of who reads archaeoastronomy as well as what they have to say about it.[2] Only then can we begin to perceive on a broader basis what we, collectively, are contributing to the fount of knowledge.

North America

I begin with the work in North America, where practically all of the research has been focused upon the US Southwest. First some statistics: I tallied up the number of papers (excluding brief reviews and notes) in our two international journals as well as the Oxford 2 volume and the present conference and categorised them by area (North American, Mesoamerican, and other American). The results are given in Table 1.

One can mull over this table and draw various conclusions. Here I will use the data to make the point that there has been a lot of work going on in North America—as much as in Mesoamerica if I use these data alone. Though I have not surveyed the non-archaeoastronomical journals and conference proceedings, I am sure that a good deal of the Mesoamerican work has been assimilated into the disciplinary journals

---

[2] In this regard the author, with advice from the organising committee, distributed a questionnaire to conference participants to help assess the value of works on archaeoastronomy as perceived by the established disciplines.

(such as *American Antiquity, Science, Current Anthropology*, and *Cuadernos de Arquitectura Mesoamericana*).[3]

Now, the US homeland is safe and easy territory to explore (much safer than Guatemala or Peru these days) and the mystique of the South-West, where many indigenous Americans still live, has a discernible force of attraction on the American scientist-citizen. Archaeologist Judge (1987) explains this attraction and accounts for the popular appeal of such studies as follows: here we are witnessing 'a "rediscovery" of the sun's path by modern Anglo populations', which is 'a function solely of our having been very effectively insulated from our environment for a long time'. Regardless of such mass psychoanalysis, much has been written lately. But what is it all about? What it used to be about was alignment hunting. 'Did you find a solstice orientation? And if so, how precise was it?' seemed to be the questions voiced at meetings. Such questions and their answers, usually dealt with out of context, were often of minimal value to the archaeologist and anthropologist and to the questions about culture and social processes that concern them. Solstice finding such as the work at Anasazi Yellow Jacket (see Malville and Walton, this volume) and seasonal and diurnal petroglyph-watching are still important activities, but now the winds of inquiry are beginning to change. These studies have departed considerably from the now antiquated (and aptly named) Old World paradigm of recording phenomena as ends in themselves. Given the similarity of the databases in British megalithic and US South-West archaeoastronomy, it is not surprising that investigators have asked the same questions, and even looked for the same alignments (Judge 1987: 13, note 6). This progressive step comes in large measure from the increased contact amongst professionals from the diverse disciplines, which breeds familiarity with different approaches and in turn facilitates the critical process so necessary for sound scholarly work. The more conferences attended, and the more works read, the more understandable are the questions and methodologies of the related disciplines. We are well past the relatively empty reportage of a decade ago. On the North American scene the prognosis is good.

Perhaps less cautious and a bit more imaginative, some investigators now begin to interpret their finds as discernible facets of 'cosmographic expression' rather than observational sightlines. Everything they say may not be subject to the rigorous tests the exact sciences can impose, but they offer us new ideas to explore and hopefully to test in other ways. Chamberlain (1984), Zeilik (1985a; 1985b; 1985c; 1986; 1987),

---

[3] A survey of the literature is needed. Data on the integration of archaeoastronomy into the standard disciplinary journals for the British and rapidly growing European archaeoastronomy ought to be assembled before Oxford 4.

Young (1987a; 1987b), Williamson (1987) and others have begun to address the question of cultural continuity from Puebloan to Anasazi to the present day historic pueblos by studying their calendars and the use of ceremonial architecture as a means of integrating anticipatory sun watching into the ritual life of these people. For example, Young, a folklorist, has sought a cultural context for the astral symbolism inferred in the rock art by placing the petroglyphs on a comparative level with ceramics and sand paintings. Still she points to an overemphasis of the petroglyphic studies relative to these other media in which astronomical knowledge might be expressed. At the same time Reyman (1987), an anthropologist and archaeologist, has discussed Puebloan astronomy as a mechanism for socio-ceremonial control in a non-egalitarian society and Wilcox (1987), an archaeologist, deals with archaeoastronomical data specifically in the context of Hohokam ceremonial systems. He argues that the metastructure of cosmological beliefs has Mesoamerican parallels. In California, research on the Chumash (Krupp 1989) and related groups (Benson and Hoskinson 1985) continues this more sure-footed trend of moving toward interpretative analysis. In sum, these investigators have boldly invaded the sanctity of those other disciplines our specialised world encourages us not to enter. Astronomy no longer generates all the hypotheses. The scientifically trained investigators who still dominate this area have begun to attempt the anthropological explanations, 'to interpret the purpose and function of these phenomena in a larger social context' as Judge (1987: 5) has challenged them to do if one ever can hope 'to see progress in the maturation of archaeoastronomy as a truly integrated discipline'.

Mesoamerica

In many parts of Mesoamerica we have considerably more than sticks and stones to work with and we also have knowledge of a hierarchy of urban structure as a motivating force that mitigates on calendrical development and the elaboration of astronomical ritual. In many instances we also have textual evidence, though we are disadvantaged in some cases by the contamination of European influence (but see Farrer and Williamson 1991).

The major development in Mesoamerican archaeoastronomy these past five years has been the realisation that things are far more complicated than they had seemed. Let me cite a few examples. Those who study astronomy in the codices are making a concerted effort to look for the empirical basis of the astronomical almanacs and ephemerides that can now be more clearly deciphered. A number of almanacs previously thought to have served as purely cyclical tables of good and bad luck days hanging in an indeterminate time frame are now being reinterpreted in a real-time framework. Some of the events portrayed in glyphs, pictures and numbers can be connected to actual conjunctions, eclipses, etc. Such tables in one part of a codex seem to be linked with those in other parts of the same codex (e.g. Bricker and Bricker 1989). At the last two Mesas Redondas de Palenque I was astounded by the large number of papers with astronomical content. Many of the monumental inscriptions and their associated architectural alignments are quite commonly referred to as political statements or expressions of the continuity and stability of royal power. What is positive about these studies is that astronomical materials are being seriously confronted by social scientist and humanist. There is actually a need for more astronomers to work with the Maya epigraphers. On the negative side, thanks to a plethora of computer discs that tell us with great facility precisely what happened in the sky, many individuals who have never seen the evening star have taken to punching up astronomical events with abandon and matching them up with inscriptions and alignments.[4]

An excellent example of the successful merger of astronomy, archaeology, and epigraphy has recently taken place at Copan, where the reading of the glyphs and archaeological excavations at Temple 22 has imposed a specific time span during which the solitary window on the west side of the building could have functioned as an astronomical sighting device. In this instance the archaeologists (Larios and Fash 1989) have made specific reference to the viability of the astronomical hypothesis for the use of the building as a device for sighting Venus. Šprajc's work (1989; see also this volume) on the greater accessibility of the evening as opposed to morning star apparitions has both strengthened this view and made even more plausible the use of this object as a predictor of the rains. Carlson's (1993) work on the existence of the Venus cult at Copan has added considerably to these studies.

Alignment studies at Dzibilchaltun (Coggins and Drucker 1988) and Chichen Itza (Milbrath 1988) have been specifically related to the iconography on the buildings and in both instances the iconographic interpretations have been employed to generate astronomically-based hypotheses. At Tikal, we now know much more about the building sequence

---

[4] For example, past appearances of Halley's Comet that were demonstrably less noticeable than that of 1986 have been accorded undue significance in the inscriptions of Palenque. Also, the dates of greatest elongation of Mercury have been matched with inscribed dates on Maya stelae across the Peten, without regard either for the fact that this phenomenon is not directly observable as the longitude difference between the sun and Mercury, nor for the fact that one must take account of the large percentage of the cycle that the planet spends near to greatest elongation.

than we did a decade ago. This gives archaeoastronomers the opportunity to reassess some of their earlier work. Now that Temple I and its adjacent structure 5D-33 have been clearly related to the foundation of the new dynasty of Ruler A (Ah Cacau) we can appreciate why they were precisely the pivotal structures in the astronomical orientation plan (Aveni and Hartung 1988). In Central Mexico, the bond between archaeoastronomy, religious studies and anthropology has been strengthened in Mexica studies thanks to Carrasco's efforts at bringing together an international working group that has focused for the past five years on ritual processes at Tenochtitlan, the basic resource materials for these studies being the chronicles of Durán and Sahagun (Carrasco 1991). Recently, the domain of archaeoastronomy has been specifically incorporated in an anthology of readings that addresses the role of mythology in the history of religions (Eisenberg *et al.* 1990). Broda's many anthropologically-based papers (e.g. Broda 1982; see also Broda 1993) place central Mexican archaeoastronomy in the context of *cosmovisión*, a world view that incorporates archaeoastronomy. Her influence has taken root in the ongoing thesis work of her UNAM students.

There are a number of areas in Mesoamerican archaeoastronomy where the codices and ethnohistoric documents cannot help us directly, and here we must proceed more by careful guesswork, never overlooking the need for rigour in weighing what data are available. For example, in Yucatan enough orientation data have been collected that we can now begin to assess them statistically. As a result, we are beginning to propose hypotheses on the evolution of the orientation calendar (Aveni and Hartung 1986). Since we last met, the study of pecked cross petroglyphs[5] has proliferated. We have doubled the number of known petroglyphs, thanks in large measure to the survey work of Wallrath and others (e.g. Wallrath 1984). This has necessitated an incipient overhaul of the old benchmark hypothesis (Dow 1967; Aveni *et al.* 1978; Chiu and Morrison 1980; Peterson and Chiu 1987) who were looking for a complete and unifying system of such markers at Teotihuacan. These new data also support the idea that different levels of explanation and multiple function can co-exist in a set of artefacts that at least at the surface seem to correspond remarkably well to a very narrow definition. It may seem difficult to those of us trained as scientists to isolate paired cause and effect, but we must face it: no single purpose or grand design can be proposed that will account for all the pecked circles, any more than for all Nazca lines (see below). It can be demonstrated statistically that some pecked circles served as counting devices, others as

---

[5] Right-angled crosses centred on single, double or triple circles pecked into rock or into the floors of buildings.

markers, the axes or lines between pairs of which indicated important astronomical or environmentally significant directions, while others may have served as sacred game boards (Aveni 1988; see also Iwaniszewski, this volume). Some may have functioned in two or more of the above ways. What unites all of these hypotheses is the four-fold way of categorising space-time which goes far beyond indicating horizon phenomena and counting days but can be shown to penetrate to certain aspects of the way these societies functioned. To borrow Brotherston's (1988) characterisation of page 1 of Codex Mendoza—essentially the Aztecs' map of their capital city—here was a 'vocabulary of ritual' showing a 'ceremonial landscape that reflected relations of political subordination. In a single space this map organises history, foundation mythology, space, time, material economy, and social (tribute) relations according to ritual principles.'

In spite of all these reassessments, mergers of different approaches, and so on, I am obliged to report that number one on the archaeoastronomy 'Top 40 of Mesoamerica' remains the controversial serpent hierophany at Chichen Itza. The number of phone calls to my office still increases dramatically around equinox time. They are usually from Club Med tour leaders wanting to know for just how long the phenomenon is 'valid' and where one best ought to stand to watch it happen. The more knowledgeable guide tells me he cannot get hold of the seminal five-page paper by Rivard (1970) describing the event, and so could I please send a copy? The hardening process that comes with age has led me to accept my hypocritical role, on the one hand as a purveyor of high, medium and low-grade archaeoastronomy to the public,[6] and, on the other, conservative critic who also unleashes his scepticism on scholarly colleagues who dare propose ideas that seem the least bit out of the mainstream. The difference between archaeoastronomers and people is that the latter are very impressed with something that simply 'works'!

South America

One measure of the sound health of a discipline is the depth of attention given to its scholarly output. Just as Ruggles and others have reassessed the Thoms' data and even re-acquired on-site data of their own, so too have other investigators critically re-examined the theories and methods

---

[6] I plead guilty to having cruised the Caribbean on more than one occasion with hosts of pilgrims on their way to the Temple of Kukulcan who, unlike the early European crusaders, took up their mission in comfortable surroundings, except for the bus ride from Playa del Carmen to the ruins.

associated with some of the earlier published New World archaeoastronomical fieldwork. Nowhere has such an effort been made with the thoroughness that we have seen in the Andean sphere. Here attentions have focused on the calendar and astronomy of the Inca, and in particular upon the alleged alignments at Cuzco.

Dearborn and Schreiber's (1989) re-examination of Zuidema's and my work has helped to bring to the surface some of the unsettled issues about the role of astronomy in Incaic Cuzco but it has also provided logical connections to some of their work at Machu Picchu. For example, they pronounce our June solstice alignments better documented than those of the December solstice, and the zenith-antizenith alignment emerges as interesting, though not fully demonstrated physically. In the scientific tradition of logical positivism, their method is piecemeal. It deals with one alignment at a time, accepting or rejecting it before passing on to the next. Their criteria for acceptability rely heavily on whether the alignment works with precision. The June solstice alignment at the Torreón of Machu Picchu (Dearborn and White 1982) certainly does work, provided one arbitrarily places a wood-framed device on the window pegs. This seems not to bother Dearborn and White at all, and they have been very patient hearing me tell them on numerous occasions why it bothers me. To see just how well it works and how much of an impression the effect can leave, see Carlson (1990). Witnessing this phenomenon is not unlike seeing the sun rise over the Heel Stone at Stonehenge, glimpsing the dagger of light at Fajada Butte, or following the descending serpent at Chichen Itza. Still, I would not trade any number of mechanical devices that make something 'work' for the words of demonstrably reliable chroniclers—words which suggest to us that the alignments in the Inca capital were very real and part of a complex social system. There are no such words, given what little is known of the history of Machu Picchu, to support the hypothesis that astronomical alignments existed there.

Scientific method relies heavily upon what Toulmin and Goodfield (1964: 237) have called the 'testimony of things'—material or physical evidence—while the social sciences make use of other kinds of evidence as well as methods that might seem foreign to the scientist.

The lesson to be learned from reading the recent literature on Native American archaeoastronomy is that as long as our interdiscipline is about both people and things it must reckon with more than one approach.[7] The root of our differences, then, as both scientists and social

---

[7] At a recent lecture I delivered to an audience of physicists at a large American university I was asked what I thought of Lévi-Strauss and his structuralism. Before I had a chance to answer, my questioner gave his own opinion

scientists, resides in the problem of what determines the credibility of archaeoastronomical arguments and this remains, as it was at Oxford 2, an issue that cuts across our interdiscipline.

Another critique (Ziołkowski and Sadowski 1989a; 1989b) leans heavily upon the calendrical interpretations of Zuidema, though it also questions the positions of the *sucanca* or sun pillars on the Cuzco horizon that he has posited. They see the calendar of the established Inca state as consisting of twelve synodic months tied to a tropical year with the count beginning at one of four different moments of the cycle (Ziołkowski and Sadowski 1989a: 167). On the other hand Zuidema, relying on Cobo's description of the *ceque* system, thinks it a more complicated affair consisting of sidereal months being counted in one part of the year, synodic in the other. Are his critics too wedded to the primacy of the solar year when they complain (Ziołkowski and Sadowski 1989b: 207) about the 'empty time' of 37 days that would be needed to fill out the native calendar of 328 days to make a full solar cycle of 365 days? The Roman, Trobriand and Maya calendars all once operated on temporal baselines that fell short of a tropical year. That the whole problem of Inca calendrics remains open is verified by the content of the papers at this conference (see Ziołkowski and Lebeuf, this volume). While Zuidema's suggestions about where the pillars were placed has been altered and while archaeological evidence is still lacking, his primary concept remains the same, namely that they were used to demarcate the planting season in a luni-sidereal-solar calendar based on an ethnohistorical description of the *ceque* system. The issue of Cuzco's sun pillars is not yet settled.

To judge by the number of books published on the subject (e.g. Morrison 1987; Hadingham 1987), the Nazca pampa has been as active a resource for current developments in archaeoastronomy as Cuzco. A recent book edited by this author (Aveni 1990) incorporates the work of a number of investigators from diverse disciplines. It tries to dispel several myths about the lines, some of which have been responsible for attracting undeserved public attention. Our research group finds that the lines were not difficult to make either in terms of workload or mathematical and technological skills, nor do we believe any blueprint or written map was necessary. There is indeed a system that incorporates

---

(paraphrasing): 'Whoever heard of numbering and classifying myths and using algebraic equations to interrelate them and discover their meaning? It's all a lot of (expletive). You can't do that with myths!' How was he so certain, I wondered, about how to deal with a body of data about which he knew nothing? I fear that there are more scientists who have not taken the trouble to learn a little anthropological theory than the number of anthropologists they have accused of failing to understand scientific method.

all straight Nazca lines: like the *ceque* system of Cuzco it consists of (interlocked) radial patterns that emanate from hydrologically significant points on the pampa. However, again like the *ceque* system, these may also have incorporated astronomical alignments at least in part. Basically, they were rather more intended to have been walked upon than looked along.

This multi-faceted explanation of the Nazca lines follows a familiar theme that by now ought to appear well embedded in the present status report. That theme was echoed in our discussions of the pecked cross petroglyphs of Mesoamerica as well as in the studies of rock art of the American South-West: the situation is far more complicated than we thought it was a few years ago. One hypothesis simply does not fit all the data and many hypotheses cannot be individually rejected. Hopefully our work on the pampa, coupled with our knowledge of pan-Andean systems of thought gleaned through reading the works of Zuidema (1982), Urton (1982), and Reinhard (1987) will have taught us, whether we examine Nazca lines or Teotihuacan petroglyphs, to go beyond framing all-inclusive hypotheses that lump together the physical evidence—the testimony of things—and human behavioural characteristics.[8]

I do not want to leave the rest of the Americas out of this report, for just as we have begun to learn at this conference that much archaeoastronomical work is going on all across Eastern Europe and Asia, so too I am obliged to report that our interdisciplinary tentacles are spreading to new places across the North and South American continent. It is unfortunate that many of our colleagues who live in the third world countries of Latin America were unable to travel here to report on their exciting activities. I want to mention specifically the work of Robiou-Lamarche (1988) in the Caribbean Antilles, Puerto Rico and the Dominican Republic, where he has published plans of pre-contact ceremonial plazas with astronomical orientations. Also the work of Magaña and his colleagues on astral myths in Surinam (e.g. Magaña 1987), and Reichel (1987) in north-east Amazonia. Though these works are traditionally classified as ethnoastronomy, Reichel and Arias (1987), Reichel (1987) and Wilbert (1981), in particular, have helped define the continuum of ethno- and archaeoastronomy by dealing with manifestations of *cosmovisión* in communal architecture.

---

[8] For example, Hawkins (1969: 27) stated: 'no matter what hypothesis is finally adopted, we must expect an almost total explanation for the lines; otherwise we have the unsatisfactory situation of explaining only part of the construction and leaving the "why?" for the remainder of the lines unanswered.'

Conclusion

In sum, archaeoastronomy in the Americas is not so much in a state of confusion as it is in a phase of hyperactivity. There is maturity in the level of argumentation about phenomena once reported as ends in themselves. While the media still distort and exaggerate our discoveries, the problem is to some extent our own for choosing the subject matter we deal with. Maya studies have emerged as the glamour discipline of the 90s and the enigmatic Nazca pampa will always attract attention. As professionals, our task should consist of thinking about what archaeoastronomy ought to plug into. To whom, to what groups of scholars, does our work have value? Exactly what does it illuminate in their fields of concern?

Conferences such as the present one are beneficial, even seminal, for we all deal with one another as peers. We are all insiders in the sense that we chose, or were magnetically attracted to, this curious interdisciplinary crack between the disciplinary floorboards that make up the house of human knowledge. But once we have talked among ourselves we must go out and do our own wiring, so to speak—to fit our own electrical outlets if we would seek to illuminate the interior of our shared domicile. I would urge publication of archaeoastronomical results in the disciplinary journals. Not only would ideas about the practice of astronomy in the ancient world gain wider readership but also archaeoastronomers would receive the dispassionate criticism from which they all can surely benefit as serious professionals. Why should we of this compound discipline demand less of ourselves than the biochemist or neurophysiologist who have walked the halls of interdisciplinary academia no longer than we? At least on the American side, I am not satisfied that the archaeologists have fully assessed our work.[9]

I believe that those of us whose tap roots grow in the scientific community should make more effort to read papers at non-scientific meetings. Though I respect my fellow astronomical colleagues I have derived more benefit, in terms of concrete feedback concerning things I did not know and would profit from knowing, when I describe my work to an audience of anthropologists or historians of religion than when I speak to astronomers, physicists, or even historians of science, many of whom are not even aware of the existence of other societies, much less other modes of thought and categories of explanation, than

---

[9]   For example, Hammond (1982: 294) states of the Maya that 'even if they lacked formal observatories, or had no structures more sophisticated than the Uaxactun grouping, [they] nevertheless used some kind of hand-held instrument for noting the precise positions and movements of heavenly bodies'.

the western European. While there is much to be gained in helping to educate our fellow scientists, I believe full integration of our interdiscipline will be gained only by educating ourselves through deeper and more prolonged contact with those in whose domains the fruits of archaeoastronomy can really matter. This means that we must learn their issues and deal with them. Since Jim Judge once quoted me, I take this as an opportunity to return the compliment with this most appropriate closure:

> When astronomers deal more with the cultural context of sites in an area and when archaeologists deal more with the ritual aspects of the same sites, we will begin to see progress in the maturation of archaeoastronomy as a truly integrated discipline.
>
> Judge (1987: 5).

Acknowledgements

I thank C. Farrer and H. Hartung for commentary on earlier drafts of this work.

References

Aveni, A.F., ed. (1982). *Archaeoastronomy in the New World*. Cambridge: Cambridge University Press.

Aveni, A.F. (1986). Archaeoastronomy: past, present, and future. *Sky and Telescope* **72**, 456–60.

Aveni, A.F. (1988). The Thom paradigm in the Americas: the case of the cross-circle designs. In *Records in Stone*, ed. C.L.N. Ruggles, pp. 442–72. Cambridge: Cambridge University Press.

Aveni, A.F., ed. (1989). *World archaeoastronomy*. Cambridge: Cambridge University Press.

Aveni, A.F., ed. (1990). *The lines of Nazca*. Philadelphia PA: American Philosophical Society.

Aveni, A.F. and Hartung, H. (1986). Maya city planning and the calendar. *Transactions of the American Philosophical Society* **76**, 1–81.

Aveni, A.F. and Hartung, H. (1988). Archaeoastronomy and dynastic history at Tikal. In *New directions in American archaeoastronomy*, ed. A.F. Aveni, pp. 1–16. Oxford: British Archaeological Reports (BAR International Series 454).

Aveni, A.F., Hartung, H. and Buckingham, B. (1978). The pecked cross symbol in ancient Mesoamerica. *Science* **202**, 267–79.

Benson, A. and Hoskinson, T., eds. (1985). *Papers from the Northridge conference on Archaeoastronomy*. Thousands Oaks CA: Slo'w Press.

Bricker, V.R. and Bricker, H.M. (1989). *Archaeoastronomy* no. 12 (supplement to *Journal for the History of Astronomy* **19**), S1–62.

Broda, J. (1982). Astronomy, cosmovisión and ideology in Prehispanic Mesoamerica. In *Ethnoastronomy and archaeoastronomy in the American Tropics*, eds. A.F. Aveni and G. Urton, pp. 81–110. New York NY: Annals of the New York Academy of Sciences 385.

Broda, J. (1993). Astronomical knowledge, calendrics and sacred geography in ancient Mesoamerica. In *Astronomies and cultures: selected papers from the third 'Oxford' international conference on archaeoastronomy*, eds. C.L.N. Ruggles and N.J. Saunders. Niwot CO: University Press of Colorado. In press.

Brotherston, G. (1988). *Cosmos* **4**, 355–56.

Carlson, J.B. (1990). America's ancient skywatchers. *National Geographic* **177**(3), 76–107.

Carlson, J.B. (1993). Venus and astrologically-timed ritual warfare in Mesoamerica. In *Astronomies and cultures: selected papers from the third 'Oxford' international conference on archaeoastronomy*, eds. C.L.N. Ruggles and N.J. Saunders. Niwot CO: University Press of Colorado. In press.

Carrasco, D. (1989). The king, the capital and the stars: the symbolism of authority in Aztec religion. In *World archaeoastronomy*, ed. A.F. Aveni, pp. 45–54. Cambridge: Cambridge University Press.

Carrasco, D., ed. (1991). *To change place: Aztec ceremonial landscapes*. Niwot CO: University Press of Colorado.

Chamberlain, von Del (1984). Astronomical content of North American Plains Indian calendars. *Archaeoastronomy* no. 6 (supplement to *Journal for the History of Astronomy* **15**), S1-54.

Chiu, B.C. and Morrison, P. (1980). Astronomical origin of the offset street grid at Teotihuacan. *Archaeoastronomy* no. 2 (supplement to *Journal for the History of Astronomy* **11**), S55–64

Coggins, C. and Drucker, R.D. (1988). The observatory at Dzibilchaltun. In *New directions in American archaeoastronomy*, ed. A.F. Aveni, pp. 17-56. Oxford: British Archaeological Reports (BAR International Series 454).

Dearborn, D. and White, R. (1982). Archaeoastronomy at Machu Picchu. In *Ethnoastronomy and archaeoastronomy in the American Tropics*, eds. A.F. Aveni and G. Urton, pp. 249–60. New York NY: Annals of the New York Academy of Sciences 385.

Dearborn, D. and Schreiber, K. (1989). Houses of the rising sun. In *Time and calendar in the Inca empire*, eds. M. Ziołkowski and R. Sadowski, pp. 49–74. Oxford: British Archaeological Reports (BAR International Series 479).

Dow, J. (1967). Astronomical orientations at Teotihuacán: a case study in astroarchaeology. *American Antiquity* **32**, 326–34.

Eisenberg, D., Lord, D., Markman, P., Markman, R., Merrill, R., Scribner, M. and White, C., eds. (1990). *Transformations of myth through time*. New York NY: Harcourt Brace.

Farrer, C.R. and Williamson, R.A. (1991). Epilogue: blue archaeoastronomy. In *Earth and sky: visions of the cosmos in native American folklore*, ed. R.A. Williamson and C.R. Farrer. Albuquerque NM: University of New Mexico Press.

Hadingham, E. (1987). *Lines to the Mountain Gods*. London: Harrap.

Hammond, N. (1982). *Ancient Maya civilization*. New Brunswick NJ: Rutgers.

Hawkins, G.S. (1969). *Ancient lines in the Peruvian desert*. Cambridge MA: Smithsonian Institution Astrophysical Observatory Special Report no. 906-4.

Heggie, D.C., ed. (1982). *Archaeoastronomy in the Old World*. Cambridge: Cambridge University Press.

Heggie, D.C. (1990). Review of 'World Archaeoastronomy'. *The Observatory* **110**, 1095.

Judge, W.J. (1987) Archaeology and astronomy: a view from the Southwest. In *Astronomy and ceremony in the prehistoric Southwest*, ed. J.B. Carlson and W.J. Judge, pp. 1–8. Albuquerque NM: Maxwell Museum of Anthropology (Papers, 2).

Krupp, E.C. (1989). Rayed disks. *Griffith Observer* **53**, 2–19.

Larios, R. and Fash, W. (1989). Architectural history and political symbolism of Temple 22, Copan. Paper presented at the VIIth Mesa Redonda de Palenque, Palenque, Mexico, June 1989.

Leaver, J.P. (1989). Review of 'World Archaeoastronomy'. *Journal of the British Astronomical Association* **99**(4), 190.

Magaña, E. (1987). Astronomía y mitos estelares de los Indios de las Guayanas. In *Etnoastronomías Americanas*, eds. J. Arias and E. Reichel de von Hildebrand, pp. 45-56. Bogotá: Centro Editorial Universidad Nacional de Colombia (45º Congreso de Americanistas, Universidad de los Andes).

Milbrath, S. (1988). Astronomical images in orientations in the architecture of Chichén Itzá. In *New directions in American archaeoastronomy*, ed. A.F. Aveni, pp. 57–79. Oxford: British Archaeological Reports (BAR International Series 454).

Morrison, T. (1987). *The mystery of the Nazca lines.* Woodbridge, Suffolk: Nonesuch Expeditions Ltd.

Peterson, C.W. and Chiu, B.C. (1987). On the astronomical origin of the offset street grid at Teotihuacan. *Archaeoastronomy* no. 11 (supplement to *Journal for the History of Astronomy* **18**), S13–18.

Reichel de von Hildebrand, E. (1987). Astronomía Yukana-Matapí. In *Etnoastronomías Americanas*, eds. J. Arias and E. Reichel de von Hildebrand, pp. 193–232. Bogotá: Centro Editorial Universidad Nacional de Colombia (45° Congreso de Americanistas, Universidad de los Andes).

Reichel de von Hildebrand, E. and Arias, J. (1987). Prefacio. In *Etnoastronomías Americanas*, eds. J. Arias and E. Reichel de von Hildebrand, pp. 7–18. Bogotá: Centro Editorial Universidad Nacional de Colombia (45° Congreso de Americanistas, Universidad de los Andes).

Reinhard, J. (1987). *The Nazca lines: a new perspective on their origins and meaning.* Lima: Editorial Los Pinos.

Reyman, J. (1987). Priests, power and politics: some implications of socio-ceremonial control. In *Astronomy and ceremony in the prehistoric Southwest*, ed. J.B. Carlson and W.J. Judge, pp. 121–47. Albuquerque NM: Maxwell Museum of Anthropology (Papers, 2).

Rivard, J. A hierophany at Chichén Itzá. *Katunob* **7(3)**, 51–55.

Robiou-Lamarche, S. (1988). Astronomía primitiva entre los Tainos y los Caribes de las Antillas. In *New directions in American archaeoastronomy*, ed. A.F. Aveni, pp. 121–41. Oxford: British Archaeological Reports (BAR International Series 454).

Saunders, N.J. (1989). The ancient sky at night. *New Scientist* **123**: 57-58.

Šprajc, I. (1989). *Venus-Maize-Rain complex in Mesoamerican world view.* Unpublished PhD thesis, Universidad Nacional Autónoma de México.

Toulmin, S. and Goodfield, J. (1964). *The discovery of time.* Chicago IL: University of Chicago Press.

Urton, G. (1982). *At the crossroads of the earth and the sky.* Austin TX: University of Texas Press.

Wallrath, M. (1984). Paper presented at the conference on archaeo- and ethnoastronomy, Mexico City.

Wilbert, J. (1981). Warao cosmology and the Yecuana roundhouse. *Journal of Latin American Lore* **7**(1), 37–72.

Wilcox, D. (1987). The evolution of Hohokam ceremonial systems. In *Astronomy and ceremony in the prehistoric Southwest*, ed. J.B. Carlson and W.J. Judge, pp. 149–67. Albuquerque NM: Maxwell Museum of Anthropology (Papers, 2).

Williamson, R.A. (1987). Light and shadow, ritual, and astronomy in Anasazi structures. In *Astronomy and ceremony in the prehistoric Southwest*, ed. J.B. Carlson and W.J. Judge, pp. 99–120. Albuquerque NM: Maxwell Museum of Anthropology (Papers, 2).

Young, M.J. (1987a). The interrelationship of rock art and astronomical practice in the American Southwest. *Archaeoastronomy* no. 10 (supplement to *Journal for the History of Astronomy* **17**), S43–58.

Young, M.J. (1987b). The nature of the evidence: archaeoastronomy in the prehistoric southwest. In *Astronomy and ceremony in the prehistoric Southwest*, ed. J.B. Carlson and W.J. Judge, pp. 169–89. Albuquerque NM: Maxwell Museum of Anthropology (Papers, 2).

Zeilik, M. (1985a). The ethnoastronomy of the historic Pueblos, 1: Calendrical sun watching. *Archaeoastronomy* no. 8 (supplement to *Journal for the History of Astronomy* **16**), S1–24.

Zeilik, M. (1985b). A reassessment of the Fajada Butte solar marker. *Archaeoastronomy* no. 9 (supplement to *Journal for the History of Astronomy* **16**), S69–85.

Zeilik, M. (1985c). Sun shrines and sun symbols in the U.S. Southwest. *Archaeoastronomy* no. 9 (supplement to *Journal for the History of Astronomy* **16**), S86–96.

Zeilik, M. (1986). The ethnoastronomy of the historic Pueblos, 2: Moon watching. *Archaeoastronomy* no. 10 (supplement to *Journal for the History of Astronomy* **17**), S1–22.

Zeilik, M. (1987) Anticipation in ceremony: the readiness is all. In *Astronomy and ceremony in the prehistoric Southwest*, ed. J.B. Carlson and W.J. Judge, pp. 25–41. Albuquerque NM: Maxwell Museum of Anthropology (Papers, 2).

Ziołkowski, M. and Sadowski, R. (1989a). The reconstruction of the metropolitan calendars of the Incas in the period 1500–1572 AD. In *Time and calendars in the Inca empire*, eds. M. Ziołkowski and R. Sadowski, pp. 167–96. Oxford: British Archaeological Reports (BAR International Series 479).

Ziołkowski, M. and Sadowski, R. (1989b). Knots and kinks. The quipu-calendar or supposed Cuzco luni-sidereal calendar. In *Time and calendars in the Inca empire*, eds. M. Ziołkowski and R. Sadowski, pp. 197–213. Oxford: British Archaeological Reports (BAR International Series 479).

Zuidema, R.T. (1982). Catachillay: the role of the Pleiades and of the Southern Cross and Alpha and Beta Centauri in the calendar of the Incas. In *Ethnoastronomy and archaeoastronomy in the American Tropics*, eds. A.F. Aveni and G. Urton, pp. 203–29. New York NY: Annals of the New York Academy of Sciences 385.

3

Space, time and the calendar in the traditional cultures of America

STEPHEN C. MCCLUSKEY

This paper is dedicated to the memory of Alexander Stephen and Alexander Thom, two Scots who enriched our understanding of traditional astronomies.

Introduction

In this paper I intend to illustrate some of the relationships between astronomical knowledge and calendrical systems in 'traditional' or 'preliterate' or 'non-state' cultures, taking my examples from the native peoples of Mexico and the south-western United States and focusing especially on the Hopi.

In looking at a number of traditional societies I have been struck by the fact that the astronomies of such societies tend to share the same primary function; they use the regular motions of the heavens to reckon the passing of time and the changes of the seasons. This should not be surprising, for as Travis Hudson (1984) has pointed out, this concern with marking the changes of the seasons seems a universal human activity; even migrating hunters and gathers must know in detail the appearance of wild plants and animals with the seasons.

Of course, keeping a calendar does not require astronomical knowledge; the behaviour of animals and the migration of birds can also be taken as signs of the coming appearance of certain wild plants or of the proper time to begin planting crops. However, the orderly, cyclical recurrence of astronomical phenomena are the most reliable indicators of the passing of the seasons; watching both the heavens and animals for signs of the seasons is documented in cultures as different as Archaic Greece (Hesiod *Works and days* 383–90, 564–73), Islamic Spain (Pellat 1961), and the present-day Maya (B. Tedlock 1992: 185–90) and Hopi (Malotki 1983: 395–405). It is scarcely an exaggeration to say that astronomy, and botany and zoology for that matter, are natural human activities.

This keeping of time differs in one important respect from timekeeping as undertaken by modern astronomers. Traditional timekeeping does not deal with the measurement of an unending flow of undifferentiated moments of time, but rather with the demarcation of special moments and days and seasons which return in a recurring cycle. Yet such noting of special days is not a characteristic of 'primitive' societies; it is more like what we do when we mark special appointments on our office calendars.[1] Astronomical phenomena provide ideal markers to punctuate the regular flow of time, noting the arrival of a special hour, marking the return of important days and delimiting the passing of the seasons.

As a historian of science looking at these astronomies I tend to focus on certain sets of problems that they raise. First, historians are interested in human activities not as static unchanging practices, but as they take place in time. A historian wants to understand the temporal dimension of astronomies, both the changes they underwent over time and, conversely, those institutions that preserved astronomies from change as they were transmitted from generation to generation. This becomes especially significant when the astronomies are relatively static while the society itself is undergoing drastic changes. Secondly, while historians of science share the concern of other students of 'traditional' astronomies to work out how these astronomies functioned at specific times and places in other cultures, we tend to see these astronomies as examples of science.

Looking at these astronomies as sciences, the most important elements for our study are the broad overarching explanatory frameworks—one sense of what Thomas Kuhn called paradigms—that 'traditional' astronomies generate and employ in order to encompass economically and make intelligible in memorable fashion the relationships between certain significant observations.

'Traditional' astronomies, like all astronomies, *must* successfully transmit their paradigms from generation to generation. Just as we are taught in secondary school that the earth is a sphere that orbits around the sun, so too is a youngster in a traditional society taught the elements of his or her astronomical paradigm. Thus we must consider the methods (such as hearing the community's myths, participating in its rituals, working with learned elders, or undergoing more formal training) by

---

[1] I differ on this point from Nilsson (1920: esp. 8–10, 355–58), who considered such 'discontinuous time-indications' to be primitive antecedents of a 'genuine system...of time-reckoning'.

which the young members of the community come to master astronomical knowledge and practice.

It is in this matter of transmission that I find the greatest difference between 'traditional' and modern science. In the absence of writing, knowledge is a fragile thing; thus in traditional sciences we find a greater emphasis on mechanisms preserving what is known and for transmitting it from generation to generation and less on mechanisms for developing new knowledge that might challenge what is already known.

One mechanism for preserving what is known is that the most general elements of 'traditional' astronomies are linked in many ways to other elements of society. The most crucial is the close involvement of 'traditional' astronomies with religion: their astronomical concepts are expressed in sacred terms; these astronomies are learned and practised as part of ritual. Modern astronomy only touches these aspects of human life tangentially. The greater interconnectedness of 'traditional' astronomies no doubt contributes to their conservatism; changes in astronomy would call forth changes in other areas. Yet this interconnectedness only sharpens the question as to how 'traditional' astronomies endure through major changes in their cultural context.

The pan-American astronomical paradigm

The most thoroughly studied 'traditional' astronomies are those of the native peoples of the Americas. These astronomies display the general characteristics of 'traditional' science: conservatism and resistance to change, close interconnectedness of astronomy with religion and society, and a strong emphasis on such practical concerns as establishing a ritual or economic calendar. Most native American astronomies also share two common cosmological themes: the four-directionality of space (whether defined by sunrise and sunset at the solstices or defined in terms of the four cardinal directions) and a duality separating summer and winter, the above and below, and by extension that which is (the present) and that which is to come (the future).

Here, I will offer a few brief comments about the above and below, which are more strongly connected to cosmological *abstractions* about time and space, before turning to the four directions, which are closely tied to empirical *observations*.

For the Hopi, as for many other peoples, the below is the place where the sun returns from west to east during the night; hence when it is night in the above it is day in the below, and when it is winter in the above it is summer in the below. In the below are made the seeds of all beings—plants, animals, and men—that live in the above; hence the crops that are growing now in the below will grow six months later in the above (Stephen 1894a: 113; 1894b).

Similarly, the Hopi saw this world in which we live as the fourth of a succession of worlds, into which their ancestors had emerged by climbing on a reed that they had planted to grow from a third world in the below.[2] In most Hopi accounts the above and below are not hierarchical; there is a simple duality in which things come forth from potentiality in the below into actuality in the above. This is sometimes extended to include a third layer in the sky.

This seems to contrast with Mesoamerican cosmologies, in which the above and below are both arranged in hierarchically ordered layers. To hazard a speculation, it seems that the more complex hierarchical ordering of Mesoamerican cosmologies reflects the more complex hierarchical ordering of their societies.

We begin our consideration of the four directions with a discussion of the number four by a Lakota wise man, Tyon, recorded early in this century (see Tedlock and Tedlock 1975: 215–16):

> In former times the Lakota grouped all their activities by fours. This was because they recognized four directions: the west, the north, the east, and the south; four divisions of time: the day, the night, the moon, and the year; four parts in everything that grows from the ground: the roots, the stem, the leaves, and the fruit; four kinds of things that breathe: those that crawl, those that fly, those that walk on four legs, and those that walk on two legs; four things above the world: the sun, the moon, the sky, and the stars; four kinds of gods: the great, the associates of the great, the gods below them, and the spiritkind; four periods of human life: babyhood, childhood, adulthood, and old age; and finally, mankind has four fingers on each hand, four toes on each foot, and the thumbs and great toes taken together form four. Since the Great Spirit caused everything to be in fours, mankind should do everything possible in fours.

This emphasis upon the fourness of things is the most common element of Native American thought. In myth and ritual, significant actions are commonly repeated four times to insure their efficacy; ceremonies are commonly four days or four nights long; animals, plants, and other natural beings are commonly grouped in fours.

---

[2] For various versions of this emergence myth see Courlander 1987: 17–33; James 1974: 2–8; Nequatewa 1967: 7–26. Compare the strikingly different version of Oswald White Bear Fredericks in Waters 1963: 1–28.

The emphasis on fourness is not limited to North America or to non-literate Native Americans. Things also appear in fours amongst the Aztec and Maya peoples of Mesoamerica. The Quiché Maya book, the *Popol Vuh*, begins by telling of the origins of the world, of

> the emergence of all the sky-earth:
> the fourfold siding, fourfold cornering
> measuring, fourfold staking,
> halving the cord, stretching the cord
> in the sky, on the earth
> the four sides, the four corners (D. Tedlock 1985: 72).

Here the poet seems to relate both the cardinal directions and the inter-cardinal (solstitial?) directions as four sides and four corners. The directions also come first, before the beginning of the world and, presumably, before the beginning of time. It is only later that the first people were formed to inhabit the fourth of a series of four worlds: Jaguar Quitze, Jaguar Night, Mahucutah, and True Jaguar.

In his fieldwork in the Peruvian village of Misminay, Urton (1981: 27–29, 41–65) noted that the villagers divided the terrestrial space around them and the celestial region above them into four quarters. In these traditional Peruvian villages the quadripartition of space is coupled with a quadripartition of the year by four *unequally spaced* Saint's days, days which are *not* determined by astronomical observations.

Such variations on the theme of the four directions suggest local variations on an underlying Pan-American cosmological paradigm. Note that Tyon and the author of *Popol Vuh* put the four directions at the beginning of their discussions of the importance of the number four and of the origins of the world. This does not seem to be an accident; I take these native accounts at face value when they make the division of space into four directions the principal source of the Native American emphasis upon fourness. In turn, this emphasis upon fourness, and the association of each of the four directions with a colour and with other beings sharing analogous fourfold classifications contributes to the preservation of the astronomical framework founded upon the four directions. Much of Native American astronomy may then be considered as different articulations of the Native American paradigm of the four directions, just as the astronomies of Ptolemy and Copernicus are different articulations of the Greek paradigm of uniform circular motion.

The Hopi and the four directions

To illustrate how this paradigm of the four directions functions in traditional astronomies, consider the specific case of the Hopi Indians of

Northern Arizona. Hopi direction concepts were something of a problem for early ethnographers. Fewkes (1892a) first noted that the Hopi orientation was skewed away from the cardinal directions and suggested that the Hopi orientation was constrained by the local topography. Years of observation suddenly dropped into place for Alexander Stephen (an Edinburgh man) as he contemplated the shrine where the Hopi had deposited prayer sticks to the sun. He wrote of his discovery in a widely quoted letter to Fewkes:

> The Hopi orientation bears no relation to North and South, but to the points on his horizon which mark the places of sunrise and sunset at the summer and winter solstices. He invariably begins his ceremonial circuit by pointing (1) to the place of sunset at the summer solstice, next to (2) the place of sunset at winter solstice, then to (3) the place of sunrise at winter solstice, and (4) the place of sunrise at summer solstice...

Stephen went on to tell Fewkes how this discovery flashed upon him in a sudden moment of insight and how the old men of the village 'one and all declared how glad they were that now I understood, [and] how sorry they had long been that I could not understand this simple fact before.' (Stephen 1893).

To the historian of science, Stephen's description of this sudden understanding of what was previously incomprehensible sounds very much like the kind of gestalt shift in perception that, according to Kuhn (1970: 111–15), accompanies the acceptance of a new scientific paradigm. Before this new insight, Stephen and his Hopi neighbours were not speaking the same language—even when they all spoke Hopi. It was only when Stephen came to comprehend the four Hopi cardinal directions in terms of the setting and rising points of the sun at the solstices that they were able to communicate.

The Hopi directions differ from the four directions known to the European tradition in several important aspects. The fundamental difference is that they do not represent geometrical idealisations of space; rather, they are empirically observed places. Since they are not abstractions, they do not provide the neat orthogonal co-ordinates with which we are familiar, but meet at an angle defined by the rising and setting of the sun on the local horizon. As distinct places they do not provide a basis for the continuous measurement of position or time; rather, they identify special places in space and the correlative special days when the sun rises or sets at those places. Finally, as physically real places they are not devoid of the 'sensible' qualities of everyday experience and have been associated with other characteristics that organise disparate elements of knowledge and ritual.

The most commonly known of these associations links each direction with a specific colour. *Kwiniwi*, the direction of summer solstice sunset, is associated with the colour yellow; *tévyuna*, the direction of winter solstice sunset, with blue or green (which, in this context, the Hopi do not distinguish); *tátyuka*, the direction of winter solstice sunrise, with red; and *hópoko*, the direction of summer solstice sunrise, with white. In Hopi art, ritual and mythology, we find repeated allusions that go beyond colours to encompass birds, plants, animals, and minerals—all organised in a taxonomy founded upon the ordering principle of the four horizontal directions, plus the above and below.[3] At first glance these proliferating symbolic associations seem superfluous growths on what could be a simple rational observational framework. Yet these associations provide a rich mnemonic device, one repeated throughout the annual ritual cycle to insure the preservation and transmission of this vital astronomical framework.

This framework is presented with striking power during the ceremony in which young boys and girls are initiated into some of the secrets of Hopi ritual. On the fourth day of the ceremony the initiates witness a dance by the Chowilawu Kachina, in which he dances four times anti-clockwise (that is, following the annual path of the sun) around a sand mosaic, each side of which is marked by the colour of the appropriate direction.

As Chowilawu dances, he displays a circular book-like object, containing six circular pages, each bearing a design which relates astronomy to the growth of crops. First is a red blossom, representing the direction of winter solstice sunrise, followed by a multi-coloured blossom flowering in the below, in the well-spring of the future. Next is a white blossom, representing the direction of summer solstice sunrise, the end of corn planting, followed by a picture of sprouting shoots of green. Last is a symbol for the moon, followed by a picture of rain clouds over growing ears of corn. The month after the solstice is the time when the first ears of sweetcorn are harvested for the Niman festival.

A Hopi child could no more be ignorant of the overall concept of the four directions than a modern child could be unaware of the notion that the Earth is a sphere. And like all theoretical frameworks it (i) summarises a wide range of observations in an economical form, (ii) makes the Hopis' observations of the sun intelligible, (iii) helps them anticipate future observations, (iv) provides a vocabulary by which they can discuss astronomical phenomena, and (v) provides a mnemonic symbolism to aid in the transmission of astronomical lore.

---

[3] Lévi-Strauss (1966: 41). The most complete listing is found in Hieb 1972 (84–89).

Markers at the four directions

For the Hopi, as for other practitioners of science, their astronomy must be grounded in observations of the natural world. The system of four directions provides the framework within which Hopi sun-watchers observe the sun's annual motion, watching the sun's set as he moves from his summer to winter house and his rise as he moves from his winter to summer house (McCluskey 1990a).

Speaking in general terms, at all four of the places where the sun pauses in his annual travels along the horizon there is said to be a sun's house, a *Tawaki*. However, only two—or in the most restricted sense, only one—of them is considered the true house of the sun. These two sun's houses at *kwiniwi* and *tátyuka* are marked by small shrines and play important roles in both religious ritual and astronomical observation.

The sun's house in the north-west is properly the house of *Huzruing wuhti*, 'hard-being woman', the deity to whom all hard substances—especially shells, coral, and turquoise—belong. The yellow of *kwiniwi*, where her house is located, complements the white shells, pink coral, and blue-green turquoise which, according to myth, decorate her house. At the summer solstice, when the sun is at *kwiniwi*, he stands directly over Huzruing Wuhti's house, then descends through a hatchway in the roof into her house (Stephen 1894a: 112–13).

Prayer sticks, or *pahos*, offered to the sun are traditionally deposited at a shrine marking this direction at the time of the summer solstice (Fewkes 1892b). Towards the north-west, towards *kwiniwi*, we find today two shrines located on a narrow promontory between First and Second Mesa called *Ponotuwi*. When I saw the shrines in July 1979 one of them contained prayer sticks deposited at the recent summer solstice, freshly coloured and with feathers, strings, and other perishable parts intact.

The sun's house proper is at *tátyuka*, at the south-east, where he comes out and stands directly above his house at the winter solstice. This direction is associated with the colour red, and so in myth it is described as red in colour, and the sun eats there from a red stone bowl (Stephen 1894a: 114). As at the summer solstice, prayer sticks for the sun are deposited at this shrine on the winter solstice (Parsons 1936: 53–60, 71–82). This sun shrine is on the top of a small rise on the promontory known as *Kwatipkya* or *Kwatipki* (Eagle point or Eagle's house). When I visited it in the summer of 1979, the shrine was clearly still being used. *Pahos* had recently been deposited there, and while the feathers and other ephemeral parts of them had deteriorated somewhat, the *pahos* themselves were only slightly faded.

This is not just Hopi ritual; the sun does, in fact, rise and set at these points near the solstices (Table 1). If we observe the two shrines at *Pono-*

Space, time and the calendar   41

Direction Marker	Azimuth	Distance	Declination	Longitude
Ponotuwi (North)	300°·10	7·2 km	23°·7	***
Ponotuwi (South)	299°·46	7·1 km	23°·2	97°·8
Luhavwu Chochomo	241°·83	127·4 km	-23°·0	258°·9
Kwatipkya	119°·03	10·4 km	-23°·2	278°·1
Kwitcala (centre)	64°·55	43–47 km	20°·2	60°·3

TABLE 1. Direction markers from Walpi Pueblo (sun bisected by local horizon). Since the sun arrives at the same declination at two different times in its annual path, once when it is moving northwards and again when it is moving towards the south, the appropriate solar longitude was selected using the Hopi convention that sunset observations are made after the summer solstice and sunrise observations after the winter solstice.

tuwi marking summer solstice sunset from the south tip of Walpi pueblo, the setting sun barely reaches the abandoned northern shrine, while it lingers for an extended period around the active shrine to the south. At the solstice itself the lower limb of the setting sun just touches this shrine; on each subsequent day it sets farther to the left until some twelve days later the upper limb touches the shrine just before the sun sets.

The shrine at the direction of winter solstice sunrise, the sun's house proper, relates to the rising sun in a way that corresponds exactly to the relation of the active sun shrine at *Ponotuwi* to the setting sun. At the solstice itself the *upper* limb of the rising sun just touches this shrine; each subsequent day it rises farther to the left until some twelve days later the *lower* limb just touches the shrine as the sun rises. Although their azimuths from Red Cape are not exactly 180° apart, this pair of shrines marks the corresponding solar declinations with a precision that exceeds the precision of our measurements of their positions. The solar longitudes, and hence the day of the solstice, correspond to within half a day.

Although these shrines mark the solstices precisely, there are two reasons why they could not be used for highly precise determination of the date of the solstice. First, they are too small to be seen from the village and second, the sun moves too slowly at the solstices for observations to determine precisely when it stops and changes direction. Astronomically these shrines confirm, rather than determine, the arrival of the sun at his house. This is consistent with their ritual function as shrines where offerings are deposited to the sun on the day he arrives at his house.

The dates of the solstices are determined by anticipatory observations of the sun's arrival at a pair of natural landmarks near the other two Hopi cardinal points that act as distant foresights. At *tévyuna*, the direction of winter solstice sunset, are the San Francisco peaks, mountains which the Hopi hold to be sacred and which can be seen throughout Hopi country. As is well known, the day of the coming winter solstice ceremony is fixed when the sun sets some eleven days before the solstice in a distant valley in the San Francisco peaks, *Luhavwu Chochomo*, some 127km distant (Parsons 1936: 442, n.1, 539–40).

In the opposite direction, *hópoko*, the direction of summer solstice sunrise, there is a series of minor irregularities on the distant horizon that mark the times to plant various crops in specific types of fields. Most striking of these is a narrow notch called *Kwitcala*, which marks the day to begin general corn planting, some thirty days before the summer solstice. This notch is formed by two precipitous cliffs at the entrance of an unnamed wooded canyon in the west side of Balakai Mesa. Since one of these cliffs is over 43km, and the other over 47km, distant from Walpi Pueblo, the notch is admirably suited for highly precise observations anticipating the solstice.

These anticipatory observations are necessary if the Hopi are to have adequate time to prepare materially for corn planting and spiritually for the solstice ceremonies. Technically, anticipatory observations of this sort indicate a Hopi concern with exact observations which are best made over a long baseline and while the sun is still moving perceptibly from day to day (Zeilik 1987).

In this complementary pattern of anticipatory observations before the solstices followed by observations on the day of the solstices, we can see elements that reflect different aspects of scientific practice. The observations in anticipation of the solstices are examples of precise applied science; they provide the exact calendar that the Hopi require in their harsh environment.

The observations at the solstices, however, meet no direct practical need. The shrines marking the sun's house are too small to be used for precise observations; even if the shrines could be seen, precise observations of the slowly moving sun would do little to fix the calendar. From the standpoint of timekeeping, the careful layout of these shrines seems superfluous; we must look beyond the calendar to justify this precision.

These shrines are more than just calendar markers; the sun's houses arc sacred places, the focus of ritual and sacred mythology. They define a sacred space set apart from ordinary space by a hierophany, the appearance of the divine sun, and in so doing specify the four Hopi directions, that fundamental paradigm of Hopi astronomy (cf. Eliade 1959: 20–76). Here we find a context in which these two sets of solstitial observations begin to make sense.

The precise anticipatory observations over distant natural landmarks establish the Hopi calendar and regulate sacred time, fixing the beginning of those sacred seasons during which the sun arrives at his houses. Observations of the sun's arrival at his houses establish a framework of astronomical co-ordinates and define sacred space, by marking the location of those sacred places from which he emerges at midwinter and into which he descends at midsummer.

The location of those sacred places is just as important to a Hopi's religious, cosmological, and intellectual framework as is the determination of the sacred times marking the turnings of the year. When the sun pauses precisely at his house, this validates the framework of the four directions which orders Hopi ritual and which frames the Hopi understanding of the sun's habitual motion.

To the Hopi, then, the four directions provide an empirically based model, combining elements of a scientific paradigm with elements of religious ritual—ritual whose dramatic impact assists the stable transmission of the Hopi articulation of a pan-American astronomical paradigm.

References

Courlander, H. (1987). *The fourth world of the Hopis*. Albuquerque NM: University of New Mexico Press.

Eliade, M. (1959). *The sacred and the profane*. New York NY: Harcourt, Brace and World.

Fewkes, J.W. (1982a). The ceremonial circuit among the village Indians of northeastern Arizona. *Journal of American Folklore* 5, 33–42.

Fewkes, J.W. (1892b). A few summer ceremonials at the Tusayan Pueblos. In *A journal of American ethnology and archaeology*, vol. 2, ed. J.W. Fewkes, pp. 24–33. Boston MA: Houghton Mifflin and Company.

Hieb, L.A. (1972). *The Hopi ritual clown: life as it should not be*. PhD dissertation, Princeton University. Ann Arbor MI: University Microfilms.

Hudson, T. (1984). California's first astronomers. In *Archaeoastronomy and the roots of science*, ed. E.C. Krupp, pp. 11-81. Boulder CO: Westview Press (AAAS Selected Symposium, 71).

Kuhn, T. (1970). *The structure of scientific revolutions*, 2nd. edn. Chicago IL: University of Chicago Press.

James, H.C. (1974). *Pages from Hopi history*. Tucson AZ: University of Arizona Press.

Lévi-Strauss, C. (1966). *The savage mind*. Chicago IL: University of Chicago Press.

McCluskey, S.C. (1990a). Calendars and symbolism: functions of observation in Hopi astronomy. *Archaeoastronomy* no. 15 (supplement to *Journal for the History of Astronomy* **21**), S1–16.

Malotki, E. (1983). *Hopi time*. New York NY: Mouton (Trends in Linguistics: Studies and Monographs, 20).

Nequatewa, E. (1967). *Truth of a Hopi*. Flagstaff AZ: Northland Press.

Nilsson, M.P. (1920). *Primitive time-reckoning: a study in the origins and first development of the art of counting time*. Lund: C.W.K. Gleerup.

Parsons, E.C., ed. (1936). *Hopi journal of Alexander M. Stephen*. New York NY: Columbia University Press.

Pellat, C. (1961). *Le calendrier de Cordoue, publié par R. Dozy, nouvelle édition*. Leiden: Brill.

Stephen, A.M. (1893). *Letter to [J.W.] Fewkes, Tewa, 29 Jun 1893*. Washington DC: Smithsonian Institution (National Anthropological Archives, box 4408(4)).

Stephen, A.M. (1894a). Legend of Tiyo, the snake hero. In *A journal of American ethnology and archaeology*, vol. 4, ed. J.W. Fewkes, pp. 106–19. Boston MA: Houghton Mifflin and Company.

Stephen, A.M. (1894b). *Letter to [J.W.] Fewkes, 14 Feb 1894*. Washington DC: Smithsonian Institution (National Anthropological Archives, file 4408(4)).

Tedlock, B. (1992). *Time and the Highland Maya*, revised edn. Albuquerque NM: University of New Mexico Press.

Tedlock, D., trans. (1985). *Popol Vuh*. New York NY: Simon and Schuster.

Tedlock, D. and Tedlock, B., eds., (1975). *Teachings from the American earth: Indian religion and philosophy*. New York NY: Liveright.

Urton, G. (1981). *At the crossroads of the earth and the sky*. Austin TX: University of Texas Press.

Waters, F. (1963). *Book of the Hopi*. New York NY: Ballantine Books.

Zeilik, M. (1987). Anticipation in ceremony: the readiness is all. In *Astronomy and ceremony in the prehistoric Southwest*, eds. J.B. Carlson and W.J. Judge, pp. 25–41. Albuquerque NM: Maxwell Museum Press (Papers of the Maxwell Museum of Anthropology, 2).

4

Some social correlates of directional symbolism

STANISŁAW IWANISZEWSKI

Introduction

Recent archaeoastronomical studies have revealed that astronomical activities are found in human societies almost universally. However, while we can assume that astronomical events are equally perceived by all humans, their cultural references demonstrate great diversity.

In general terms, any new cultural element (invented or borrowed) may be rejected or gradually incorporated to the pre-existing cultural matrix. While new elements are fitted into the matrix, the older ones have to be modified in order to accommodate them. The basic assumption here is that any cultural element related to astronomy is correlated with other specific cultural variables and that there are some universal tendencies in the development of concepts of astronomical events.

Two different approaches may be suggested. The first is the study of different social attitudes to specific astronomical objects and phenomena. Based on a world-wide sample analysis some generalisations on the nature of the relations between astronomical events and sociocultural systems may be deduced. The further goal of this approach should lead to the development of astronomical systems typologies.

The second approach is the study of the astronomy-based cultural paradigms in the context of a specific social system's trajectory. A series of different paradigm states should be observed at different stages of social evolution in a given society. This approach should contribute to our understanding of the change in astronomical systems.

In this brief paper I shall illustrate both approaches. A single common topic, concerning four-directional symbolism, will be examined here from both methodological positions.

Cardinal directions and social correlates

In several works Eliade observed that man's relations with the cosmos can be defined after a fixed point (a house, village, city or territory) has been established. Some reference points in a space must be chosen to define the sacred area within the profane space. These points form the framework within which man's relations with the divine are ensured. Anthropology offers numerous examples of how people use space to express distinctions of age, sex, rank, ethnic provenance, and so on. Following Kus (1983) we can say that the human use of space includes the ordering of the quotidian and the cosmological. Cosmovisional mapping justifies the quotidian patterning of space. Thus, any study of the cultural importance of the cardinal directions should reveal some mutual sociocultural and astronomical relations.

	G	Ht	F	Hs	A	Ag
RS_0	0	0	0	1	0	0
RS_1	3	4	1	0	10	7
RS_2	0	0	2	0	18	0
SS_1	2	4	0	0	10	8
SS_2	0	0	1	0	21	0
$\sim RS_0$	1	0	0	0	0	0
$\sim RS_1$	6	1	1	0	8	6
$\sim RS_2$	0	0	1	0	15	0
$\sim SS_0$	1	0	0	1	0	0
$\sim SS_1$	7	1	2	0	8	5
$\sim SS_2$	0	0	2	0	12	0

TABLE 1. Relationships between the three social variables and the sun-derived nomenclature for east and west. Columns denote mode of subsistence (Murdock variables 7–11): G – gathering, Ht – hunting, F – fishing, Hs – animal husbandry, and A – agriculture. In five cases the specialisation by sex in the performance of the dominant economic activity is observed. We deal here with female gatherers, male hunters, fishermen and animal husbandmen and both female and male agriculturists. Ag means agriculturists with equal or irrelevant sex participation. These data were provided by the Murdock variables 54–62. The letters in rows refer to the east/west nomenclature. RS/SS denotes the cases when the terms for east/west are derived from the labels for the rising/setting of the sun. The absence of such connotations is marked by $\sim$. The second variable denotes prevailing types of settlement pattern (Murdock variable 31): 0 – fully migratory/nomadic bands, 1 – semi-nomadic, semi-sedentary or impermanent settlements, and 2 – compact, permanent or complex settlements.

	G	Hs	Ht	F	A
~SS$_0$	1	1	0	0	0
~RS$_0$	1	0	0	0	0
RS$_0$	0	1	0	0	0
	A	Ht	G	F	Hs
RS$_1$	7	3	1	0	0
SS$_1$	7	3	2	1	0
~SS$_1$	6	1	5	1	0
~RS$_1$	6	1	4	1	0
	A	F	G	Ht	Hs
SS$_2$	15	1	0	0	0
RS$_2$	13	1	0	0	0
~RS$_2$	10	1	0	0	0
~SS$_2$	10	1	0	0	0

TABLE 2. The data from Table 1 after conversion into percentages and reordering to place the greatest frequencies on the left.

My paper starts with Brown's (1983) analysis of nomenclature for the cardinal directions. His conclusions indicate that for much of human history people were not interested in the development of special terminology for cardinal points. Brown considers that in recent times people's interest in cardinal points has increased considerably and that languages have frequently developed labels for east/west by extending terms for the rising/setting of the sun. He also suggests that both terms were encoded before the north and south terminology was developed and that the latter has no prominent referents associated with celestial phenomena.

To examine the relationship between the nomenclature for the four cardinal directions and social variables, I used Brown's sample of 127 globally distributed languages and Murdock's (1967, with later modifications) list of ethnographic societies. In the present study I examined the terminology for east and west and three social variables: basic subsistence economy, settlement pattern, and types of specialisation by sex in the performance of the dominant economic activity. Table 1 presents the sample data of 85 languages and ethnographic societies. The data were then tabulated in percentages and ordered in Table 2.

The following conclusions may be drawn from Table 2. First, the cultural salience of east-west directions increases notably in semi-sedentary and sedentary societies. The association of the east/west terminology with solar events is higher in agricultural semi-sedentary and

	Specialisation by sex	Equal or irrelevant specialisation by sex	No data
S	71	6	15
~S	49	18	11

$$\chi^2 = 9.56, p < 0.01$$

TABLE 3. Relationships between east/west terminology and types of economic specialisation by sex. S denotes sun-derived east/west nomenclature and $\sim S$ refers to the lack of such a terminology.

	Labour specialisation		
	Female	Male	Equal sex or irrelevant
S	21	50	6
~S	21	28	18

$$\chi^2 = 11.57, p < 0.01$$

TABLE 4. Relationships between the gender of economic specialisation and east/west terminology.

	G_f	Ht_m	F_m	Hs_m	A_f	A_m
S	5	8	4	1	16	37
~S	15	2	6	1	6	19

$$\chi^2 = 15.83, p < 0.01$$

TABLE 5. Relationships between the sun-derived east/west nomenclature and basic economic activities performed by female or male specialists.

sedentary societies. Second, semi-sedentary societies of gatherers and both semi-sedentary and sedentary agriculturists have a relatively high degree of non-astronomical referents for the east/west nomenclature.

One of the measures of social complexity is the presence or absence of labour specialisation by sex (see variables 54–62 in Murdock). Since the data presented in Tables 3 and 4 are statistically significant we can conclude that societies with male predominance in activities related to the basic subsistence economy are more likely to develop terms for east/west using solar connotations. In societies in which the social division of tasks by sex is irrelevant, the cultural salience of solar referents for the east/west terminology is less pronounced.

The data of Table 5 offer some more details. It seems that agricultural societies in which only one sex dominates in performing agricultural activities are more likely to develop the nomenclature for east/west using solar connotations. When this specialisation is not clearly established the cultural salience of non-astronomical referents is greater. It may be observed that male hunters use solar connotations more frequently than the female gatherers.

In conclusion, the terms for east/west tend to be defined by the rising/setting of the sun in complex and sedentary societies, with agriculture as the basic subsistence economy and with the clearly established task specialisation by sex. Solar referents are less often encoded into east/west terms is smaller and less sedentary societies in which the gathering of plants and of small fauna is dominated by females.

Cardinal directions in Late Preclassic Uaxactun

The Mayan site of Uaxactun is located in the central-north-eastern part of the Peten district in Guatemala. It was excavated by the Carnegie Institution of Washington between 1926 and 1937 and by the National Tikal Project in 1983–85. The ceramic sequence at Uaxactun begins with a Mamon ceramic complex (600–300 BC) and continues through Chicanel (300 BC–AD 250), Tzakol (AD 250–550), and Tepeu (AD 550–889). These periods roughly correspond to the Middle Preclassic, Late Preclassic and Protoclassic, Early Classic, and Late Classic periods respectively.

The site consists of several clusters of architectural remains dispersed in the tropical rain forest. Groups A, E, and H were occupied in the Preclassic period. At the end of the Protoclassic Group H was abandoned and during Early Classic Group E gradually declined in importance as Group A expanded. Groups B, C, D were also occupied in the Early Classic. During the Late Classic Groups A and B became connected by a raised causeway and the importance of Groups C and D diminished.

Uaxactun started as an independent polity, but came under heavy Tikal influence around AD 378. It recovered its autonomy in the Late Classic era (Matthews 1986). It is assumed that local ruling families occupied each of the architectural groups. The expansion of Group A and the decline of Group H at the beginning of Early Classic is interpreted in terms of the struggle for a dominant position at Uaxactun. The peak of the struggle is associated with the transition to Classic times (Valdés 1988: 12).

Group E (Fig. 1) is well known in the archaeoastronomical literature. Aveni and Hartung (1989) noted that the assemblage could have served as an 'astronomical observatory' when the observer was placed at the top of the first structure of pyramid E-VII sub. The height of this position was estimated to be some 3·3–3·5m above ground level, so the observer's eye level here is comparable in height to the top of the East Mound platform (4·6m). In other words, the platform was altered to the height of

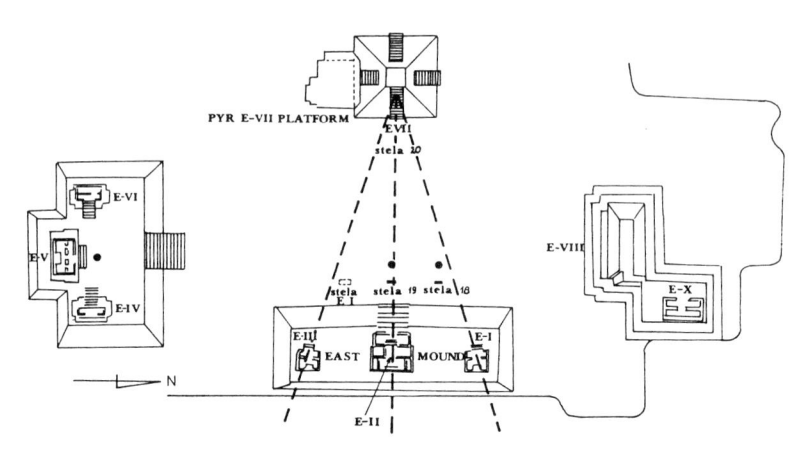

FIG. 1. Group E, Uaxactun (after Ricketson and Ricketson 1937).

the natural horizon. When the later E-I, E-II and E-III construction phases proceeded, the view towards the eastern horizon became obstructed.

Archaeological evidence (Laporte 1989; Rosal 1987; Valdés 1988; 1989) indicates that the east-west reference line was included in Group E from the very beginning. The first structures built in the first phase of Chicanel, E-VII sub 1 and East Mound (E-XVI-1), were of the same height (c. 3m), and an observer at the top of the pyramid could observe the natural horizon over the top of the platform. Very soon this situation changed. During Chicanel 2 the height of East Mound was elevated to about 4·6m, corresponding to the height of natural horizon as seen from the top of E-VII sub 1. By the end of Middle Chicanel (fifth stage, c. AD 1) a much greater pyramid, E-VII sub 2, was built over E-VII sub 1 and the observer's location lost its privileged position at the top of the pyramid. However, at the top of the first pyramid structure astronomical observations could be carried out without any problems. With the beginning of Tzakol 1 (around AD 250), the earliest versions of the three temples E-I, E-II and E-III were constructed over East Mound and the direct observation of horizon events was blocked, though solar solstice phenomena could still be perceived. Final versions of E-I, E-II and E-III were finished during Tzakol 2 (AD 300–78), and no major architectural activity has been detected at the site after this period. The site was used exclusively for ceremonial purposes.

The large-scale and open form of the group of buildings and the easy access to the centre from different sides indicate, according to Chase (1985: 37), that it was dedicated to public activities that could have involved much of the local population. No burials of important individuals were found in Group E (e.g. Coggins 1980: 731–32) and it may be assumed that the place was dedicated to perform group-oriented ceremonies.

Schlak (cited by Aveni and Hartung 1989: 455) defined these ceremonies as decapitation rites. More information on the on the so-called 'stela cult' provided by Hammond (1982) and Justeson and Matthews (1983) leads to the conclusion that in Group E specific ceremonies involved the erection of stone monuments, sacrificial burials of decapitated individuals, and ritually cached offerings, and that they were carried out at regular 360-day intervals. Since most of the solar orientations of the Preclassic tend to refer to solstitial points (Aveni and Hartung 1986), I suggest that these ceremonies were performed at solstices.

While Group E served for ceremonial purposes, Group H (see Fig. 2) was transformed into a seat of the political elite (Valdés 1989). The symbols of ancestors, masks of solar gods, and iconographic representations of bloodletting rites define the place as a seat of the local ruling family.

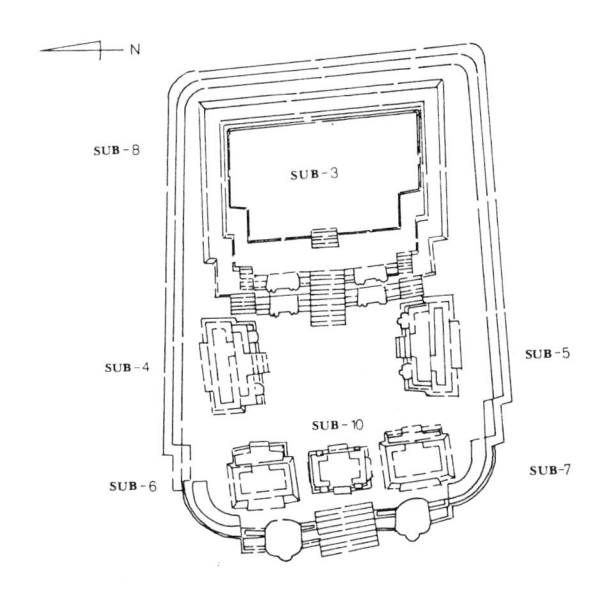

FIG. 2. Group H, Uaxactun (after Valdés 1988).

Here the programme of directional symbolism also started with the establishment of the east-west axis. On a huge platform other buildings were gradually constructed, of which Sub 2 and Sub 3 were the first. The images of ancestors on the walls of Sub 2 were later covered by the huge masks of Sub 3 representing *axis mundi*, or the vertical symbolic division of the universe. The symbols of rulership were placed between the superior and inferior planes alluding to the ruler as a mediator between the sacred and the human.

Two later palaces (Sub 4 and Sub 5) on both sides of Sub 3 established the north-south reference line. Images of a lunar deity covered the walls of the north building while masks of Venus gods appear on the southern one. Finally, during the subsequent stages of architectural activities the platform under the buildings was extended to the west. The access to the whole group led from the west through a staircase composed of nine steps and of masks and other symbols of political elites. Valdés (1989) concludes that Group H was a seat of the local ruling family that performed family-oriented rituals involving auto-sacrifice and bloodletting.

This evidence indicates that in the Chicanel period two different symbolic systems related to cardinal directions coexisted. In Group E these directions seem to be defined by solstitial points. Solstice directions could create a framework for the further ordering of space. On the climatological-agricultural level, the date of winter solstice could mark the beginning of the hostile hot and dry season, and the lack of agricultural activities.[1] On the social-religious level this date could refer to the rites that reinforced the sacrosanct status of the elite.[2]

Other examples from Late Preclassic times sites indicate that symbolic values attached to the four directions were commonly shared by Maya elite groups, with minor differences.

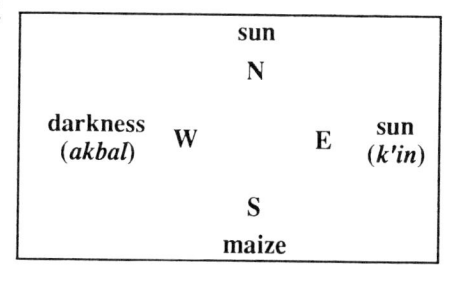

	sun	
	N	
darkness (*akbal*) W		E sun (*k'in*)
	S	
	maize	

FIG. 3. The programme of directional symbolism derived from the Late Preclassic (100 BC–AD 100) sites in Belize (after Hammond 1987).

---

[1] Rains start in May and continue until October. Archaeological data and analogies from other sites indicate that in maize agriculture planting and sowing takes place between March and June. Harvesting occurs between October and January.

[2] It has been argued (Justeson and Mathews 1983: 586) that the stela cult rites were related to important events at the level of the elite. The Late Classic ruling family of Group A used several winter solstice alignments with the abandoned Group E buildings, presumably to get support of the (deified) founders of Uaxactun.

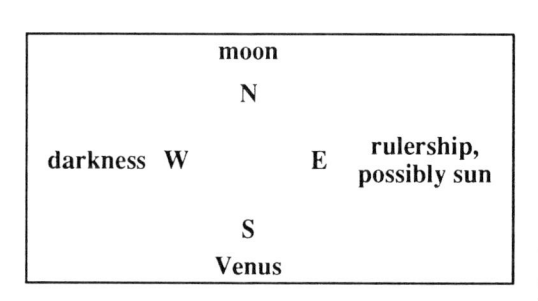

```
            moon
             N
darkness  W      E   rulership,
                     possibly sun
             S
           Venus
```

FIG. 4. The programme of directional symbolism assumed at Uaxactun during the same period.

The programme of directional symbolism derived from the Late Preclassic (100 BC–AD 100) sites in Belize is shown in Fig. 3; that during the same period in Uaxactun in Fig. 4. The assumed Uaxactun pattern seems to be repeated some five hundred years later on the walls of Tomb 12 at Río Azul (Carlson and Stuart 1986; Bricker 1988). During the Oxford 3 conference Carlson presented new evidence of the existence of the four-directional symbolism linked to the elites. In the Late Classic structure 66c of Group 8N-11 at Copan, a ceremonial throne with panel reliefs has been recently discovered. The relief presents the Uaxactunian symbol set.

Conclusions

Late Preclassic Uaxactun represents a sedentary agricultural society. It is composed of at least two different social ranks (Gibson 1986: 167) and there is advanced craft specialisation. This society developed collective representations attached to the cardinal directions. At Group E they were defined by the solar solstitial referents and related to the justification and legitimisation of the existing social order. With the emergence of a ruling family a new programme of directional symbolism tended to ensure its elevated position amongst other competing lineages. One of the new methods was to associate the ruler with the world's axes. Being a mediator between the divine and the community, and being supported by the ancestors, the ruler could intervene in earthly social order. The association of the ruler's seat with the centre of the world defined by the four directions was a further step towards the legitimisation of his power. At the end of this process, his position was strengthened by the association with the image of the rising sun.

In both Group E and Group H, this symbolism starts with the establishment of the east/west axis. Thus it is probable, as the recent discussion on the meaning of directional glyphs suggests (Closs 1988; Bricker 1988; Coggins 1988), that among the Maya languages the labels for east and west were encoded before the terms for north and south were invented. The nomenclature for east and west was derived from the solar movements.

I have argued that sedentary, agricultural and somewhat stratified societies tended to develop a terminology for cardinal directions, *ergo* they worked out concepts of the world's quadripartition. However, Brown (1983) concluded not only that the terms for east/west preceded those for north/south but that they also tended to be expressed in terms of solar movements.[3]

Our Uaxactunian example shows that at least three centuries after the sedentary and agricultural style of life had been established, four-directional symbolism started to be encoded in alignments of stable ceremonial structures. Two different sets of quadripartite space division observed at the site possibly reflect distinct cosmovisional patterns. While the Group E type based on solstitial directions seems to be, in my opinion, a more general pattern correlated with an agricultural, sedentary but rather simple society, the Group H symbolism reflects a dynamics resulting from more stratified society. Possibly the upper social stratum 'mastered' and transformed a traditional pattern in order to legitimise their elevated positions. By that time, concepts of spatial division also changed: space was no longer divided by the solstitial lines but distributed into four quadrants. The east and west quadrants probably corresponded to the zones defined by the extension of the solar arc. The emphasis placed on the solstitial directions on one hand and the Maya cosmogram relating east/west with the sun on the other, supports Brown's other conclusions. In the light of the aforementioned discussion on the meaning of terms for north and south among the Maya, it seems more probable that they did not refer to solar movements. Thus, for example, the term for north may mean 'above' but not necessarily 'the sun above' or 'the sun at zenith' (but see Bricker 1988 and Closs 1988).

Finally, we note that while the first of our two approaches is useful in establishing more general (and static) sociocultural correlates of astronomical, calendrical and cosmovisional activities, the second offers insights into the nature of their further transformations by associating them with social dynamics.

References

Aveni, A.F. and Hartung, H. (1986). Maya city planning and the calendar. *Transactions of the American Philosophical Society* **76**(1), 1–87.

---

[3] More detailed study is needed. Examples from Mesolithic and early Neolithic Europe suggest that the development of four-directional cosmovision is more closely related to the tendency towards a sedentary style of life than to agriculture. Again, the east/west orientations preceded the north/south ones, and the east/west alignments usually fell within the solar arc (Iwaniszewski 1991).

Aveni, A.F. and Hartung, H. (1989). Uaxactun, Guatemala, Group E and similar assemblages: an archaeoastronomical reconsideration. In *World archaeoastronomy*, ed. A.F. Aveni, pp. 441–61. Cambridge: Cambridge University Press.

Brown, C.H. (1983). Where do cardinal direction terms come from? *Anthropological Linguistics* **25**(2), 121–61.

Bricker, V.R. (1988). Comment on Closs (1988).

Carlson, J.B. and Stuart, D. (1986). Early Classic Maya four-directional cosmology: tomb no 12 at Río Azul, Peten, Guatemala. Paper presented at the Second Oxford International Conference on Archaeoastronomy, Mérida, January 1986.

Chase, A.F. (1985). Archaeology in the Maya heartland. *Archaeology* **38**(1), 32–39.

Closs, M.P. (1988). A phonetic version of the Maya glyph for North. *American Antiquity* **53**(2), 386–411.

Coggins, C.C. (1980). The shape of time. Some political implications of a four-part figure. *American Antiquity* **45**(4), 727–39.

Coggins, C.C. (1988). Comment on Closs (1988).

Gibson, E.C. (1986). Inferred sociopolitical structure. In *A consideration of the Early Classic period in the Maya lowlands*, ed. G. R. Willey and P. Matthews, pp 161–74. Albany NY: State University of New York at Albany (Institute for Mesoamerican Studies publication no. 10).

Hammond, N. (1982). A Late Formative period stela in the Maya Lowlands. *American Antiquity* **47**(2), 396–403.

Hammond, N. (1987). The sun also rises: iconographic syntax of the Pomona flare. *Research Reports on Ancient Maya Writing* **7**, 11–24.

Iwaniszewski, S. (1991). From the Mesolithic to the Neolithic times: archaeoastronomy of the Funnel Beaker Culture (TRB) in Central Europe. Paper presented at the second conference on *Current Problems and Future of Archaeoastronomy*, Székesfehérvár, Hungary, October 1991. To be published in the Bulletin of the Institute of Astronomy, Hungarian Academy of Sciences, Budapest.

Justeson, J.S. and Matthews, P. (1983). The seating of the *tun*: further evidence concerning a Late Preclassic Lowland Maya stela cult. *American Antiquity* **48**(3), 586–93.

Kus, S.M. (1983). The social representation of space: dimensioning the cosmological and the quotidian. In *Archaeological hammers and theories*, eds. J.A. Moore and A.S. Keena, pp. 277–98. New York NY: Academic Press.

Laporte, J.P. (1989). El Grupo B, Uaxactun: arquitectura y relaciones sociopolíticas durante el Clásico Temprano. In *Memorias del segundo coloquio internacional de Mayistas, vol. 1*, pp. 625–46. Mexico City: Universidad Nacional Autónoma de México.

Matthews, P. (1986). Maya Early Classic monuments and inscriptions. In *A consideration of the Early Classic Period in the Maya Lowlands*, eds. G. R. Willey and P. Mathews, pp. 5–54. Albany NY: State University of New York (Institute for Mesoamerican Studies, Publication 10).

Murdock, G.P. (1967). Ethnographic atlas: a summary. *Ethnology* **6**(2), 121–61.

Ricketson, O.G. and Ricketson, E.B. (1937). *Uaxactun, Guatemala, Group E 1926–1931*. Washington DC: Carnegie Institute of Washington (Publication 437).

Rosal, M.A. (1987). El Preclásico del Grupo E, Uaxactún, Peten. Paper read at the second International Colloquium of Mayanists, Campeche, August 1987.

Valdés, J.A. (1988). Breve historia de la arquitectura de Uaxactun a la luz de nuevas investigaciones. *Journal de la Société des Américanistes* **74**, 7–23.

Valdés, J.A. (1989). El Grupo H de Uaxactun: evidencias de un centro de poder durante el Preclásico. In *Memorias del segundo coloquio internacional de Mayistas, vol. 1*, pp. 603–24. Mexico City: Universidad Nacional Autónoma de México.

II

NEW HORIZONS

5

Moon Man and Sea Woman: the cosmology of the Central Inuit

SUSAN M. PEARCE

Amongst the Hudson Bay Inuit, the Powers that made Earth and Mankind are called *Ersigifaunt* ('those we fear'), or *Mianerisannt* ('those we keep away from and regard with caution'). The powers are threefold: the Woman in the Sea, who also has dealings with Birdland; the Spirit of the Weather, the storm and the snow drift, who lives between the sky and the flat surface of land and sea; and the Moon Man, who lives on the moon in the sky. The Sea Woman and the Moon Man in many ways carry equal weight, and the Weather Spirit is employed by both to execute punishments connected with the weather. The Polar Inuit of North Greenland say that the object of the whole system is 'to keep a right balance between mankind and the rest of the world' (Rasmussen 1929: 56–63). This paper will try to unravel the nature of Central Inuit cosmology, using a structuralist approach as the most useful analytical technique, and then endeavour to show how the Inuit view of the relationship between the heavenly bodies, the earth and humankind perpetually sustains and reinforces the ideological basis of Inuit society.

In this paper the term 'Central Arctic' is interpreted fairly loosely, and is intended to include the Inuit communities of Baffin Island, the islands and peninsulas immediately to the west of Baffin, the communities of the high Arctic islands, and those to the south in Hudson Bay and its islands (Fig. 1). Although this is a vast area of tundra, sea and pack ice, encompassing some four million square kilometres, the Inuit groups within it, up to about AD 1600, shared a reasonably homogeneous lifestyle, modified a little since that date as individual groups adapted to particular local and changing circumstances (and now, of course, largely destroyed). This lifestyle was characterised by the Thule culture, which spread rapidly across the Arctic from west to east in the centuries following AD 1000, and which is expressed in a common language,

Inupik, spoken in a range of local, but broadly mutually intelligible, variants from western Alaska to eastern Greenland.

On the evidence of the anthropological record (Boas 1964; Rasmussen 1929), the traditions of these groups, which enable us to achieve a purchase on their cosmology, differ in some important respects; however, their broad agreement, reflecting a shaped cultural history, is sufficient to allow them to be discussed as a coherent entity. The most important myths revolve around the Sea Woman, sometimes called Sedna, and her relationship with Belowsea Land, with her father and with her suitor, the fulmar, who takes her to Birdland (Boas 1964: 175-83; Pearce 1987); the stories of the Goose Maiden, her husband Itiqtaujaq, and his experiences of Salmon Father and Birdland (Kleivan 1962; Pearce 1987); and the stories about the Moon Spirit. The principal ritual revolves around the Beginning-of-Winter Festival. During the brief Arctic summer there is little or no darkness, and during the long and difficult winter there is little or no daylight, with the Moon providing the principal illumination, and so assuming corresponding importance. The Inuit have little in the way of star lore, because between sunlight, moonlight and bad weather the stars and planets are seldom very visible. The cosmological structure naturally reflects these (rather odd) physical conditions.

The tundra and the Arctic sea is an environment rich in food to patient and resourceful hunters who know how to fear the Powers and endure the bad times, but its severe conditions impose a restricted lifestyle upon those who live within it. Since Mauss's classic paper (1906) it has been recognised that Central Inuit life was organised around the major dichotomy of land and sea, which produced a clear seasonal distinction embracing caribou hunting on the land of the interior from temporary camps during the summer, and sea-mammal hunting—principally seal and walrus and sometimes small whale—on and near the ice from permanent coastal settlements during the winter. The implication of this summer / land / caribou : winter / sea / sea mammals dichotomy is profound and permeates social practice. Amongst some communities, for example, sea mammal meat must not be eaten at the same times as venison, and walrus hide must not be taken into the interior because this is the domain of the caribou (Boas 1964: 187). To this, two further important elements should be added. The sea is ruled, as we have seen, by the Sea Woman, and women generally are regarded as belonging with the sea and the winter, while men belong more to the land and the caribou. Similarly, raw materials and the artefacts made from them participate in the structure, so that women's sewing equipment is made from sea mammal ivory or bone, and seal hunting equipment is similarly made of bone or ivory, but caribou hunting arrows have antler heads (McGhee 1977; Pearce 1989).

The cosmology of the Central Inuit

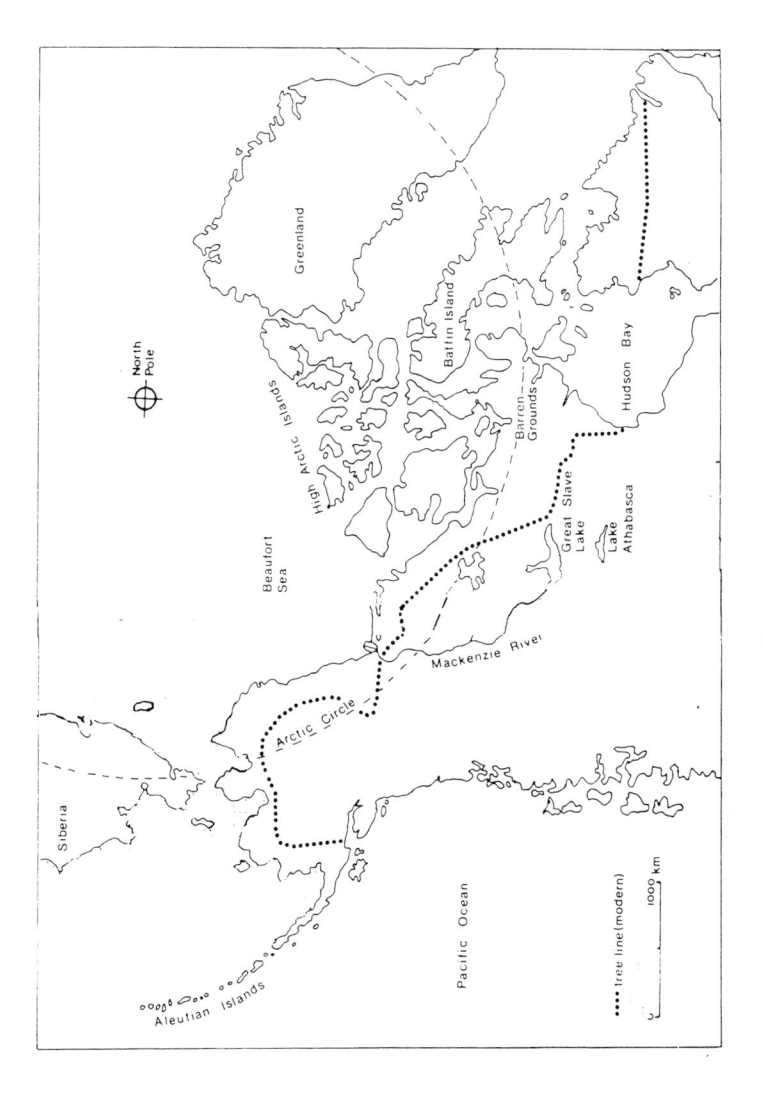

FIG. 1. The lands of the 'Central Arctic' region.

The land/sea pair, and all that it implies, is completed by the integration of the sky into the physical world, and its mythological equivalent, the Land Above, into the cosmological structure. The Land Above has two aspects: the Land of the Moon Spirit and Birdland. As we have seen, the Inuits' view of the heavens concentrates upon the Moon and, to a much lesser extent, the Sun. The Moon Man or Spirit rules over a generally happy land in the sky, from which he sometimes visits the earth, driving across the ice of the clear sky and the more difficult snow fields of the clouds in his dog sledge. The land of the Moon Spirit seems to be the world above in its winter aspect: its summer equivalent is the Land of Birds, and, as in many other mythologies, birds tend to parallel the lives of humans. Human progress through life is woven into this triple pattern of Land, Sea and Land Above.

Amongst the Cumberland Sound Inuit new-born babies are regarded as little birds and, in a point where poetic cosmology and utilitarian function meet, their first dress is the warmest clothing available, the skin of a large bird like a snowy owl turned inside out with its feathers still attached to make a sort of sleeping bag. This feather dress is kept by each individual as his/her most powerful personal amulet, and it is worn on the end of the hood, allowed to hang down so that the amulet protects the back, at the great festival which marks the beginning of Winter. People retain their bird nature through life. The same festival features a tug-of-war between the ptarmigans, those born during the winter, and the ducks, those born during the summer.

At death, a distinction is made between the dead body and the insubstantial spirit. Dead bodies are wrapped in caribou skin, and among the Igloolik Inuit this must be done by a widow and young girl so that the fertility of married women is not endangered. The body must be buried as soon as possible (graves are very shallow because the permafrost prevents excavation to any depth). In Igloolik, men are buried facing the east and the dawn, women facing the unlucky south from which the winds bring storms and bad weather. Later, the grave is visited with gifts of caribou venison, never sea-mammal meat. Stores of venison are cached in the Autumn in a very similar way, and it seems clear that a correspondence is perceived between dead bodies and caribou. In desperate times, the Inuit resort to cannibalism, usually eating the bodies of those who have already died of hunger, so this whole area is characterised by a grim species of gallows humour.

Spirits go both to Belowsea Land and to the Land of the Moon Spirit: both are regarded as happy places, that is, places where large quantities of food are freely available, but the Land of the Moon is usually regarded as the better of the two. When the happy souls with the Moon play the Walrus skull ball game, those here below see it as a display of

Northern Lights. The Igloolik Inuit stress that the souls in the Land of the Moon include (or perhaps are only) those who were drowned, murdered, committed suicide or died in child birth, that is those who would otherwise be regarded ritually unclean or taboo. The importance of the link between Moon Spirit and the system of taboos is a subject to which we shall return.

Two elements in social practice bring together all of these states of being and make them transparent to each other. The first belongs within the temporal mode and stands at the crisis of the turning year when the short summer ends and the ice floes begin to form and clash together, showing the beginning of winter. In most communities this is marked by a festival, like that already referred to at Cumberland Sound, of which we happen to have a detailed description. Here, the men gather in the middle of the settlement early in the morning. Most run sun-wise shouting through the houses, but those born by abnormal presentations wear women's clothes and run in the opposite direction. At the climax of the festival, two gigantic figures—the *quailetetang*—appear, wearing grotesque clothing, including masks and women's jackets, and carrying a harpoon (men's gear) and an *ulu* (moon-shaped knife, women's gear). They pair off the women and men (not in married couples) and the pairs will spend the next night and day together. They go down to the beach and invoke the north wind which brings fine weather, warning off the stormy southern wind. The men then pretend to attack the *quailetetang*, and they both act as if they have been killed. They are brought back to life with fresh water, and then tell individuals what their futures hold. There is no need to elaborate upon the Other World themes which this ritual embodies, or to stress its relationship to the whole system of Inuit cosmology.

The second act of fusion focuses upon the person of the Shaman. The Shaman of each community can control both the spirits of the dead and the spirits of animals and natural features. His Helping Spirits conduct and preserve him on long, dangerous journeys to the Land of the Dead, both in the Sky and Belowsea. He plays a major part in the Beginning-of-Winter Festival by defeating the Sea Woman who might otherwise overwhelm the community. In his own person, he acts as a bridge between this world and the Otherworld, and in fulfilling this role one of his most important functions is to be summoned in cases of sickness in order to discover the transgressions of social rules which have caused the disorder.

In cosmological terms, it is clear that Land, occupying the middle band of the scheme, represents This World or, put another way, human norms. Here belong men, summer and light, the caribou, equipment made of caribou products, venison, and buried bodies. Both the sky and the sea can be experienced from Land, but always have an Other World character. This triple nature is reflected by the tripartite structure of the territories which make up the Otherworld. Belowsea Land is inhabited by the Sea Woman, the sea mammals with skins, meat and ivory, the spirits of the dead, and women in one of their aspects. Birdland, in the sky, holds birds, new babies, and women in their other aspect. The Land of the Moon Spirit, and his dead spirits, represents the central band, or Land itself in its Otherworld aspect. Here, too, belongs the Weather Spirit, a kind of incarnation of the power of the greater spirits to affect the condition of Land, Sea and Sky.

Within this structure, there are complexities. Each of the Otherworlds, being complete worlds in their own right, also contain elements which belong with the norms which Land represents. In Belowsea Land, the Sea Woman lives with her father, and the myth to which we have already referred, and which exists in a number of variants, tells how her fingers, cut off by her father, became the sea mammals, while his feet, gnawed off by her dogs, seem to have become the caribou (Pearce 1987). Amongst the people of Igloolik the father seems to be an evil spirit, unhelpful and dangerous to humans. Similarly, the Weather Spirit has an evil companion, but there are hints that the Weather Spirit itself is a child or some kind of half-animal (Rasmussen 1929: 73). Birdland is inhabited both by bird-women and the male fulmar.

In the Land of the Moon Spirit, also called the Land Above, the Land of the Dead in the Sky, and the Land of Day, the Moon Spirit lives with his sister in a house with two living areas but one common entrance. The origin myth of the Moon Spirit (Rasmussen 1929: 81) describes how he became blind and was given (probably supernatural) sight by a loon, and how the two children kill their grandmother. They come to a land where the people have no arms and no genitals, and show them how to complete themselves. Meanwhile, the girl discovers to her horror that she has been committing incest with her brother (an ever-present problem in Inuit life, as we shall see). She runs from him, but both are caught up in the sky, where he becomes the moon and she the sun. This story seems to represent the turning away from abnormalities and aberrations which the Moon Spirit in his role as Life-Maintainer represents. The Sister-Sun plays no further role in myth or morals, no doubt because the sun plays a relatively limited role in Inuit life.

The Moon is, however, accompanied by an evil female spirit called *Ululiarnaq* ('the one with the woman's knife', or *ulu*). She is never identified with the sister, although the suspicion of a link inevitably comes to mind. She looks grotesque and acts grotesquely in order to make humans within the Moon Spirit's house laugh; but the moment they do so, she disembowels them. Perhaps this may best be interpreted as a general

The cosmology of the Central Inuit 65

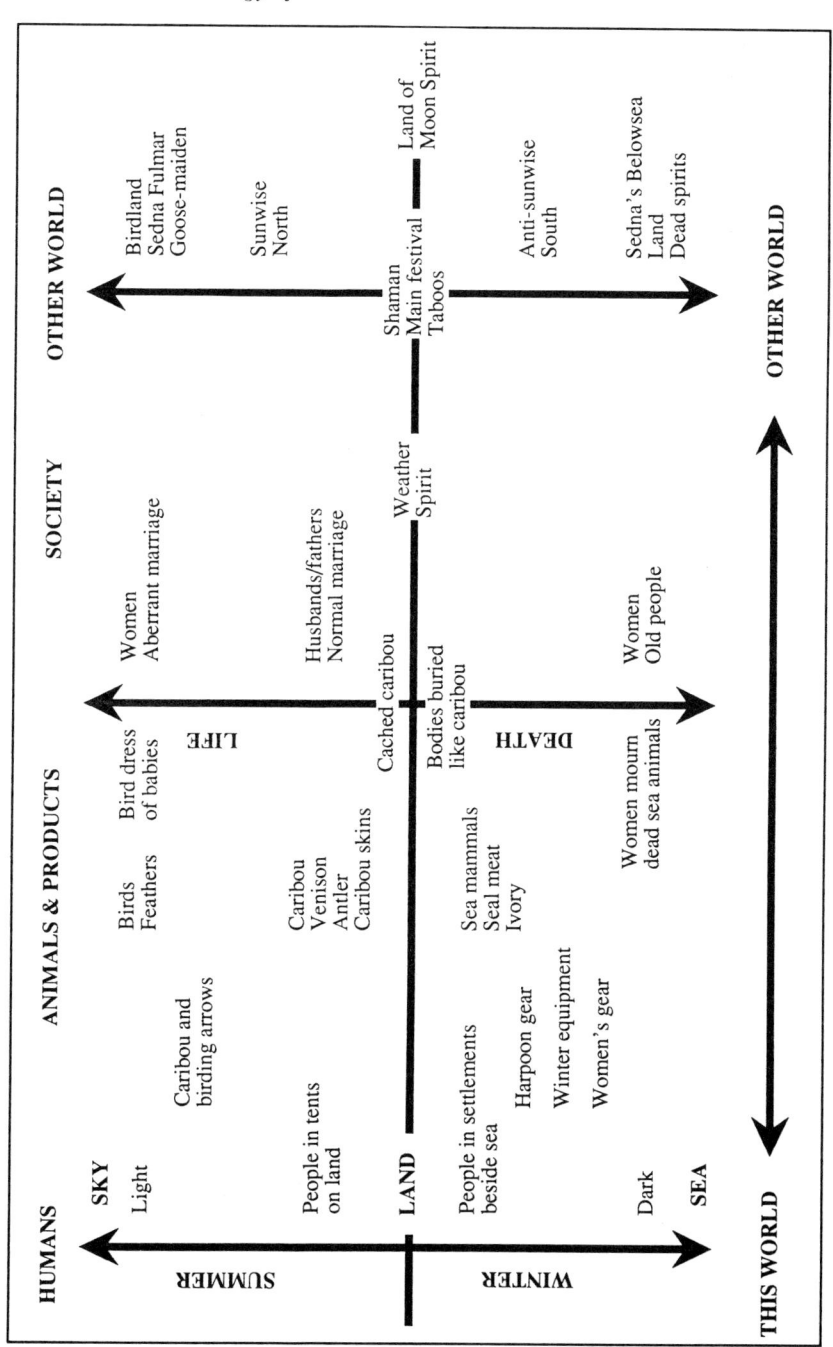

FIG. 2. Structuralist plot of the Central Inuit cosmological and social scheme.

threat to all normal forms of social life, because humour of all kinds, formal and informal, is a very characteristic element in Inuit social relationships. Clearly, however, a firm structure is maintained in which each of the two Otherworlds has a main power, a male in the Sky balancing a female in the Sea, paired with a secondary character where the sexes are reversed. The structure is reflected by the Weather Spirit and its companion who are, as we would expect, of lesser powers, weaker and of rather indeterminate character. The structure of Birdland is less clearly worked out than that of Belowsea Land and the Land of the Moon, but nevertheless holds male birds, Sedna for important episodes in her life, and other bird women. The whole cosmological scheme can be set out in a structuralist plot of the kind shown in Fig. 2.

Amongst the Central Inuit, at any rate, the Moon Spirit appears as wholly kind and helpful. Through his link with the tides, he can distribute seals along the coast, and when he is not in the sky, he is out hunting sea mammals. He offers protection to hunters, who often go out by moonlight in the winter. He brings fertility to women, where his cyclical nature clearly matches theirs, and sometimes he impregnates them himself. Most importantly, he is the moral guardian of mankind and this means that he endeavours to maintain the network of taboos which surround the Inuit in daily life, and punishes those who transgress. One of the most widespread of all Inuit stories describes how the Moon purifies a homeless boy called Kagjagjuk of the taboos accumulated within him, and he becomes a mighty hunter (Rasmussen 1929: 88–89). The moral, of course, is that those who obey the rules, succeed. There is a clear link between his moral guardianship and his role of providing winter food-animals through the tides, and of encouraging women's fertility because many of the taboos revolve around the prohibition of women to do many things while they are menstruating. There are also several stories showing how the Moon Spirit befriends women or brings them into a correct frame of mind (Rasmussen 1929: 85–88). It is the Shaman who is called in to elucidate taboo-breaking which has resulted in sickness or death, and it is clear that the idea of the Moon Spirit fulfils the very earthly ideological role of reinforcing the system of taboos and the social organisation, including the powerful position of the Shamans, which they underpin.

This analysis of lunar power play is true as far as it goes, but it is also rather simplistic and superficial. A more fundamental analysis suggests that the cosmological structure as a whole both mirrors Inuit life as it is led on tundra and sea, and provides or reflects a system of checks and balances which enables the community to absorb, or at any rate to tolerate, the tensions inherent in its (as in every other) form of social organisation, and so to continue to reproduce itself. This emerges when the myths which embody the cosmology are related to social practice. A viable hunting group numbers some thirty or forty souls. Briefly, men are expected to hunt, and women to sew the waterproof skin boots and clothing: without both food and clothes, human beings die. Young people are allowed to marry as soon as they can perform these adult tasks, but a man is still expected to hunt for his own parents—presumed elderly and fairly helpless, that is around forty years old—and also his parents-in-law. The couple begin life with her parents, and it is not until these are dead that the man is finally his own master. Marriage is in many ways an economic, rather than a sexual, arrangement and both parties are free to enter into any liaisons which may appeal to them, so brother-sister (or at least, half-sibling) incest is, although forbidden, an ever-present possibility, and fatherhood is a matter of hope rather than of secure knowledge. Both divorce and re-marriage are very simple for both sexes. The wife's mother is a powerful figure, who may, if she wishes, command a divorce.

The story of Sedna and her father can scarcely be called one of domestic content, but it should be noted that Sedna found her Fulmar suitor and Birdland bitterly disappointing, and that father and daughter remain together at the end of the story. The implication seems to be that girls serve their own interests best by putting their fathers before their husbands, and in this the myth reflects the normal arrangements of early marriage, and relates dangerous social aberrations to Birdland. Why Birdland should be a source of tension is hinted at in the story of Itiqtaujaq and the Goose Maiden. Women belong with Birdland and, as all sexual motifs in the story show, this is a danger to men; equally threatening is the woman's tendency to take her son, who may or may not be her husband's son, but on whom he must largely rely in his old age, to Birdland with her.

The myths reflect the struggles of men to retain powers over wife, daughters and sons, all of whom are necessary to his economic wellbeing, now and in the future. They concentrate upon the problems of the male side because hunting prowess fades much more quickly than does the skill to sew, and the crucial importance of sewing is probably one of the reasons which stand behind the great sexual freedom women enjoy, even though it undermines the position of men. Similarly, the power of older women, of mothers and mothers-in-law, is recognised in custom and mythologically linked with Belowsea Land; their kinship with the large sea mammals is recognised by their obligation to mourn as for human kin when such an animal is killed. Equally, of course, the group as a whole needs its men to be successful hunters and its women to be fertile, and these things are encouraged by the beneficent powers of the Moon Spirit. We can see that Belowsea Land, principal source of food, identified with winter and women, is linked with the social weight of the old. Land, place of men, caribou and summer and social norms, is linked with good family relationships. The Land Above is ambivalent: containing dangerous but necessary Birdland, from which comes little birds and babies, and the happy home of the Moon Spirit from whence comes the help which supports the group.

Viewed from the outside, Central Inuit cosmology is mixture of natural elements drawn from the world the Inuit inhabit and the social customs which they have found make life in that world possible, developed into an ideological scheme which reinforces and constantly recreates society. Viewed from the inside, the cosmology describes the mystical unity of all things in which balance and fear mingle, and the heart of man finds comfort. To judge between the value of these opposed views would be a difficult task.

References

Boas, F. (1964). *The Central Eskimo*. Lincoln NE: University of Nebraska Press.

Kleivan, I. (1962). The Swan Maiden myth among the Eskimos. *Acta Arctica* **13**, 6–47.

McGhee, R. (1977). Ivory for the Sea Woman: the symbolic attributes of a prehistoric technology. *Canadian Journal of Archaeology 1, 141–49*.

Mauss, M. (1906). Essai sur les variations saisonnieres des sociétés Eskimos. *L'Année Sociologique* **9**, 39–130.

Pearce, S.M. (1987). Ivory, antler, feather and wood: material culture and the cosmology of the Cumberland Sound Inuit, Baffin Island, Canada. *Canadian Journal of Native Studies* **7**(2), 307–21.

Pearce, S.M. (1989). Objects in structures. In *Museum studies in material culture*, ed. S.M. Pearce, pp. 47–59. Leicester: Leicester University Press.

Rasmussen, K. (1929). *Intellectual culture of the Igloolik Eskimo. Report of the fifth Thule expedition 1921–24*. Copenhagen: Gyldendalske Boghandel, Nordisk Forlag.

6

Time-reckoning in Iceland before literacy

THORSTEINN VILHJÁLMSSON

Introductory remark

The history of science in Iceland is a field of study which has received little attention to date. Many of the fundamental works are old and outdated, and authors with basic scientific knowledge have been sadly absent from the arena. The present paper is a partial effort to remedy this situation.

The subject matter of the paper belongs to archaeoastronomy in the sense that it deals with astronomy and time-reckoning in an illiterate society without the knowledge and instruments of classical or late medieval times. However, it lacks any strictly archaeological input at present. Owing to adverse climate and the limited powers of the small and scattered medieval population, only extreme luck would produce significant archaeoastronomical findings in Iceland. Indeed, an effort was made recently to see if archaeology could bring us somewhat closer to the twelfth-century astronomer Star-Oddi, but the archaeological study turned out not to be particularly successful in this respect (Ólafsson et al. 1992).

The Vikings

Traditionally, the Vikings originating in Scandinavia in the early Middle Ages are associated with violence and brutal force. However, the views of modern scholars paint a less monochromatic picture (e.g. Foote and Wilson 1984; Jones 1986; 1990; Graham-Campbell 1989). The present paper relates to one aspect of this, namely the knowledge and science of the Vikings and their immediate successors in Iceland and other Scandinavian countries.

Many of the activities of the Vikings required and produced knowledge of time-reckoning and of what we would nowadays classify as astronomy. For example, their extensive travelling and trade must have involved some knowledge of astronomy. The necessity of such knowledge is generally recognised in the case of coastal navigation, but also holds for inland travel through previously unknown areas, such as the vast lands of Eastern Europe.

Inland travel and coastal navigation is one thing, but regular transoceanic traffic is quite another. Yet such traffic was required to support the Scandinavian settlement of Iceland and Greenland, around the years 900 and 1000 respectively, at a time when the people of Europe knew nothing of the compass or the sextant. Even with good luck the oceanic voyage would take about a week, and without it land might not be sighted for several weeks. The navigational methods used included both terrestrial and celestial observations (Einarsson 1970: 57–63; Schnall 1975, ch. 4; Marcus 1980: 100–18; McGrail 1989: 59–63). There is hardly any doubt that the knowledge written down on vellum in Iceland in the twelfth and thirteenth centuries derives to a high degree from these observations and this experience.

The need for a calendar

Turning inland, both textual sources and common sense reasoning indicate needs for certain kinds of knowledge in the community settling in the novel environment of the large island of Iceland. For instance, the Icelandic summer is short, making it a matter of primary concern to utilise summer time as well as possible.

In 930 the Icelanders decided to establish the *Althingi*, a kind of parliament where an important part of the population gathered once a year for purposes of legislation and justice. Those who went there would spend two to five weeks away from home at a precious time of the year. The farms were scattered at long distances and the landscape often barely passable. Therefore the method previously known to the Scandinavians, to summon conventions like the *Althingi* by message, was not viable. It was much more convenient to have a simple and reliable calendar to help people know when to start from home so as to arrive at the same time as the others, and also to date the parliament at the time of summer when the loss of domestic labour was least harmful.

To understand the need for a calendar we may also look at the agriculture itself and its annual cycle. Certainly, the caprices of Icelandic weather and nature are such that the calendar may often be a bad guide for action. In deciding when to let cattle and sheep out on grass or when to start hay-making it is better to observe the actual signs of nature than the calendar. But there are certain kinds of annual operation where the calendar proves superior: for example, in determining when to sow the grain, something which people had tried with little success in the first centuries of settlement in Iceland. Another good example is that of deciding when to let the ram to the ewes. It is important to do this at the right time in the winter so that the lambs have the best possible prospect of growing in the short summer, without too much risk of interludes of bad weather in the spring just after they are born. When the individual farmer makes his decision on this at some point around Christmas time, he has no clear natural signs of a terrestrial nature to go by.

The calendar reform of 955

In the brief history of Iceland called *Íslendingabók* (The Book of the Icelanders, *Libellum Islandorum*), written by Ari the learned in the period 1122–33, we have a report on a calendar reform of about 955:

> This was when the wisest men of the country had counted in two semesters 364 days or 52 weeks—then they observed from the motion of the sun that the summer moved back towards the spring; but there was nobody to tell them that there is one day more in two semesters than you can measure by whole weeks, and that was the reason.
>
> There was a man called Thorsteinn the black...a very wise man. When they came to the Althing he sought the remedy that they should add a week to every seventh summer and try how that would work...
>
> By a correct count there are 365 days in a year if it is not a leap year, but then one more; but by our count there are 364. But when in our count a week is added to every seventh year, seven years together will be equally long on both counts. But if there are 2 leap years between the ones to be augmented, you need to add to the sixth.[1]

According to this, people started by counting 52 weeks or 364 days in the year. When they realised the insufficiency of this they tried the remedy of intercalating one week every seventh year (*sumarauki*), thus making the average year 365 days. The method chosen may seem strange to us but it is a natural consequence of the important role of the week in the original calendar.

---

[1] Benediktsson 1968: 9–11; see also Beckman and Kålund 1914–16: 65–66, esp. the manuscript Gl. kgl. sml. 1812 4to.

So far the interpretation of the text seems straightforward. However, the text continues to describe the relation and adaptation of the Icelandic calendar to the Julian one, which must have been gradually introduced in Iceland in the eleventh and twelfth centuries, following formal Christianisation of the country in the year 1000. The text says that if there are two leap years between the years to be increased by a week, then the sixth year (instead of the seventh) should be increased. This is plainly wrong and would yield a worse approximation than the more simple rule of intercalating a week every sixth year. I have found it impossible to make sense of this except by assuming the Latin meaning of the numerals. Thus '*septimo quoque anno*' actually means 'every sixth year' by our count (see, e.g., Ginzel 1958: III, 66). In this way Ari's text can be interpreted so as to coincide with practice in his time, as seen from almost contemporary Easter tables (Beckman and Kålund 1914–16: 69–71; Beckman 1916: v–ix). Also, he would escape Occam's razor, since his formula would otherwise be more complicated than necessary for its accuracy.

It is interesting to consider the possible methods which Thorsteinn the Black might have used in determining his intercalation (Einarsson 1968). His farm was favourably located in the country to utilise the so-called mountain circle method (Vilhjálmsson 1989: 93), i.e. to follow the annual motion of sunrise and sunset near the horizon where he would have suitably distant mountains and other reference points in the landscape to make fairly exact observations possible. At high latitudes the points of sunrise and sunset move so fast that this method could easily be used to determine the length of the year to within a day.

Twelfth-century knowledge of the solstices and equinoxes

The story of independent Icelandic determination of the length of the year and its subsequent adaptation to the Julian calendar ends in the twelfth century. However, for the same reasons of co-ordination and timing the Scandinavians also needed to know the dates of the solstices and equinoxes. One of the sources on this matter is attributed to a twelfth-century farm labourer, Star-Oddi Helgason, who is described as a man well versed in time reckoning and wise in many other respects (Vilmundarson and Vilhjálmsson 1991: 459).

The text attributed to Oddi, showing his astronomical capabilities, is called Oddi's tale (Beckman and Kålund 1914–16: 48–53; Ólsen 1914). It only covers a couple of pages in print. The text is clearly of Icelandic origin and there are no parallels in medieval literature so far as I know. The tale comprises three sections treating different aspects of the annual motion of the sun. The first gives the time of the summer and winter

No. from leap year	Summer or winter solstice	Date	Sun's direction	Time
0	S	Jun 15	SE	09:00
	W	Dec 15	N	00:00
1	S	Jun 15	SW	15:00
	W	Dec 15	E	06:00
2	S	Jun 15	NW	21:00
	W	Dec 15	S	12:00
3	S	Jun 16	NE	03:00
	W	Dec 15	W	18:00

TABLE 1. The date of the solstices and their direction according to the first section of Oddi's tale, together with the corresponding time, given in modern terms.

solstices, from a leap year and through the subsequent three years until the cycle is supposed to close so that the story repeats itself. The second section describes how 'the solar motion increases in sight' from winter solstice to summer solstice and then 'decreases' to the next winter solstice. The third section tells us the direction of dawn and nightfall through the year. We shall here focus only on the aspect of time-reckoning, i.e. on the timing of the solstices and equinoxes.

The first section of Oddi's tale not only informs us about the date of the solstice but also the 'direction' of the sun at solstice accurate to one eighth of the circle (see Table 1). This is an indirect way of giving the time of the day to within three hours (Ólsen 1914: 13; Þorkelsson 1926: 46–47). The dates given do not coincide with the ones we are used to, the difference being due to the accumulated error of the Julian calendar in the twelfth century. Thus the dating in the text adheres to the real solstices rather than following the ecclesiastical decree of dating the solstices almost a week later (Jun 21 and Dec 21). Another Icelandic author in the thirteenth century tried to do away with this discrepancy by supposing that these later dates were valid 'in the middle of the world' (Beckman and Kålund 1914–16: 121, 175), thus betraying that he was describing an Icelandic observation and believing his own eyes better than the church.

The interval between the solstices in Oddi's tale is incorrect. The interval from winter to summer solstice is not generally the same as that from summer to winter; the difference amounted to about eight hours in the year 1100 and three hours in the year 1200 (calculations based on Moesgaard 1975: 172), thus exceeding the accuracy indicated by Oddi. We conclude therefore that this aspect of the text is more likely an exercise in the recently accepted Julian calendar than a report on Oddi's own observations.

The main subject of the second section of the Oddi's tale is that of the 'increase of solar motion' from winter to summer solstice and the subsequent decrease. This section implies that Oddi places the equinoxes at equal intervals from the solstices, thus dividing the year into four equal parts. This assumption for the timing of the equinoxes would have been several days in error in the twelfth century. It seems to me that in points like this, Oddi was led more by a conception of symmetry and simplicity than by exact observation.

Origins and usefulness of Oddi's knowledge

In summary, Oddi's results may quite well be some kind of crystallisation of knowledge and motivation gained in the Viking Age. Besides this, he was living at a time and place close to the flourishing literary and cultural interest shown in the sagas.

However, as always in cases like this, we should beware of focusing too strongly on the person. There are many complicated reasons for Oddi's tale being known to us. One of the most important ones is that other people have found this kind of knowledge useful and considered it worthwhile to copy the text onto precious vellum again and again.

It is not too hard to imagine how the knowledge described in Oddi's tale would have been of use to his contemporaries in their daily fight for a living in circumstances where the optimal use of your powers may be of crucial importance. We have spoken here mostly of the first section of the tale, which was mainly useful as an exercise during the gradual introduction of the Julian calendar in Iceland. A clear understanding of this calendar has been necessary in order to adapt the old Icelandic calendar to it, which people did so well that we still have vestiges of the old calendar going strong in the country. The other parts of the tale seem mainly to have been useful for navigational purposes.

In all probability, the knowledge in Oddi's tale has not been gained through foreign books but through independent observations, maybe spanning several generations, and perhaps with a little help from oral (e.g. mathematical) information from the continent.

Imported books replace original observations

After the days of Star-Oddi in the twelfth century it seems that Iceland experienced a decline in the field of independent observations of nature. This coincides with increased literacy from around 1100 onwards and with all kinds of literary activities: the import of continental books, their reading and translation, the writing of old poems and new sagas and so on. At this time it gradually became easier to obtain valuable books (Vilhjálmsson 1990: 41–49) instead of, for example, looking firsthand at the sky and trying to make sense of the phenomena as Oddi had done. We have thus missed the chance to see what a full-blown astronomy of the North would have looked like.

Conclusions

In summary, the settlement of Iceland around year 900 demonstrates that the Scandinavians had reached a level of knowledge and skill sufficient to maintain regular ocean traffic and a co-ordinated society. A considerable part of this knowledge related to astronomy and other fields of natural science. This kind of knowledge was further developed by independent observations in Iceland after settlement, of which we have here dealt with one of the best examples. However, the advent of literacy and literary interest from the twelfth century onwards made it easier to obtain knowledge from books than from first-hand observations.

References

Beckman, N. (1916). Inledning [Introduction]. In *Alfræði íslenzk: Islandsk encyklopædisk litteratur: II. Rímtöl [Encyclopaedic literature on the calendar]*, by N. Beckman and Kr. Kålund, pp. i–cxciv. Copenhagen: S.L. Møllers.

Beckman, N. and Kålund, Kr. (1914–16). *Alfræði íslenzk: Islandsk encyklopædisk litteratur: II. Rímtöl [Encyclopaedic literature on the calendar]*. Copenhagen: S.L. Møllers.

Benediktsson, J., ed. (1968). *Íslenzk fornrit I: Íslendingabók, Landnámabók fyrri hluti [The Book of Icelanders and the Book of Settlements]*. Reykjavík: Hið íslenzka fornritafélag.

Einarsson, T. (1968). Hvernig fann Þorsteinn surtur lengd ársins? [How did Thorsteinn the Black find the length of the year?]. *Saga* 6, 139–42.

Einarsson, T. (1970). Nokkur atriði varðandi fund Íslands, siglingar og landnám [On the discovery of Iceland, sailings and settlement]. *Saga* 8, 43–64.

Foote, P. and Wilson, D.M. (1984). *The Viking achievement: The society and culture of early medieval Scandinavia*. London: Sidgwick and Jackson.

Ginzel, F.K. (1958). *Handbuch der mathematischen und technischen Chronologie, I–III*. Leipzig: J.C. Hinrichs (Nachdruck der Ausg. 1906–1914).

Graham-Campbell, J., ed. (1989). *The Viking world.* London: Windward.
Jones, G. (1986). *The Norse Atlantic saga: being the Norse voyages of discovery and settlement to Iceland, Greenland, and North America,* 2nd edn. Oxford: Oxford University Press.
Jones, G. (1990). *A history of the Vikings,* 2nd edn. Oxford: Oxford University Press.
McGrail, S. (1989). Ships, shipwrights and seamen. In *The Viking world,* ed. J. Graham-Campbell, pp. 36–63. London: Windward.
Marcus, G.J. (1980). *The conquest of the North Atlantic.* Woodbridge: Boydell.
Moesgaard, K.P. (1975). Elements of planetary, lunar and solar orbits, 1900 BC to AD 1900, tabulated for historical use. *Centaurus* **19**, 157–81.
Ólafsson, G., Vilhjálmsson, Þ. and Sigurðsson, Þ. (1992). Fornleifar á slóðum Stjörnu-Odda [Archaeology in the tracks of Star-Oddi]. *Árbók Hins íslenzka fornleifafélags [Yearbook of the Icelandic Archaeological Society],* pp. 77–123.
Ólsen, B.M. (1914). Um Stjörnu-Odda og Oddatölu [On Star-Oddi and Oddi's tale]. In *Afmælisrit til Dr. Phil. Kr. Kålunds bókavarðar við safn Árna Magnússonar 19. ágúst 1914* [a Festschrift, without editor], pp. 1–15. Copenhagen: Hið íslenska fræðafélag.
Schnall, U. (1975). *Navigation der Wikinger: Nautische Probleme der Wikingerzeit im Spiegel der schriftlichen Quellen.* Oldenburg: Gerhard Stalling Verlag.
Vilhjálmsson, Þ. (1989). Af Surti og sól: Um tímatal o.fl. á fyrstu öldum Íslands byggðar [On the calendar in the first centuries of settlement]. *Tímarit Háskóla Íslands [The University Journal]* **4**, 87–97.
Vilhjálmsson, Þ. (1990). Raunvísindi á miðöldum [Icelandic science in the Middle Ages]. In *Íslensk þjóðmenning, VII, [National Icelandic culture],* ed. F.F. Jóhannsson, pp. 1–50. Reykjavík: Þjóðsaga.
Vilmundarson, Þ. and Vilhjálmsson, B., eds. (1991). *Íslenzk fornrit, XIII: Harðar saga o.fl. [Icelandic medieval literature: the saga of Hörður and other sagas].* Reykjavík: Hið íslenzka bókmenntafélag.
Þorkelsson, Þ. (1926). Stjörnu-Oddi [Star-Oddi]. *Skírnir* **100**, 45–65.

7

The geometry of pastoral stone octagons: the Basque *sarobe*

ROSLYN M. FRANK AND JON D. PATRICK

Introduction

This report concentrates on findings resulting from archival research and a field study of stone octagon complexes associated with pastoralism in the Basque Country (Euskal Herria). The fieldwork was conducted primarily in Gipuzkoa and Nafarroa, two of the seven Basque provinces composing the modern-day Basque Country (Fig. 1). In the rural zones, where transhumant shepherding has survived to this day, stone octagons called *sarobe* in Basque (Euskera) have continued to be utilised and surveyed into the twentieth century. The octagons, constructed originally within the limits of the common lands, represented the multipurpose space within which the shepherds were required to build their huts and corrals. Traditionally, each stone octagon had its own name.

According to custom and, later, to written law codes, although during the daylight hours the animals and their keepers had access to pastures of the common lands, each night the shepherd group had to return with its flock to the confines of the group's own stone octagon to bed down. The members of each transhumant pastoral collective inhabited their lowland winter *sarobe* for six months. At the beginning of May they would drive their sheep and other livestock up to the summer *sarobe* in the highlands, only to journey back down to the winter stone octagon in the valley at the beginning of November.

The *sarobe*, initially thought to be unique to the Basque geographical zone, form part of a larger and more widespread body of knowledge and social practices associated with transhumant pastoralism in other parts of Europe. For example, a similar parcelling of grazing lands is found

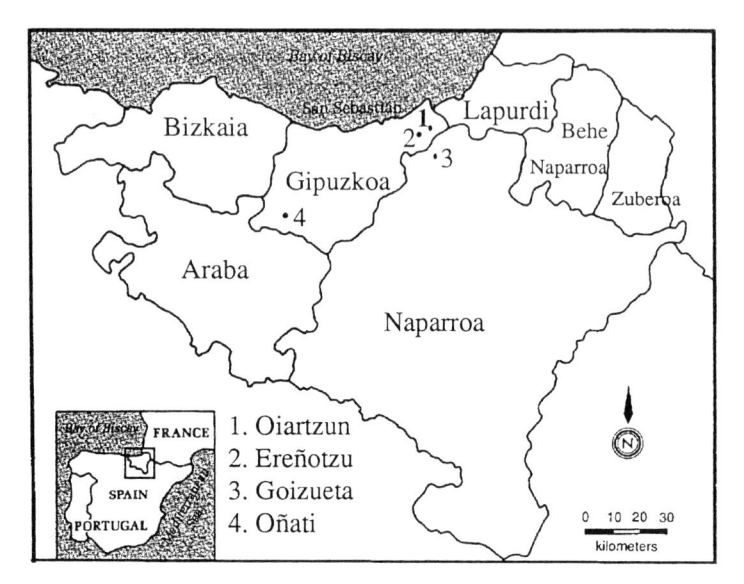

FIG. 1. Map of the Basque Country, showing the seven provinces.

across the northern region of the Iberian peninsula, in Santander, Asturias and Galicia. Evidence for their existence extends north into France and south into Spain as far as Extremadura (Caro Baroja 1973: 173–82; Haristoy 1883–84: 551 ff.; Lecuona 1959: 297–300). Also, there are indications that smaller stone enclosures, along with septarian units such as the 21ft 'perch', may have structured the practices of transhumance and the laying out of traditional land holdings in Ireland, Wales, England, Scotland and the Orkney Islands (Clauston 1932; Hadingham 1975; Hibbert 1831; MacKie 1977). Historically, in Euskal Herria the stone octagons are intimately linked to transhumant practices of the Basque-speaking shepherds of this zone, while the absolute *terminus ante quem non* of the design of the octagons themselves is quite uncertain.

Units of measurement

The 'geometric foot' and the *vara*

The basic unit of measurement employed in the *sarobe* of Gipuzkoa is the 'geometric foot' (g.ft). In practice, this unit is equivalent to one third of the *vara de Castilla* or 0·2786m. In 1568, the latter replaced the *vara de Toledo* as the official standard and ruler for Spain and its colonies abroad (*Nueva Recopilación* 1919: ley I, tít. XIII, lib. V). The length of the *vara de Castilla* coincides with that of the bar standard of Gipuzkoa. These rods date back at least to the Middle Ages and are derived from the same metrological traditions (Martínez Gómez 1816 [1795]).

The *gizabete* and the *toise*

In addition to being used in multiples of three, as in the *vara*, the geometric foot was the basis of the 7g.ft unit of measurement called *gizabete* in Basque and *toise* in French. It was this 7g.ft measuring rod that was employed by the French at the end of the eighteenth century to carry out the measurements of the earth's quadrant that led to the invention the modern-day decimal metric system. However, in France the 7g.ft *toise* unit had been converted earlier into six *pieds du Roi* (King's feet or royal feet). Consequently, at the time of the invention of the decimal metric system, in France, the *toise* was viewed as a unit of measurement within a sexagesimal system.

That the 7:6 ratio between the two standards was common knowledge is clearly documented in the historical record. For example, in his report dating from 1736, the Basque surveyor Villarreal de Berriz (1973 [1736]: 116–17) states:

> In a case I have a royal foot of Paris, done in bronze, extraordinarily precise and having measured it and compared it many times with the foot of Castille, I find that 6 feet of Paris, which is the *toise*, are equal to precisely 7 feet of Castille.[1]

Furthermore, the widespread awareness of the 7:6 ratio is demonstrated by the fact that Jorge Juan, the Spanish astronomer who measured the arc of the meridian in Peru, used it in his report on the 1752 expedition (García Franco 1957: 82). Thus, in reality, the decimal metric system itself resulted from surveys conducted using the *toise* standard of 7g.ft, a point overlooked by modern historians of science. The latter have always viewed the *toise* merely as being six 'feet' in length. As a result, they have failed to recognise that these 'feet' were 'royal' and that the *toise* unit of measurement was linked to an earlier septarian number system still being utilised by the Basques to lay out their stone octagons at the time of the creation of the metre.

---

[1] 'En un estuche tengo un pie real de París en bronce muy exacto, y aviéndolo medido, y cotejado muchas veces con el pie de Castilla, hallo, que 6 pies de París, que es la *toysa*, hace 7 pies justos de Castilla.' (Translation by RMF.)

The *hogeitabatoin/postura* unit of measurement

In addition, the seven-foot shepherd's staff called *gizabete* appears to be related to septarian units of measurement once found commonly throughout much of Europe. The Basques, for example, traditionally laid out their land holdings using a 21g.ft. pole, referred to as *hogeitabatoin* in Basque and *postura* or *pértiga* in Spanish. The latter term is related to the expression 'perch' in English. Also, it should be noted that the old Irish acre 'was measured by a perch of seven yards or 21 feet in length' (McKerral 1944: 47).

The *gorapila* 'knot'

In keeping with the traditional septarian measurement system, Basque surveyors worked with poles such as the 7g.ft *gizabete*, the 14g.ft *hamalauoin* and the 21g.ft *hogeitabatoin*. They also employed a 49g.ft unit called *gorapila* to lay out the stone octagons. In Euskera the term *gorapila* itself means 'knot' and knotted ropes were used for surveying purposes. Septuagesimal units based on the number 70 and its multiples, such as the 4900g.ft *gizalan*, were also utilised to measure surface areas.

The Basque sarobe

Cultural emplacement of the *sarobe*

Evidence for the standard design for the *sarobe* was first encountered in the law codes of Gipuzkoa. The legal records specify the design, construction and celestial orientation of stone pillars utilised to lay out the Basque pastoral octagons. Although in the rest of Europe stone-circle building was abandoned many millennia ago, in Euskal Herria the stone octagons have survived as functioning entities into the twentieth century. The sites have been fundamental organisational and structural components within the cultural assemblage of the Basque shepherd group. Therefore, they reveal specific socio-cultural contextualisations that have connected them intimately to the social architecture of Basque shepherd society (Ott 1981).

Each centre was astronomically aligned and linked to the theme of cosmic order. It represented and reaffirmed the identity of the people who built it, acting as the seat of government and the locus of religious rites, and, consequently, serving as the source for law, order and unity (Barandiarán 1973a; 1973b; Caro Baroja 1973; 1974; Galbete 1953; Leizaola 1977; Ott 1981; Peña 1901). The same standard units of length and septarian values employed to lay out the stone octagons served to structure the ritual space where public debates were held and judicial and political processes were conducted. Concomitantly, the *makila* staff standards provided the conceptual framework through which legal and political procedures were conceived to determine 'just measures'.

The configuration of stones at the site of Gerediaga is typical of the open-air assembly points linked to these social practices. In Gerediaga, in front of a hermitage, there was a large stone cross and a circle of 28 stones which 'were the seats of the representatives of the fourteen republics' belonging to that pastoral territory or *merindad* (Caro Baroja 1974: 371). In the centre stood a large rock serving as a table upon which the decisions of the assemblies were written. The stone configuration at Gerediaga, in addition to many other well-known judicial assembly points such as Gernika, was located next to a sacred oak tree. Reminiscent of similar structures among the Greeks and other European peoples, these stone figures hark back to the origins of Basque political society.

In summary, amongst the Basques the common septarian system of measures was enmeshed in a cosmological network of social practices and beliefs rather than functioning merely at an instrumental level, as is true in the case of the decimal metric system today in other parts of Europe (Lasa 1964; Balzola 1917 [1853]). Hence, the metrological standard, geometry and septarian number system should be understood as fundamental elements within the Basque cultural assemblage, as tools for structuring and giving meaning to the relationships and modes of thought found in the social architecture of the people in question. To examine these features in isolation is to fail to recognise that such a number system is a cultural product.

The nature of the *sarobe*

The Basque stone octagons are connected to pastoral practices that can be traced back to AD 853 in the written record of the Iberian peninsula (Caro Baroja 1973: 180). Documentary and archaeological evidence from Euskal Herria demonstrates the evolution of the *sarobe*. They appear to have been established initially as inhabited enclaves and clearings located inside communal woodlands and pastures. Over a period of many centuries many of these *sarobe*, which were exploited collectively by the community, passed into private hands. This transition was particularly common in the case of *sarobe* located in the valley lowlands, since their situation made them more desirable for agricultural purposes (Arregi 1980; Ormachea and Zabala 1988; Ugarte 1987).

Even though the *sarobe* often appear on maps in the form of a circle, in Gipuzkoa they were laid out in the shape of an octagon following the traditional septarian counting method documented in social practice

(Larramendi 1745: 280) and in Basque law codes (e.g. the *Cuadernos de Fueros de Guipúzcoa*, tít. 20, cáp. 2:268, for the years 1583 and 1696 in the *Recopilación de leyes y ordenanzas...* 1983 [1583] and the *Nueva Recopilación* 1919). The earliest known written record of instructions for the design of the *sarobe* is found in a law code from 1484, recently discovered and transcribed by researchers working on the collection of manuscripts housed in the Archivo General de Gipuzkoa in Tolosa (Carmen Alvarez, Begoña Irazu, Ana Otaegi and Gabriela Vives, pers. comm.).

The *sarobe* of Gipuzkoa were constructed with eight stones, four of which were placed at each of the cardinal points while the other four marked the inter-cardinal positions. The centre was marked by the placement of a stone called *haustarria* ('ash-stone'), under which ashes from the yuletide logs along with pieces of clay tile from the participating households were deposited. It is said that by means of this practice the centre stone could be distinguished from natural stones.

Each cardinal point stone was set on a radius of one *kordel* ('cord') and laid out using a special knotted rope divided into twelve *gorapila*, each of 49g.ft. Thus, the one-*kordel* radius yields a radius of 588g.ft. Each of the four inter-cardinal stones was placed at a distance of nine *gorapila* from the adjoining cardinal stone, giving a total perimeter of 72 *gorapila* (Frank 1980a; 1982). It is important to note that the legal records require that the 72-*gorapila* perimeter be constructed by using the chord distances of nine *gorapila*. Consequently, a ratio of 1:6 is achieved between the radius and perimeter. As a result of this legal requirement, while the radial distances from the *haustarria* to each of the four cardinal stones was set at twelve *gorapila*, the distance from the centre stone to each inter-cardinal stone was not twelve, but rather 11·49 *gorapila* (563 g.ft). Hence, in fact, the octagon was not set out as a 'circle'. Rather, the surveyors needed to exploit some other aspect of the geometry to achieve the correct layout of a perimeter of 72 *gorapila* (Fig. 2).

Furthermore, the law codes specify that the '*remate*' of the figure also must measure 72 *gorapila* and be carefully verified during the field surveying. Although the meaning of the term '*remate*' is somewhat unclear in this context, other references to it suggest that it was the apex created by the intersection of the tangents to a circle. In the case of the design of the *sarobe*, the only figure that we have been able to deduce as comparable to the *remate* and which conforms to the dimension of 72 *gorapila* is the perimeter of the rectangle formed by the contiguous halves of the adjacent sides (see Fig. 2). The specification of 72 *gorapila* in the perimeter and in the *remate* suggests that the number 72 also had special significance in the Basque number system.

The geometry of pastoral stone octagons: the Basque sarobe 83

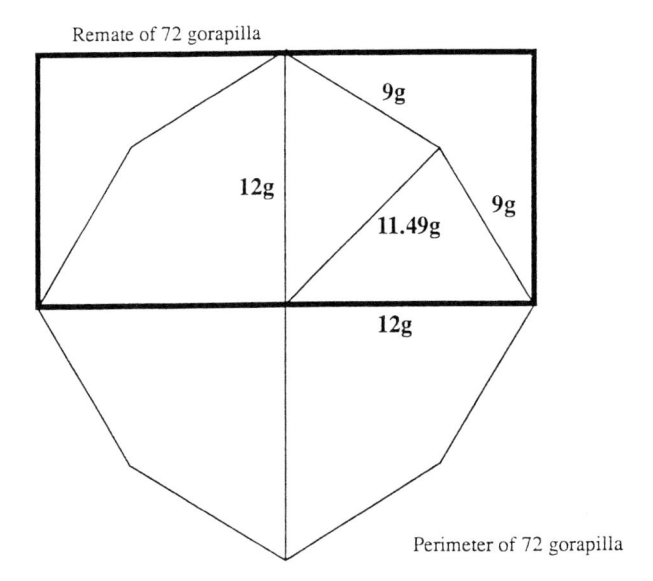

FIG. 2. Diagram showing the dimensions of the *sarobe* of Gipuzkoa.

In addition to the standardised design and dimensions of the stone octagons, the top surface of the *haustarria* itself was employed to map the location of the perimeter stones. On the top of the *haustarria* a small centre circle was frequently carved with eight radial lines pointing to the outer boundary stones. In one case a cross-like figure indicating the four cardinal directions was incised into the flat rock face.

Often, three or four smaller *lekukoak* or 'witness' stones were placed in the ground immediately next to the *haustarria*, although the exact configuration of these stones is not described in the law codes. From the sites we have seen, the actual practice varies greatly. At one location, no witness stones were discovered while the centre stone in question was a 1m-high pillar with a 20cm-square top. On this surface was the small centre circle with its eight radial lines pointing to the eight boundary stones that were still in place. In another case, local researchers discovered the *haustarria* with a broken piece nearby, indicating an original height of more than 1m. They also found that four small stones had been packed about the base of one side of the centre stone (Luix Mari Zaldua and Luis del Barrio, pers. comm.). Other field evidence for the use of 'witness' stones indicates that they are found only on three sides of the base of the outer stones and that they were placed at some small but unspecified distance away from them. In documents dating from the end of the seventeenth century which relate to land conflicts over the *sarobe* of the 'community of blacksmiths' and the local town council, there is a set of instructions for correctly laying out the octagon. It includes the placement of the four 'witness' stones next to the *haustarria* as well as the placement of three 'witness' stones next to each of the outer stones (Lecuona 1959: 297–300).

In Gipuzkoa, the standard *sarobe* measuring one *kordel* in radius is sometimes encountered next to or adjoining a double *sarobe* measuring two *kordel* in radius. From the data we have encountered in the other provinces of Euskal Herria, there is clear evidence that the *sarobe* octagons were laid out with radii of 6, 9, 12, 18 and 24 *gorapila*, i.e. 0·5, 0·75, 1, 1·5 and 2 *kordel*. The traditions governing the construction of *sarobe* with different radii, ranging from approximately 82m to 327m, are still unknown (Fermin Leizaola, pers. comm.).

The extent of the *sarobe*

Although there is no comprehensive field study of the Basque *sarobe* yet available, a review of materials from regional researchers in Euskal Herria gives a picture of the data that could appear. The study by Ugarte (1987) of the municipal district of Oñati, which covers about 150km^2, produced a list of 28 sites still extant and 71 referenced in archival sources. A study of the plans of land titles of the same district revealed 39 potential sites. An extrapolation of these figures suggests that at least a few hundred sites still exist and perhaps as many as five hundred or more could be identified in some way.

For example, a map showing the *sarobe* of Goizueta, a district of Nafarroa of about 100km^2, provides significant information concerning the large number of sites still identifiable in that zone only 130 years ago. Dating from 1863, this detailed map is a copy of an earlier 1792 map of the same zone. In addition to showing the location of each and every tree in the zone, it lists 113 *sarobe* by name, and indicates that about eighty of them were still extant with their centre stone in place. In the case of approximately thirty of the sites named and identified, the stones could not be found at the time of the map survey. Local researchers working in the area in recent years have been able to find only a few of the centre stones of these sites.

In documents relating to a 1773 law suit found in the archives of Real Chancellería de Valladolid, there are plans showing 22 *sarobe* in the Hernani district of the province of Gipuzkoa. Over the past decade, local researchers have been able to locate about half of these sites. While their original date of construction is unknown, the name of one of the 22 sites mentioned in the 1773 law suit can be traced back six centuries through documentary sources to the year 1189 (Luix Mari Zaldua, pers. comm.).

In another case, Inaki Arbelaitz, Juan Manuel Lekuena and Nicolas Etxebeste, a team of Basque researchers working in the municipal district of Oiarzun, spent many years conducting painstaking fieldwork along with an extensive review of toponymic data, without locating even one centre stone. However, their efforts recently paid off in the discovery of a very unusual *haustarria*. At present, the site and the stone are undergoing further analysis.

The dangers to which the *sarobe* sites are exposed is exemplified by the experience of researchers who spent two years looking for the centre stone of one site. They eventually came across it in a ditch where it had been left after being pushed aside by heavy equipment in a road-widening project. Hence the *haustarria*'s original location could not be identified. This particular centre stone was unique, being round with an outer circle around its rim as well as an inner circle along with the expected eight radial lines (Luix Mari Zaldua and Luis del Barrio, pers. comm.).

With respect to the overall state of repair of the sites visited, while a few are well-preserved with all eight outer uprights and the centre stone still in place, others have only their centre stone intact. In many other cases, few or no stones can be located even though the perimeter of the site is sometimes well demarcated by a fence, earthen wall or tree line. The outer dimensions of these sites could be recovered to within about five metres, but not more accurately.

The Ereñotzu sites

The best-preserved complex of sites that we have observed to date is located in the hills above Ereñotzu in the district of Hernani, about 10km south of Donostia (see Fig. 3). These sites are instructional for a number of reasons. First, all eight sites shown on the plan are still known by their original names: Listorreta, Pagota, Muniskue (Mulisko), Alkatxurain, Akola, Añua, Zaragueta and Arlotegi. The traditional practice of naming each octagon facilitates the identification of many other sites on the basis of local toponyms (Lekuona 1989; Ormachea and Zabala 1988) since the term *sarobe* is appended to the name, as is the case with the above sites, e.g. Listorretako Sarobe, the 'Sarobe of Listorreta'. In this municipal district it is clear from the toponymic record alone that at least twice as many sites once existed. Through recourse to medieval documents and law suits we have references to the existence of a total of 38 sites, including the eight mentioned above, which can be dated back to the Middle Ages (Ayerbe 1986-87). This fact indicates that the Ereñotzu sites themselves must have been constructed even earlier.

Second, as the plan shows (see Fig. 3), the sites are represented as circles and not as octagons, demonstrating that an awareness of their original geometry is no longer present in the mind of the modern mapmaker. Additionally, some sites show boundary perturbations in the form of either disjoint boundaries, e.g. the south-west corner of the Alkatxurain *sarobe*, or a pronounced oval shape, as in the case of the Akola *sarobe*. Furthermore, this complex reveals that sites can overlap each other, a factor for which no explanation has been forthcoming.

Also, large and small *sarobe*, i.e. those measuring two *kordel* (24 *gorapila*) and those measuring one *kordel* (12 *gorapila*) in radius, can be found in the same locality and at the same altitude. Previously, the reason given for the existence of these two different-sized *sarobe* was that they marked off the summer and winter habitats of the shepherds and their livestock. The dimensions of the small *sarobe* were explained by saying that they corresponded to the needs of the sheep: the small *sarobe* were said to be located in the highlands and used only during the summer months. The dimensions of the large *sarobe* were said to correspond to the size needed by the flock during the winter months in the lowland pastures (Iturriza 1884 [1775]). The fact that both types of *sarobe* are found together in the highlands, at the same altitude and adjoining each other, clearly makes this explanation no longer tenable.

Finally, six of the eight sites on the plan (Muniskue, Alkatxurain, Akola, Añua, Zaragueta and Arlotegi) are within one hundred metres of either megalithic dolmens or cromlechs. In this respect, the Basque scholar Barandiarán frequently stressed the link between the distribution of megalithic structures and the practice of transhumance by the pastoral population of Euskal Herria. He repeatedly noted that the megalithic structures are concentrated where there is abundant pasture, good soil, a nearby water source and shelter from inclement weather. These conditions also coincide with those that characterise the location of the stone octagon habitat areas of Euskal Herria (Barandiarán 1953). Undoubtedly, once complex sites such as those at Ereñotzu have been excavated and carefully carbon-dated, we will have a much better picture of the antiquity of the stone octagons and their possible relationship to the megalithic pastoral practices of the region. It should be recalled that the zones in which the stone octagon sites are found have been

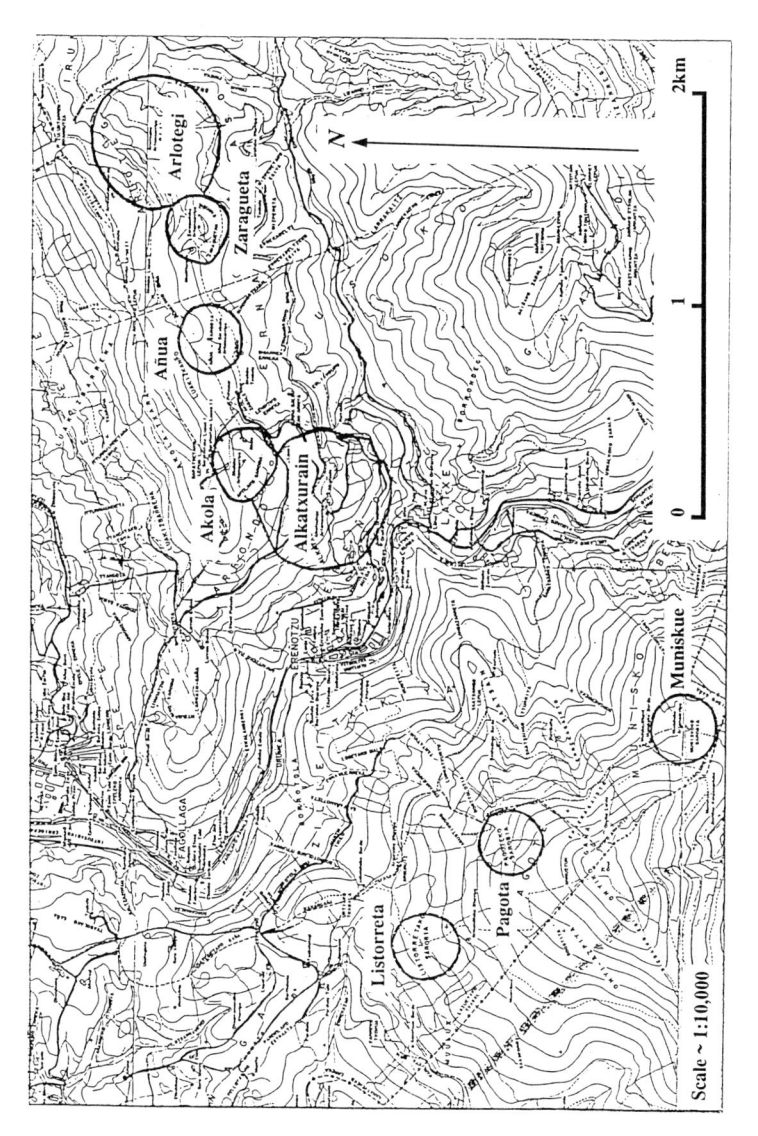

FIG. 3. Diagram of theEreñotzu sites.

occupied since time immemorial by Basque-speaking pastoral groups. Consequently, their transhumant mode of existence has not undergone the linguistic and cultural rupture which has been associated with pastoralism in other parts of Europe.

Implications for archaeoastronomy

Until now, investigations of western European megalithic archaeoastronomy have focused almost solely on the disposition of archaeological remains. This approach assumes that in western Europe, because of cultural and linguistic rupture brought about by the incursion of Indo-European speakers into the zone, no traces of the earlier indigenous pre-Indo-European cultural assemblage survive in the present-day practices and knowledge base of populations inhabiting areas where megalithic monuments are found. Such a position has been summarised by Ruggles (1984: 14):

> In some countries, archaeological studies benefit from evidence other than just the present disposition of archaeological remains: in Mesoamerica, for instance ...there exists ethnohistoric evidence (accounts by Spanish invaders of practices current when the sites were in use), ethnographic material which is clearly relevant (present-day practices by descendants of the groups being studied) and first-hand accounts (Maya and Aztec codices). However, none of these other sources is available in British work.

In the case of research in the Basque country, these other sources are abundantly available. In the zones we have studied, the transmission of the cultural practice from generation to generation has taken place in Euskera, a language that is recognised as being pre-Indo-European. Consequently, although at this time we cannot state that the stone octagons themselves can be dated back to the megalithic period, in Euskal Herria they clearly have functioned as an integral part of the Basque cultural assemblage and the practice of transhumance.

Significantly, the results of linguistic and serological studies of the Basques have uniformly demonstrated that they belong to the most ancient pre-Indo-European substratum of western Europe (Bernard and Ruffié 1976; Frank 1980b). In contrast to the situation of modern Indo-European-speaking populations of western Europe, the Basque language and people have not suffered the same kind of cultural and linguistic rupture with their own past. Hence, in Euskal Herria many strands of evidence are available to the researcher other than just the present disposition of archaeological remains, particularly for the investigator who speaks Euskera.

Conclusions

While references to the *sarobe* are found in the earliest medieval records, their *terminus ante quem non* is quite uncertain. Given their employment by transhumant pastoral groups to set out clearings within the common woodlands and pastures, the Basque stone octagons may date back to much earlier pastoral practices. Likewise, the numerical importance of the numbers 7 and 72 and the use of the seven-'foot' staff and 21-'foot' perch by this pre-Indo-European pastoral group should make it worthwhile to search for the presence of similar cultural artefacts among other transhumant traditions in Europe. Such an approach could contribute significantly towards building better foundations for the archaeoastronomical studies of these regions. In summary, the fact that such a tradition could be preserved so effectively and for so long suggests that the Basque cultural and linguistic assemblage deserves to be investigated in greater detail by archaeoastronomers.

Acknowledgements

We would like to express our thanks to the many individuals in Euskal Herria whose help and collaboration made this fieldwork project possible and most particularly to the following researchers: Inaki Arbelaitz, Abel Ariznavarreta, Luis del Barrio, Alvaro Carredono, Ignacio Carrion, Nicolas Etxebeste, Fermin Leizaola, Jose Manuel Lekuona, Josu Tellabide and Luix Mari Zaldua. Also, we would like to extend our special thanks to Antonio Aranburu and the Udaletxea of Hernani. We are particularly indebted to the brothers Jesús and Ceferino Etxegia for their aid on our hunting expeditions through the forests and mountains of Goizueta and Erasun in search of the elusive *haustarria*. Finally, we would like to express our thanks to the Diputación Foral de Gipuzkoa and most especially to Julián Peña for sharing with us his knowledge of the aerial photographic record of Gipuzkoa. The co-operation of all these individuals and their willingness to work many long hours with us, often under difficult conditions, was a fundamental element in the collection of data used in this paper.

References

Arregi, G. (1980). Auzo. In *La etnia Vasca: Euskaldunak*, ed. E. Ayerbe, vol. 3, pp. 601–56. San Sebastián: Etor.

Ayerbe, M.R. (1986–87). Sobre el hábitat pastoril y las pasturación de ganado en el Valle del Urumea (Guipúzcoa). *Acta Historica et Archaeologica Mediaevalia* (Barcelona) **7-8**, 311–20.

Balzola, P. de (1917 [1853]). *Tablas de correspondencia de todas las pesas y medidas de Guipúzcoa y las principales del extranjero con las del sistema métrico.* San Sebastián: Imprenta de la Provincia.

Barandiarán, J.M. (1953). Aspectos sociográficos de la poplación del Pireneo Vasco. *Eusko Jakintza* **7**, 3–26.

Barandiarán, J.M. (1973a). Contribución al estudio de los establecimientos humanos y zonas pastoriles del País Vasco. In *Obras completas*, vol. III, pp. 275–344. Bilbao: Editorial de la Gran Enciclopedia Vasca.

Barandiarán, J.M. (1973b). Vida pastoril vasca. Alberques veraniegos, trashumancia intrapirenaica. In *Obras completas*, vol. V, p. 398. Bilbao: Editorial de la Gran Enciclopedia Vasca.

Bernard, J. and Ruffié, J. (1976). Hématologie et culture: le peuplement de l'Europe de l'ouest. *Annales, Économies, Sociétés et Civilisations* **31(4)**, 661–76.

Caro Baroja, J. (1973). *Estudios Vascos.* San Sebastián: Editorial Txertoa.

Caro Baroja, J. (1974). *Ritos y mitos equívocos.* San Sebastián: Editorial Txertoa.

Clauston, J.S. (1932). *A history of Orkney.* Kirkwall: W.R. MacIntosh.

Frank, R.M. (1980a). The Basque stone circles and geometry. *Archaeoastronomy* (Center for Archaeoastronomy) **3**, 29–33.

Frank, R.M. (1980b). *En torno a un mito: el Euskera y el Indoeuropeo.* San Sebastián: Hordago.

Frank, R.M. (1982). The Basque Nautical League and terrestrial geometry. *Archaeoastronomy* (Center for Archaeoastronomy) **5**, 24–29.

Galbete, V. (1953). Algunas medidas empleadas en el antiguo Reino de Navarra. *Príncipe de Viana* (Pamplona) **14**, 395–400.

García Franco, S. (1957). *La legua náutica en la Edad Media.* Madrid: Consejo Superior de Investigaciones Científicas, Instituto de Marina.

Hadingham, E. (1975). *Circles and standing stones.* New York NY: Walker and Company.

Haristoy, P. (1883–84). *Recherches historiques sur le Pays Basque* (2 vols.). Bayonne: E. Laserre.

Hibbert, S. (1831). Memoir on the Tings of Orkney and Shetland. *Archaeologia Scotica (Transactions of the Society of Antiquaries of Scotland)*, **3**, art. XIV.

Iturriza, J.R. (1884 [1775]). *Historia general de Vizcaya comprobada con autoridades y copias de escrituras y privilegios fehacientes... escrita... en Berriz, año de 1775.* Barcelona: Subirana.

Larramendi, M. de (1745). Introducción. In *Diccionario trilingüe de castellano, bascuene y latín*. San Sebastián: Bartholomé Riego y Montero. Two volumes.
Lasa, J. (1964). Las luchas en torno a los seles y caseríos de Albitxuri. In *Homenaje a Don José M. Barandiarán*, vol. 1, pp. 157–88. Bilbao: Diputación de Vizcaya.
Lecuona, M. de (1959). *Del Oyarzun antiguo. Monografía histórica*. San Sebastián: Imprenta de la Diputación.
Leizaola, F. (1977). *Euskalerriko artzaiak*. Donostia: Etor.
Lekuona, A. (1989). Sarobeak eta Herri-Lurrak. *Oiarzun* (Gipuzkoa), 21–22.
McKerral, A. (1944). Ancient denominations of agricultural land in Scotland: a summary of recorded opinions, with some notes, observations and references. *Proceedings of the Society of Antiquaries of Scotland* **78**, 39–80.
MacKie, E.W. (1977). *The megalith builders*. Oxford: Phaidon Press.
Martínez Gómez, V. (1816 [1795]). *Manual de comercio en que se halla la descripción de las monedas, pesas, y medidas que se usan en los reinos de España*. Madrid: Imprenta de la Viuda de Barco.
Nueva recopilación de los fueros, privilegios, buenos usos y costumbres, leyes y ordenanzas de la M.N. y M.L. provincia de Guipúzcoa (1919). San Sebastián: Imprenta de la Provincia.
Ormachea H., A. and Zabala U., A. (1988). Espacios ganaderos en la Vizcaya del antiguo régimen. In *Veinte-cinco años: Facultad de Filosofía y Letras, II: estudios de Geografía e Historia*, ed. L.M. Villar, pp. 401–28. Bilbao: Universidad de Deusto.
Ott, S. (1981). *The Circle of Mountains: a Basque sheep-herding community*. Oxford: Clarendon Press.
Peña, V. de la (1901). De los seles. In *Derecho consuetudinario de Vizcaya*, pp. 92–95. Madrid: Asilo de los Huéfanos del Sacrado Corazón de Jesús.
Recopilación de leyes y ordenanzas de la M.N. y M.L. provincia de Guipúzcoa (1983 [1583]). San Sebastián: Diputación Foral.
Ruggles, C.L.N. (1984). *Megalithic astronomy: a new archaeological and statistical study of 300 western Scottish sites*. Oxford: British Archaeological Reports (BAR British Series, 123).
Ugarte, F.M. (1987). Los seles el Valle de Oñate. *Boletín de la Real Sociedad Bascongada de los Amigos del País* **23**(3–4), 447–510.
Villarreal de Berriz, P.B. (1973 [1736]). *Máquinas hidráulicas de molinos y herrerías y gobierno de los árboles y montes de Vizcaya*. San Sebastián: Sociedad Guipuzcoana de Ediciones y Publicaciones de la Real Sociedad Vascongada de los Amigos del País y Caja de Ahorros Municipal de San Sebastián.

8

The moon and Indo-European calendar structure

EMILY LYLE

One recent shift in the study of Indo-European culture has been in the direction of an increased awareness of its underlying dualism (see, e.g., Littleton 1987: 212 and Mallory 1989: 140–41), and it will be useful to look at the implications of this shift for our understanding of the organisation of time, especially as it relates to the moon and the calendar year. There are three possible dualities to be perceived in the lunar cycle and cultures that incorporate it into a dualistic system may select one or more aspects from this range. These dualities are:

- waxing moon and waning moon;

- maximum moon (the half containing the full moon) and minimum moon (the half containing the crescents); and

- the period of the moon's visibility and the period of its invisibility.

I have suggested that Indo-European culture made use of all three (Lyle 1991: 71), but I am concerned here only with the last pairing, which contrasts two periods of greatly unequal length. The short period of invisibility is familiarly known as the dark moon, and I call the long period of visibility the 'light month' following the German scholar, W.H. Roscher, who employed the term '*Lichtmonat*' for the 27/28-day month of which he found indications in Classical antiquity (Roscher 1903: 5, 68–71). The actual lengths of light month and dark moon vary inversely (with a longer light period involving a shorter dark period and vice versa) within the total period of the synodic month of 29/30 days. The longest possible period of the light month is 28 days but this is of rare, or relatively rare, occurrence;[1] the marginal existence of the 28th-night moon is caught rather interestingly by the Rindi on the island of Sumba in Indonesia at latitude 10° S who call it the 'flickering/contested moon (i.e., open to debate whether still visible or not)' (Forth 1983: 54).

Eliade (1955: 86–88) has brought out the importance of the lunar cycle for religious thought about death, seen as corresponding to the disappearance of the moon, and rebirth, seen as corresponding to the appearance of the young crescent. Where this kind of thinking is current, death cannot be divorced from regeneration, and so the period of dark moon may be seen as a time of gestation preceding birth. Darkness is known to have preceded light for the Indo-Europeans in the case of the 24-hour day, which began at sunset (Schrader and Nehring 1917–29: 2.505), and it is probable that the lunar sequence was ordered in the same way although I know of no direct evidence on this point.

Each whole—the 24-hour day or the synodic cycle—consists of two complementary parts, those of darkness and light. In both cases, the two parts have a natural existence, but calendars impose cultural structuring on time even more than they follow the lines of nature (see Piccaluga 1987) and, when we come to the Indo-European calendar year, we seem to find the cultural imposition of a complementarity of darkness and light which has no basis in the natural year.[2] It seems that the lunar

---

[1] The degree of rarity has not yet been established but Lis Brack-Bernsen informs me (pers. comm., Oct 1990), that only one of the forty light months that can be traced in the Babylonian records of 652–262 BC is of 28 days (see Sachs and Hunger 1988: 78–81). Bradley E. Schaefer writes (pers. comm., Nov 1990): 'Under extraordinarily good conditions, it is possible to have a 28-day light period. For example, the observations in [Bortle 1990] lead to this.' Roger W. Sinnott of *Sky and Telescope* (pers. comm., Dec 1990) draws my attention to the point that the observation with binoculars by John E. Bortle of the young crescent on the night of Apr 25–26, 1990, and the naked-eye observation by Stephen W. Bieda, Jr. (see Bieda 1991) of the old crescent on the night of May 22–23 provide a record of a 28-night period of visibility. He adds the following explanatory comments: 'Moreover, the fact that they were at different locations makes no difference in this case. Since Bieda was both west and south of Bortle, it would have been *easier* for him to see the April 25–26 young moon than it was for Bortle. See also Jeffrey Jones' photograph [Bortle 1990]. The fact that Bortle had to use binoculars is explained by the clouds at his site. April 25–26 was an ideal night for naked-eye sightings of the moon; there are many cases of naked-eye sightings when the moon was less than 20 hours old.' I am most grateful to all my correspondents for their information. Since the moon is occasionally visible on the 28th night, it may be worth keeping in mind the light month as well as the sidereal month when considering the 28 lunar mansions in the Arabic and Chinese traditions.

[2] This assumes that the Indo-European calendar was not constructed by people living inside the Arctic Circle.

cycle may have provided the germ of the idea for the structuring of the year calendar and I want to explore that possibility here.

If the dichotomy between light and darkness was to be represented in the year calendar, there were two possible models: the day and the month. If the 24-hour day was taken as model, winter would correspond to night and summer to day, i.e. darkness and light would be represented by periods of roughly equal length. On the other hand, if the synodic month was taken as model, only a short period would correspond to dark moon and most of the year would correspond to the light month. It can be suggested that the short period marked by reversal which is found in Indo-European tradition (for example the Saturnalia, when slaves were waited on by their masters, and the Christianised festival of the twelve days of Christmas, when a mock king ruled) corresponds to dark moon and is an indication that the Indo-European ritual year was modelled on the month.

I should sketch in a little background to this idea. Indo-European calendars, such as those of the various Greek cities or the Gaulish Coligny calendar, work with a lunar year of 354 or 355 days and intercalate an additional month when required to keep the months in line with the seasons (see Samuel 1972; Duval and Pinault 1986). Where does the period of reversal fit into this pattern? There have been two answers to this question, both of which are compatible with a lunar model for the year. One is that the twelve-day period for which there is traditional evidence was outside the lunar year and was an intercalation giving approximate solar years of 366 and 367 days. This explanation interested Frazer who explored the idea in *The Golden Bough*, but he pointed out himself that, if this intercalation had been practised, it is rather strange that we find whole month intercalation in the historical calendars (Frazer 1913: 342). An alternative answer to the question of the location of the twelve-day period is that it was placed inside the lunar year as a supplement to a period of 342 days consisting of twelve schematic light months of 28 days plus six festival days. This is my own suggestion and a detailed model of a hypothetical early calendar including these features may be found in Lyle (1991: ch. 6). I will not pursue this here, but will consider the design of the structure, whether this consists of a solar year incorporating the twelve reversal days or a lunar year incorporating them. Whichever year is seen as a totality consists of twelve days followed by twelve months. It is quite clear that the days come first since it was held possible to foretell the weather of the twelve months to come by observations on each of the twelve days of Christmas. The twelve-day period is a miniature of the year.[3]

---

[3] Granet (1950: 107–8) finds a pattern like this in Chinese calendrical tradition.

There is an additional point about an Indo-European way of conceptualising time which makes it possible for us to realise how imperative it might have been for the year to have a period which was treated ritually as 'dark' although it was not this in actuality. Some of the Vedic hymns indicate that the sun, after completing its day journey from east to west, rose again into the sky with a dark side turned towards the earth and retraced its journey back to the east invisibly during the night (Lyle 1991: 93–95, 163). In other words, the period of darkness covers a reverse movement back to the starting point. With such a view of the temporal process, the year would have *had* to have a period of 'darkness' in which this reverse movement could take place.

Cyclic time is often contrasted with linear time in a simple opposition, as if there were nothing problematic about the notion of cyclic time in itself. However, there are actually two modes of cyclic time: the constant circling as of a freely turning wheel, which is the more familiar idea, and the tracing and retracing of a path, as in the partial rotary motion of a steering-wheel or in to-and-fro pendulum movement (see Tuan 1978: 7, 14; Lyle 1991: 62–66, 92–104). Both the circling and reversing movements return periodically to the starting point at which a new cycle commences, but it is worth noting which process is involved since it may give vital clues to the mode of thought. The reverse movement of the sun corresponds to the unlucky anti-clockwise or 'widdershins' turning associated with death, and it is this that seems to be represented in the 'dark' period of the year.

It can be suggested that, since the totality of human experience includes life and death, both were projected onto the light and dark periods of the day and the month and both were represented (in a way that followed the lunar model of unequal lengths of time) in the ritual year. It is against a background of this sort, I think, that old world intercalary periods can be most fully understood. It seems likely that there was a convergence between the practical solution of the problem set by the movements of the sun and moon and the human perception of a pattern of darkness and light which was much older than any luni-solar calendar. If there was such a convergence, there would have been no difficulty about incorporating the extra days of a luni-solar year into the conceptual system. Whenever the desire for a means of correlating the lunar and solar years arose, it would have been possible—and perhaps inevitable—to think of the extra days as the time of death and the months as the time of life.

Old world calendars have been studied much more in terms of history of science than in terms of the total cultures that employed them (cf. Piccaluga 1987: 7), but we cannot afford in the old world, any more than in the new (where fine integrated calendar studies have been produced), to neglect the gods whose stories were interwoven with the ritual year.

For example, the to-and-fro movement of reversing time already referred to appears to be expressed in the Indian myth of the churning of the ocean in which the *devas* led by Indra and the *asuras* led by Yama, god of death, pull alternately on a snake wound round the world mountain. The complementary forces of light and darkness are, I have argued (1991: 62–65, 105–18), expressed in a pair of opposed gods throughout the Indo-European world and beyond. This makes it seem unwise to cut Indo-European structuring off too sharply from that of other old world peoples, and there is a particularly close parallel with the Egyptian gods, Horus and Seth, which invites us to look at Egyptian calendar structure in the light of what has been said above.

Both the posited Indo-European ritual year calendar and the Egyptian civil calendar have the same design: a short period extraneous to the months followed by a period of twelve months (Parker 1950: 51, 53; Piccaluga 1987: 10). Since the Egyptian calendar has twelve schematic months of thirty days and an intercalary period of five days, the superficial differences are considerable when the two calendars are looked at simply as day counts. However, looked at in terms of a structure of darkness and light, the correspondence between the two becomes obvious, and I suggest that the two sets of days extraneous to the sets of months are alike, not simply in both being liminal periods, but specifically in being tied analogically to the dark moon. Since it appears that the Egyptians had a lunar year calendar before they had a solar year one (Parker 1950), it may be that there is a shared calendar structure lying behind the Egyptian and Indo-European calendars. At any rate, in the Indo-European structure which is our main concern, it can be claimed that the moon both fills the greater part of the year with twelve months and also supplies the model in its dark phase for a short complementary period preceding the months of the year.

References

Bieda, S.W. (1991). A lunar double whammy. *Sky and Telescope* **81**, 5–6.

Bortle, J.E. (1990). April's old and young moon. *Sky and Telescope* **80**, 215.

Duval, P.-M. and Pinault, G. (1986). *Recueil des inscriptions gauloises, vol. 3: Les calendiers (Coligny, Villards d'Héria)*. Supplement to *Gallia* no. 45. Paris: Éditions du Centre National de la Recherche Scientifique.

Eliade, M. (1955). *The myth of the eternal return*. London: Routledge and Kegan Paul.

Forth, G. (1983). Time and temporal classification in Rindi, eastern Sumba. *Bijdragen tot de Taal-, Land- en Volkenkunde* **139**, 46–80.
Frazer, J.G. (1913). *The golden bough*, 3rd edn. *Part VI: The Scapegoat*. London: Macmillan.
Granet, M. (1950). *La pensée chinoise*. Paris: Albin Michel.
Littleton, C.S. (1987). Indo-European religion: history of study. In *The encyclopedia of religion*, ed. M. Eliade, pp. 7.204–13. New York NY and London: Macmillan and Collier Macmillan.
Lyle, E. (1991). *Archaic cosmos: polarity, space and time*. Edinburgh: Polygon.
Mallory, J.P. (1989). *In search of the Indo-Europeans: language, archaeology and myth*. London: Thames and Hudson.
Parker, R.A. (1950). *The calendars of ancient Egypt*. Chicago IL: University of Chicago Press.
Piccaluga, G. (1987). Calendars: an overview. In *The encyclopedia of religion*, ed. M. Eliade, pp. 3.7–11. New York NY and London: Macmillan and Collier Macmillan.
Roscher, W.H. (1903). Die enneadischen und hebdomadischen Fristen und Wochen der ältesten Griechen. Monograph no. 4 in *Der Abhandlungen der philologisch-historischen Klasse der königlichen sächsischen Gesellschaft der Wissenschaften*, vol. 21. Leipzig: Teubner.
Sachs, A.J. and Hunger, H. (1988). *Astronomical diaries and related texts from Babylonia, vol. 1: Diaries from 652 BC to 262 BC*. Vienna: Verlag der Österreichischen Akademie der Wissenschaften.
Samuel, A.E. (1972). *Greek and Roman chronology*. Munich: Beck.
Schrader, O. and Nehring, A. (1917–29). *Reallexikon der indogermanischen Altertumskunde*, 2nd edn. Berlin and Leipzig: Gruyter.
Tuan, Yi-Fu (1978). Space, time, place: a humanistic frame. In *Timing space and spacing time, vol. 1: Making sense of time*, ed. T. Carlstein, D. Parkes and N. Thrift, pp. 7–16. London: Arnold.

9

Some remarks on the moon cult of Teutonic tribes

EMÍLIA PÁSZTOR

Introduction

In the sixth century, most of the Carpathian Basin was under the jurisdiction of two warlike Teutonic tribes. The eastern part, called Tiszántúl, had been occupied by the Gepids since AD 300, but the Longobards came into North Pannonia from the delta of Elbe only after the death of Theodoric the Great. The great war between the Goths and Byzantines brought Sirmium, a town important even for the Romans, into the hands of the Gepids. But taking possession of Sirmium was a disastrous political blunder on their part, for the Romans only had to await their chance to drive them from there, and from the whole Carpathian Basin as well. This opportunity soon arose in the person of Auduin, leader of the Longobards, who formed a league with emperor Justinian against the Gepids in about 545–46. War broke out the next year, under the pressure of which Thurisind the Gepid leader asked for a two-year armistice. At the end of this, the tribes were preparing for another battle. Prokopius (1966), a contemporary, writes about this battle in his 'Bellum Gothicum'. He tells an odd story of something that happened to both armies at the same time. The two camps were not far from each other but still out of sight, when a helter-skelter flight broke out and without any cause the soldiers ran away, with only the chieftains and their escorts remaining. They then concluded another two year armistice.

What could be the event that was observed by both tribes at the same time although they were out of sight of each other? What made both armies take to flight, panic-stricken? The phenomenon (or phenomena) must have caught the eye and is likely to have taken place after dusk when the chieftains and their escorts were carousing in their tents and getting so drunk (on mead) that they could neither understand nor follow their fighters. Sevin (1955), who has dealt with the history of the Gepids, chooses the year 549 as the date of the battle (although there has never been a written proof of this) and claims (despite previous assertions), quoting Ginzel (1914), that a solar eclipse could not have been observed in that year from the Carpathian Basin. He also rejects the idea of the flashing of a bright comet, for according to him it could not give enough light to inspire such fear. He agrees nonetheless that Prokopius's description refers unambiguously to some kind of natural phenomenon.

In this paper, I shall consider a possibility not considered by Ginzel, namely that a lunar eclipse occurred and caused headlong fear amongst the armies. Evidence in support of this idea comes from Roman accounts and from Teutonic mythology. Astronomical and historical evidence then seem to confirm the date of 549 as that of the battle.

The historical and mythological evidence

From both Caesar's (1917) and Tacitus's (1943) accounts it is well known that the Teutonic people laid great emphasis on the importance of lunar cycles when they wanted to attend to a variety of matters. In Caesar's opinion the Teutons reckoned amongst their gods only those whom they could see or whose help they could experience materially, namely *Sol*, *Vulcanus* and *Luna*: the Sun, Fire and the Moon. Although Caesar's information is not entirely correct—it is known that the Teutons also venerated other gods—his emphasis shows that he regarded these three as the most important. He mentions the role of the moon in two other cases when he says that, before the decisive battle with Ariovistus, the Teuton chieftain postponed fighting because 'according to Teutonic custom the mothers of families decided by several omens and oracles whether it was worth having a brush with the enemy or not, and they declared that Fate would not let the Teutons win the day if they took the field before new moon'. They carved magic symbols on their weapons that they should help them to get the victory during the battle. The moon may also be found amongst the magic symbols on two lance-heads that have come to light in Kovel and Müncheberg (Krüger 1983: 322).

From the second comment of Caesar it can be supposed that the Teutons considered the course of the moon the basis of their early calendar: 'they measure the progress of time by the number of nights but not days'. This is also confirmed by Tacitus, who lays further emphasis on the importance of lunar cycles: 'they assemble together unless some unexpected incident occurs on a definite day at new or full moon, for they believe this to be the most blessed time to start attending to their business or affairs. They do not count by days as we do, but by nights. An appointed date always begins with the night.'

An approximate picture of old Teutonic time reckoning can be obtained from Ginzel's (1914) investigations. He found that the days of the year were counted using the cycle of lunar phases. The name of the month also refers to this: *Monat*. A leap month was applied to synchronise the calendar to the movement of the sun. The week with seven days began to gain ground before the adoption of Christianity. Later, when time reckoning developed further (with the use of the so-called rune calendar which became general from the thirteenth century), it was still a basic problem to correct the difference between the computed and the actual time of new moon, i.e. the start of the new month.

Let us now see what we can tell about the moon from old Teutonic mythology. There are gaps in our knowledge of this, for the Teutons were converted to Christianity at an early stage and this caused deep changes in their life. In Scandinavia, to which Christianity spread later, mythology kept old customs in existence longer. It survived on the island of Iceland up to the so-called Icelandic Renaissance in about the thirteenth century (see Vilhjálmsson, this volume). Here pre-Christian cultural heritage was shielded with solicitous care. The beginnings of Teutonic religion go far back into the distant past. Scandinavia can be said to be lucky in this respect since its rock engravings or pictures preserve traces of Indo-European invaders' funeral and fertility cults from the second half of the second millennium BC. Amongst the several symbols representing ships, ploughing, carts, people, and animals there are numerous symbols which are termed solar by experts. Their different forms and shapes may represent a difference in style or in the various outward forms of the sun at different times (time of day, month, season and so on) but I think it is quite plausible that some of the many kinds of symbols may refer to the moon.

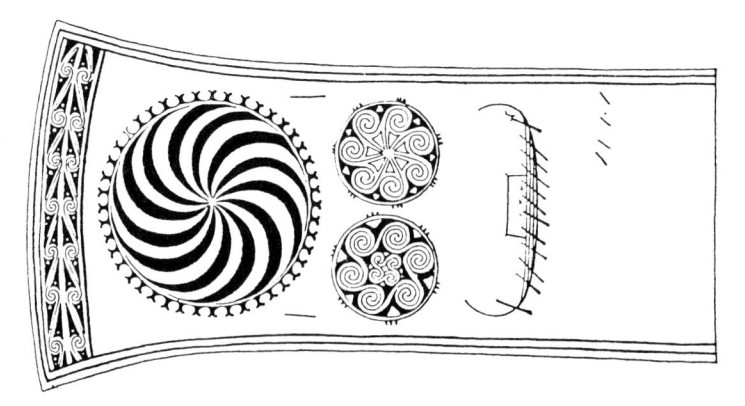

FIG. 1. Stone picture originating from the island of Gotland, probably dating from the fifth or sixth century.

Abb. 43. Sternornamente.

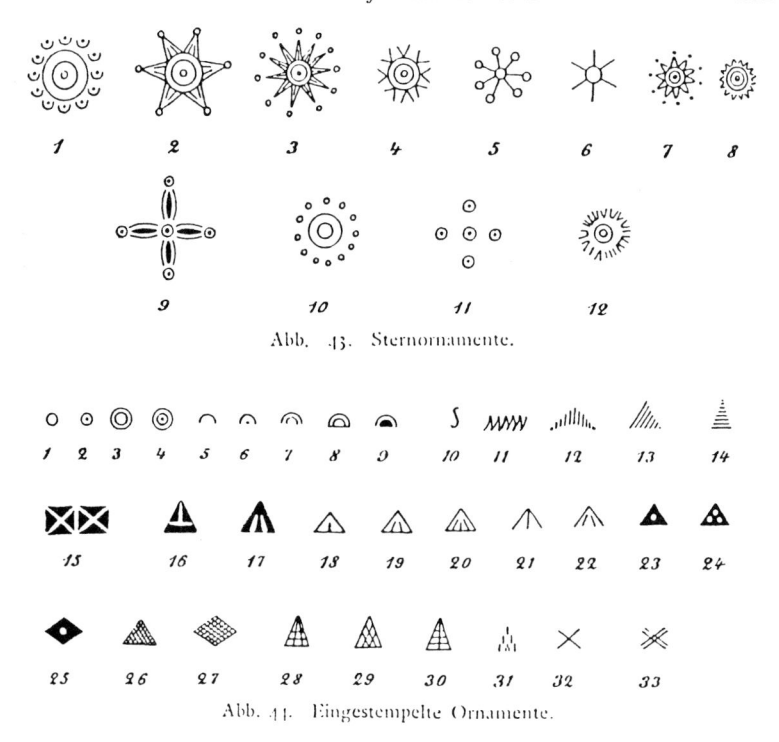

Abb. 44. Eingestempelte Ornamente.

FIG. 2. Typology of Teutonic fibulae from the third to seventh centuries. After Åberg (1919: figs. 43, 44).

An example is a stone picture originating from the island of Gotland (Broby-Johansen 1979; Nordén 1923). It probably dates from the fifth or sixth century and depicts a larger spiral with two smaller circles beneath it and, below these, a ship (Fig. 1). The rightmost of the two small circles has seven whorls and seven hatches of triple bundles while the left one contains six of each. Hoppál et al. (1990) consider these symbols to be histograms belonging to the sun and moon respectively.[1] Could the left circle, then, be a lunar symbol?

Åberg (1919) collected and made a typology of Teutonic fibulae from the third to seventh centuries whose feet were decorated with astral symbols (which he named star symbols). Most of these fibulae originate from German lands but there are some from Scandinavia. He took the most

---

[1] According to Hoppál et al., the number seven belongs to the sun as well as being a sacral number because the sun is able to shine into the seven corners of the world: primitive cosmogony symbolises space using the four cardinal points and the trinity 'underneath–halfway–above'. The sacral number of the moon in astral mythology is three (for its phases) and its multiples.

decorative symbols to be astral and treated all the others as stamps or seals (Fig. 2). Amongst these can be found the solar symbols well known from the rock engravings, and also ones which are the same or similar but divided into two. In my opinion these symbols represent the sun and the moon.

In Scandinavian mythology, the Moon (named *Mani*) is the Sun's brother. He has an important role, for he directs the motion of the stars. This role is depicted, for example, in a stone picture from the church at Håggeby, Uppland, Sweden (Fig. 3) in which two chargers with a crescent on their heads struggle against each other.[2]

The archaeological finds of the Carpathian Basin are no match for the Teutonic or Scandinavian ones (Pulszky 1897). They are relatively few and generally contain hardly any pictures or symbolic representations. Luckily, there are exceptions amongst the richly furnished graves, such as a wonderful necklace with lunulae encrusted with small red stones from a princess's tomb at Bakodpuszta (Fig. 4). A few more gold bracelets exist whose origin is still debated because they entered museums as stray finds at the end of the nineteenth century. They appear Teutonic. Their common characteristic is that the ends of the bracelets are finished with two crescents turning opposite each other (Fig. 5). The smallest one is from Transylvania and, oddly, its diameter is so small that it hardly fits even a girl's hand.

FIG. 3. Stone picture from the church at Håggeby. After Schlette (1971).

---

[2] The horse occupies a special position in Indo-European mythology because of its role in the economy and migration of the old Indo-European peoples. It is the symbol of the depths of the Earth and the Waters, light, the moon, sexuality, death, magic and so on. The Christian priests knew well enough that they had to make use of pagan symbols if they were to make people adopt Christianity. Thus the horse, as the symbol of Hel, the goddess of darkness (as well as the nether world), was admitted into the church even though Christianity regarded the sacrificial feast of horse-flesh as desecration.

FIG. 4. Necklace with lunulae encrusted with small red stones from a princess's tomb at Bakodpuszta. It is about 14 cm across. After Fettich (1951: pl. II). Reproduced by courtesy of Archaeologia Hungaria publishers.

One can find references both in the old or poetic Edda and in the new or epic Edda to how the old Teutonic people of North Europe imagined the start and the end of the Universe and what role they intended for the Moon and the Sun at that time (Dömötör 1965). On his adventurous journey King Gylfi got to know that at the end of the world wolves will carry off the Sun and the Moon. One of these wolves (called Ragnarök in old Icelandic), foreshadowing the world's end, could be the wolf Fenrir who was one of the three monsters from the nether regions and who was begotten by the god Loki. According to every oracle, Fenrir was created to destroy the gods, so he was fastened with a magic chain. One can see a dog- (or wolf-)like creature amongst the astral symbols on a disc-shaped brooch from Tangendorf (Krüger 1983: 237). Might this be Fenrir? He certainly appears on a carving in Oslo (Rácz 1983), where he is struggling with the god Odin on doomsday. He is also seen devouring Odin on a stone relief from the Isle of Man (Wilson 1980). Houzeau (1880) showed that Roman and Greek authors mention a number of times the custom of making a great clamour during a lunar eclipse, believing that the sounds would expel the dog-like demons fastened to the Moon. Dog- or wolf-like creatures who begin to eat the Moon are also found in the mythology of Mid-Eurasian peoples (Tokarev 1988).

Dating the battle

From all this evidence I think it is most likely that a lunar eclipse caused the postponement of the battle. It could easily have thrown the crowd into panic because, in accordance with their ancient beliefs, it might have been seen to herald the start of the world's end, and undoubtedly to indicate at least the failure of a disastrous battle.

The historical sources do not tell the exact time of this collision between the Gepids and the Longobards. However, limits can be given for the time of possible lunar eclipses. It follows from the chronology of the historical events that the postponed battle took place in 548, 549 or 550. Oppolzer's (1962) tables show that no solar eclipses could be seen in the Carpathian Basin in these years. There was a lunar eclipse each December, but is hardly credible that the tribes wanted to struggle in the depths of winter. This leaves Jun 25, 549 and Jun 15, 550. On the first date the eclipse started at 23:53 and was total from 01:19 to 01:47. In 550, conjunction began at 01:56, and totality lasted from 03:02 to 04:08. Fig. 6 shows the motion of the moon in relation to the Earth's shadow in the two cases. The eclipse of 549 is the most likely candidate. The phenomenon starting at about midnight and culminating between one and two o'clock is more likely to have been seen by more people, whereas in the other case it took place after three o'clock, when the most nervous and even the sentries might already have dozed off. Moreover, the second eclipse was less conspicuous, for it took place next to the horizon and before its ending the sky was growing light. The date of 549 is also suggested by historical events, for the alternative of 550 would imply

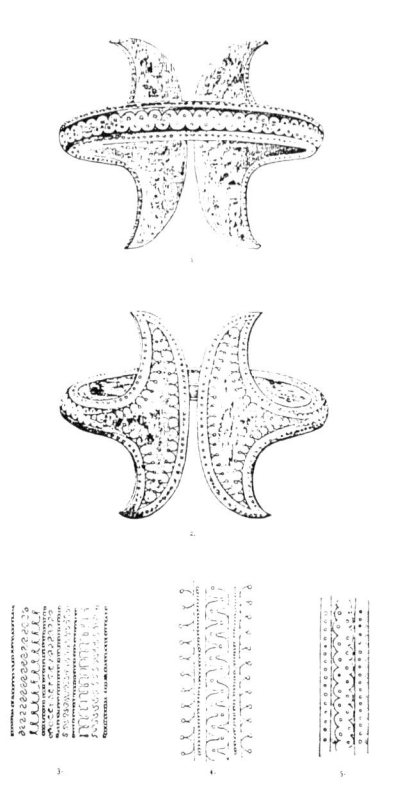

FIG. 5. Gold bracelets, apparently Teutonic, with their ends finished with two crescents turning opposite each other. After Hampel (1880: pl. XXXIII; 1894: no. 186).

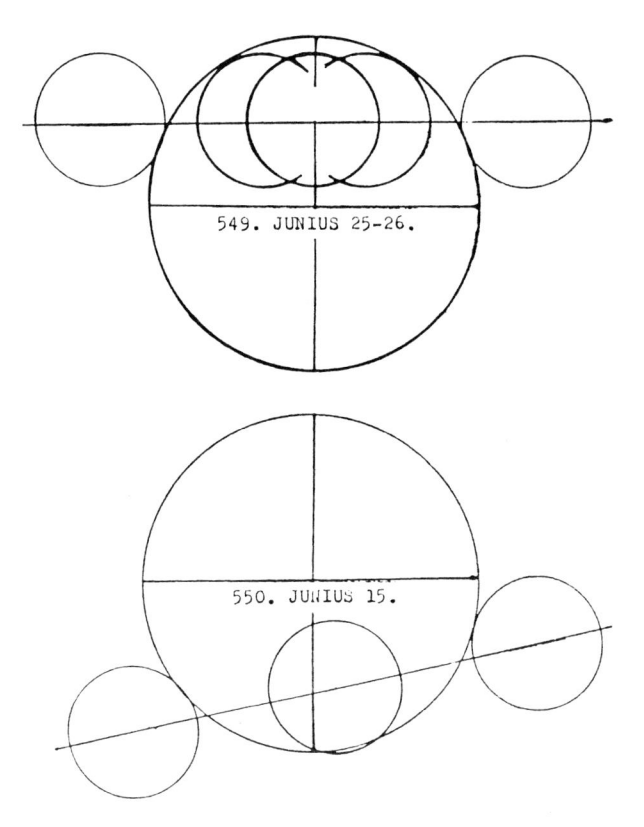

FIG. 6. The motion of the moon in relation to the Earth's shadow during the lunar eclipses of Jun 25, 549 (upper diagram) and Jun 15, 550 (lower diagram).

that the postponed battle was not fought until 552, by which time the two Teutonic armies were both in Busta Gallorum as auxiliary forces of the Eastern Roman Army.

References

Åberg, N. (1919). *Ostpreussen in der Völkerwanderungszeit*. Uppsala.
Broby-Johansen, R. (1979). *Északi sziklarajzok*. Budapest: Gondolat.
Caesar, Julius (1917). *Gallic wars*, trans. H.J. Edwards. London: Heinemann.
Dömötör, T., ed. (1965). *Germán, kelta regék és mondák [German and Celtic tales and sagas]*. Budapest: Móra Kiadó.

Fettich, N. (1951). Régészeti tanulmányok a késöi hun fémmüvesség történetéhez [Archaeological studies on the history of late Hun metallurgy]. *Archeologica Hungarica* **31**, whole issue.

Ginzel, F.K. (1914). *Handbuch der mathematischen und technischen Chronologie, 3*. Leipzig: Hinrich.

Hampel, F. (1880). Tomb found at Puszta-Bakod. *Archeologiai Értesitö* **14**, 55–62.

Hampel, F. (1894). *Régibb középkor emlékei Magyarhonban [Remains of the old Middle Ages in Hungary], 1*. Budapest: Országos Régészeti és Embertani Társulat Kiadványa.

Hoppál, M., Jankovich, N., Nagy, A. and Szemadán, G.V. (1990). *Jelképtár, Curiositas/III [Symbol collection]*. Budapest: Helicon.

Houzeau, J.Z. (1880). *A csillagászat történelmi jellemvonásai [Historical characteristics of astronomy]*. Budapest: Természettudományi Könyvkiado-Vállalat.

Krüger, B., ed. (1983). *Das germanen Handbuch 2*. Berlin: Akademie Verlag.

Nordén, A. (1923). Felsbilder der Provinz Ostgotland in Auswahl. Hagen i.W. and Darmstadt: Folkwang Verlag.

Oppolzer, R. (1962). *Canon of eclipses*. New York NY: Dover.

Prokopius (1966). *Gotenkriege*, ed. O. Veh. Munich: Ernst Heimeran Verlag.

Pulszky, F. (1897). *Magyarország archeologiája [Hungarian archaeology] 2*. Budapest.

Rácz, A. (1983). *A vikingek öröksége [Viking heritage]*. Budapest: Képzömüvészeti Kiadó.

Schlette, F. (1971). *Germanen zwischen Thorsberg und Ravenna*. Leipzig: Urania Verlag.

Sevin, H. (1955). *Die Gebiden*. Munich: Selps Verlag.

Tacitus (1943). *Germania—Agricola*. Budapest: Parthenon.

Tokarev, S.A., ed. (1988). *Mitológiai Enciklopédia [Encyclopaedia of mythology] 1*. Budapest: Gondolat.

Wilson, D.M., ed. (1980). *The northern world*. London: Thames and Hudson.

10

Astronomical knowledge in Bulgarian lands during the Neolithic and Early Bronze Age

TSVETANKA RADOSLAVOVA

1 Introduction

Archaeological studies in Bulgaria, particularly during the last twenty years, have shown that the lands forming present-day Bulgaria were the cradle of one of the earliest civilisations in Europe. It grew up between the fifth and the second millennia BC, and then mixed gradually with the cultural entity of the Thracian settlers. Besides their technical achievements, the ancient inhabitants of Bulgaria were acquainted with methods of time measurement. This knowledge was a result of systematic observations of the celestial bodies which at the same time appeared as objects of veneration and therefore were directly relevant to the adopted ritual practice.

Many written sources, most of them coming from Greek authors, testify to the astronomical knowledge of these ancient people. Unfortunately, they were not themselves literate, which creates the main difficulty when studying their cultural and historic heritage. Authors such as Jordanes (*Getica*, 69–74), Strabon, and Herodotus (*History* IV, 94) inform us that the Thracians were acquainted with the stars and their positions in the sky; with the twelve signs of the Zodiac; with the sun's and the planets' motions; and with the alterations in the lunar disc. In the opinion of these authors, this knowledge was being acquired from Egypt. Doubtless the ancient civilisation in Bulgarian lands had been keeping in touch with the great Mediterranean civilisations; however, the astronomical knowledge of Babylonians and Egyptians could not have been used directly in the Balkan peninsula, because of the different geographical and climatic conditions. Furthermore, the tribes inhabiting Bulgarian lands had come mainly from the north-east and had certainly retained traditions and religious beliefs differing from those of their new neighbours. Thus, while cultural exchange certainly existed, the originality of the Balkan civilisation as a whole and of the astronomical practice in particular can in no way be denied.

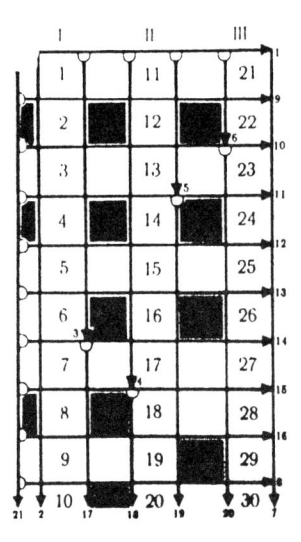

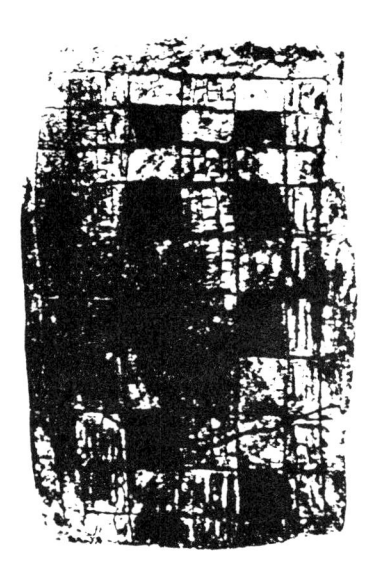

FIG. 1. The design on the base of the Slatino model furnace: (a) reconstruction; (b) photograph of the original. After Chohadjiev (1984).

2 The model furnace from Slatino

The most ancient piece of evidence supporting this assertion is a splendid find representing a model of a furnace, excavated from a burnt lodging in the village of Slatino in western Bulgaria. The object was dated to the fifth millennium BC and was interpreted as a lunar calendar, a precursor of the more sophisticated calendrical systems of Babylon and Egypt (Chohadjiev 1984).

The model furnace is rectangular in form, with a base 17 cm × 11 cm and a height of 11 cm. It represents an exact copy of the real thing, a practice repeated in several artefacts found in Bulgaria, which are thought to be connected with the cult of home and fireside. The object is completely covered by ornaments carved in clay. It has a snake-like handle; in prehistoric times the snake was considered to be a protector of the home.

The base of the model furnace is also decorated. However, this decoration has nothing in common with the rest of the ornamentation; indeed, it is quite different from the traditional decoration of early Neolithic pottery. It might be supposed that it fulfilled a special function of some importance, being depicted on a cult object and in a position preventing it being seen by just anyone. The design on the furnace base (Fig. 1) is composed of carvings ordered in a three column-table. Each column consists of ten cells so that the total number of cells is 30.

This table has been interpreted as a calendar based on the lunar phases. The argument is based upon the arrangement of the cells as well as the appearance of the carvings within them. Cells 1 to 11 contain a small number of carvings, and are interpreted as being related to the period between a new moon and a full moon. Cell 12 is empty, presumably warning about the beginning of the next period, that of the full moon marked in the next six cells, within which the carvings are most numerous. Cells 19 to 22 denote the waning of the moon; their carvings are similar to those in the first cells. In cell 23 they disappear again, now warning that the moon will soon disappear. The last seven cells, which are the least carved, are probably related to the disappearing moon and the moonless phase. This interpretation implies that the appearance of a thin new crescent marked the beginning of a month, as it did for the majority of Mediterranean peoples, for whom the synodic month served as a basis of their lunar calendrical systems.

A careful examination of the table suggests that it was probably also used to keep track of the relationship between the lunar months and the seasonal (solar) year, with which the management of agriculture needed to keep track. Five lunar signs carved on the front part of the furnace may have denoted the difference of nearly five days between twelve synodic months and the solar year. To the left of the three columns of carvings are further columns of cells, some of which are painted in red ochre. These number respectively three, five and four, and may represent the numbers of months in the three seasons of a year (like the three seasons of the Egyptian calendar): the three winter months starting from winter solstice, the months when agriculture is carried out (from sowing to harvest); and the autumn months.

Another, similar archaeological find, also interpreted as a seasonal agricultural calendar, is worthy of mention here. This is the so-called 'cult scene' of Ovcharovo (Todorova 1983). It consists of 26 cult objects excavated from a burnt lodging in northern Bulgaria dating to the beginning of the fourth millennium BC. The objects are made of clay and, like the Slatino furnace, represent small-scale copies of everyday objects: tables, chairs, four figures of priestesses and three ornate altars. It is supposed that a system of symbols is encoded within the scene, intended to denote important dates within the agricultural calendar.

It appears that the ancient calendars, although necessarily linked with religious concepts and rituals, simultaneously served specific practical, and particularly agricultural, needs. If, in fact, the neolithic calendar of Slatino was an 'improved' lunar calendar, i.e. one adapted to the solar year, then its inventors must have been observing the motions of the sun as well as the changing phases of the moon. However, the interest in the moon must have been the primary one.

3   The rock designs of Bajlovo

We now turn to a site that may help us to outline the process by which a calendar evolved, starting with lunar phase observations and arriving at calendrical records possibly connected with the sun. Surrounding the village of Bajlovo in western Bulgaria is a complex of caves. Two of these, including the rock in front of them and some of the neighbouring rocks, are carved, the carvings forming circles or parts of circles (Stoytchev 1988). These carvings are overlaid by drawings from a later epoch. According to the archaeologists, the cave complex was being used as a sanctuary from the third to the first millennia BC.

Originally, the sanctuary was thought to be connected with an ancient solar cult, the carvings representing solar symbols. This presupposes that the sun could have been observed from it, but in fact the site offers a good view of the night sky and not of the solar motion. The form of the rock carvings also supports the hypothesis that they were related to the moon rather than to the sun. The majority of them are not full circles but half and quarter ones. Thus the view prevailed that the carvings were images of various lunar phases. Support for this interpretation is found

Astronomical knowledge in ancient Bulgarian lands 111

FIG. 2. Rock designs at Bajlovo. After Stoytchev (1988).

in the fact that the ratio of the different carvings' shapes correlates with the length of occurrence of the different lunar forms during a synodic month. This fact suggests that certain groups of lunar images may have been used for calendrical purposes.

Testing such a hypothesis involves deciphering the supposed calendrical records, a problem that is still being tackled. This task is complicated by the fact that the carvings are produced in three different ways, varying in their degree of complexity. This suggests that there were several periods in the use of the sanctuary, a supposition corroborated by the occurrence of the above mentioned rock designs. The latter (Fig. 2) are of a particular interest. They were made by laying bat dung upon the carvings. The designs are thought to represent a story in the form of a succession of ritual scenes. The last one contains groups of five and ten dashes, thought to represent 'small' and 'great' weeks respectively, where each dash stands for a day. Some of the rock designs are accompanied by a great cross interpreted as a record of the emergence (heliacal rise) of a bright star.

The most spectacular design is that of a human face surrounded by two concentric circles, with lines radiating from the inner circle. A similar image was discovered earlier in the so-called 'solar room' in a famous cave ('Magura') in northern Bulgaria, whose walls are covered with splendid drawings also made with bat dung. The two images are almost identical and may be solar symbols designed to note an important solar event such as vernal equinox. The 'story in pictures' at Bajlovo may have been used for recording the number of days between significant astronomical events and the arrangement of important feast days amongst these events (Stoev *et al.* 1989). In other words, the designs may have formed a calendar catering for the needs of the adjacent community. The task of their registration was probably entrusted to a priest attending the cave sanctuary.

4 Further evidence for a solar cult

Evidence is accumulating to suggest that lunar observations gradually declined in importance, perhaps because they were insufficient for the purposes of a practical calendar, and, bit by bit, priority was given to the sun. Some of the earliest archaeological evidence for a solar cult in south-eastern Europe was uncovered in northern Bulgaria at the village of Poljanitsa (Nedelchev 1988), where cult objects were found dating from the fourth millennium BC. They are similar to those from Slatino and Ovcharovo and represent clay models of twelve houses in the form of columns of rectangular section. Three storeys and a ridged roof are marked on. The roofs of two of the houses are richly ornamented by solar symbols, jagged circles with encrustations. These 'solar houses' are supposed to symbolise sunrise and sunset. Solar symbolism can also be found in the number twelve, characteristic of the lunar-solar calendrical systems. Some investigators consider these objects to be models of pillars rather than of houses, speculating that such a system of pillars could have been used, for example, for determining the day of vernal equinox.

Further evidence for a solar cult is found in the cult complex of Paleocastro, considered to be the most ancient temple of the sun in Thrace (it dates to the beginning of the second millennium BC). Paleocastro is a

Thracian fortress: an imposing rocky complex situated at the foot of the Sackar Mountains in south-eastern Bulgaria. The whole eastern valley of the rocky chain is exposed to the rising sun; at this place more than 160 'suns' have been carved in the rock. They vary in diameter from 25cm to 90cm. Some of them are carved in relief; others are incised. They are often augmented by small rays. The rocks are of granite and gneiss with a high alloy of mica and shine brilliantly when lit by the beams of the sun. Nearby, a massive statue of a man was found, with a stone sun in place of his head. The statue is cut out of a monolithic stone block. It lies on the ground, but in times gone by it must have been upright with its face turned to the east. The statue is thought to be an anthropomorphic image of the solar god.

There are many written records, dating from later times, attesting the solar cult in ancient Thrace (Fol *et al.* 1981). The stone suns of Paleocastro might have been precursors of the solar symbols described by Herodotus in his story about Macedonian dynasty. These symbols were placed on the ground so that the sunlight could touch them. The Bithynians too, when discussing important problems, sat with their faces turned towards the sun, so that it could help them to take wise decisions or to predict the future. The solar cult in ancient Thrace was closely related to the idea of immortality of the soul. This idea came from the Dionysiac religion and was suitably adapted to the Thracian religious doctrine, relevant to the state institution of the Thracian society. The principal figure there was the king-priest, who combined political with religious power and was endowed with godlike attributes. The most important element in Thracian cult ceremonies was the annual resurrection of the king, who was at the same time a high priest and a personification of the sun.

The question follows of where these sacred Thracian sites were located. There exist a number of written sources shedding light on this subject. Pseudo-Erathostenus (*Katest*, 24) tells that Orpheus—the legendary Thracian singer—ascended Pangaeus every day before sunrise in order to address his prayers to the sun, the greatest amongst the Gods. Herodotus mentions a Dionysian sanctuary in a mountain region inhabited by the Thracian tribe Satrae. The Roman historian Macrobe describes another temple, in the form of a rotunda with an open roof, situated on the Zilmisos hill.

The archaeological data have confirmed the ancient legends. During the last twenty years, many sites have been found all over the country, and mostly in south-eastern Thrace. There are also many other Thracian megalithic monuments: dolmens, rock-cut tombs and niches, fortresses and sacrificial stones. A strong rock cult obviously reigned in this vast area. Its roots may be sought in very ancient religious beliefs, probably coming from the Island of Crete. It seems that this cult became a component of Thracian religious doctrine, unifying the solar and the chthonic forces in the Thracian rock sanctuaries.

Many facts indicate the relationship between solar cult practice and the Thracian sanctuaries. But was this practice their only function? Engaging in the solar cult necessitated a certain knowledge of the sun's motion. This suggests the idea that the sanctuaries themselves, by being suitably located, could have served a parallel purpose as solar observatories. This idea has rapidly gathered momentum amongst Bulgarian astronomers and archaeologists. It has resulted in a number of studies, bearing mostly upon astronomically significant orientations identified in Thracian rock sanctuaries (for a review see Radoslavova and Stoev 1990).

It is instructive to examine how these orientations are being outlined. In most cases natural features and peculiarities of the rocks are used. Yet, traces of human activity have been detected all over the sites. It seems that the ancient architects, taking into account the particular natural silhouette of the rocks, skilfully shaped them in order to obtain adequate functional structures. In some places the orientations were marked by means of artificial structures applied either independently or in combination with natural ones. Occasionally these were elements of the site itself, but in other cases external elements were used, such as ground or stone mounds. At present there is no evidence for the use of local horizon features; this possibility is still under consideration. Another possible way of marking the directions was by using a system of wooden pillars (recall the pillars from Poljanitsa). While no such pillars have, of course, survived, a number of holes hollowed in the rock may have been used as pillar foundations and would consequently mark their positions. The orientations themselves are of three kinds:

- towards sunrise on the day of autumnal equinox;

- towards sunrise on the day of winter or summer solstice; and

- along cardinal directions (N–S and E–W).

It is difficult to generalise because only three of the numerous sanctuaries have been studied to date. The cardinal lines (Fig. 3) are of obvious interest; however, their significance is unclear in the light of our current knowledge of Thracian beliefs.

The analysis of the astronomical context of the Thracian rock sanctuaries in Bulgaria is still at its beginning and it would be unwise to launch into speculative inferences. It is, however, certain that solar observations were carried out at the Thracian rock sanctuaries, and they have to be seriously considered from this point of view.

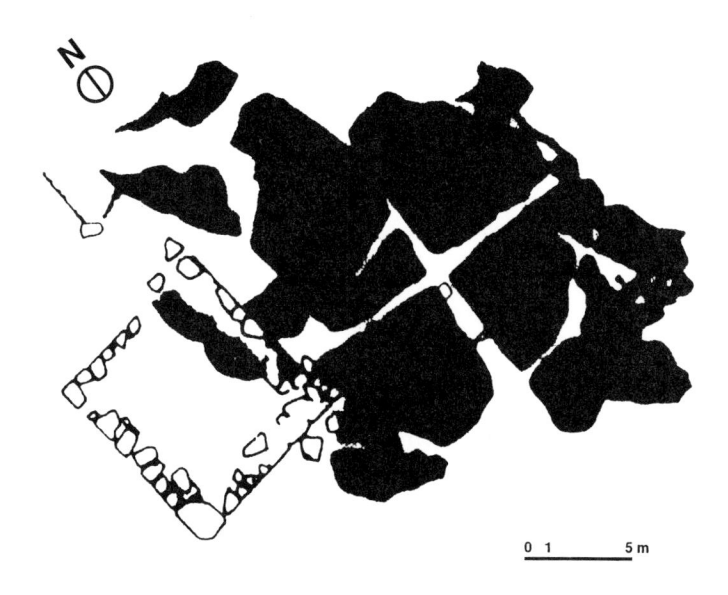

FIG. 3. The Thracian rock sanctuary at Kabile, near Yambol.

Conclusions

There exists a great deal of archaeoastronomical evidence in Bulgaria that still awaits detailed study. The objects discussed in this paper are situated in a rather limited historical and geographical area, range widely in time, from the fifth millennium BC to the second half of the first millennium BC (which marks the decay of the Thracian civilisation as a self-dependent cultural entity), and have themselves by no means been fully studied yet.

Even on the basis of this small sample of the evidence potentially available in Bulgaria, we may easily perceive several particular stages in the evolution of astronomical knowledge in our lands, as well as clearly defined trends towards continuity and succession. The opportunity of studying this evolution is a unique and rare chance for Bulgarian archaeoastronomy.

References

Chohadjiev, S. [Чохаджиев, С.] (1984). Археологически данни за календар от ранния енеолит [Archaeological data on the early Neolithic calendar]. *Археология [Archaeology]* **2–3**, 1.

Fol, A., Venedicov, I., Marazov, I. and Popov, D. [Фол, А., Венедиков, И., Маразов, И. и Попов, Д.] (1981). *Тракийски легенди [Thracian legends]*. Sofia.

Nedelchev, N. [Неделчев, Н.] (1988). Слънчевите къщи от Поляница и Вору-Монофациу в соларните религии на Балканския полуостров [The sun houses from Polyanitza and Voru-Monofaciu in solar religions at the Balkan Peninsula]. *Интердисциплинарни изследвания [Interdisciplinary Studies]* **15**, 104.

Radoslavova, Ts. and Stoev, A. (1991). Astronomical traces in ancient rock monuments in Bulgarian lands. In *Colloquio Internazionale Archeologia e Astronomia*, eds. G. Romano and G. Traversari, pp. 176–79. Rome: Giorgio Bretschneider Editore (Supplementi alla RdA, 9).

Stoev A., Guerassimova-Tomova, V. and Stoytchev, T. (1989). Complèxe de grottes près du village Bajlovo. In *Proceedings of the tenth international conference on speleology*. Budapest.

Stoytchev T. [Стойчев, Т.] (1988). Тракийско скално светилище близо до сло Байлово свързано с култа към Пуната [A Thracian sanctuary dedicated to a lunar cult near Bajlovo]. *Археология [Archaeology]* **2**, 32.

Todorova, H. [Тодорова, Х.] (1983). *Овчарово, разкопки и проучвания [Ovcharovo, excavations and studies]*, vol. IX. Sofia.

11

Four approaches to the Borana calendar

CLIVE RUGGLES

Introduction

The calendar of the Borana, a group of nomadic cattle-herders whose territory straddles the border between Ethiopia and Kenya, was studied as part of anthropological work undertaken in the 1960s and published in 1973. Several years later, technical inconsistencies were discovered in the account of the Borana's astronomical observations. A possible resolution was suggested, which led to the idea that Borana time-reckoning stemmed from a calendrical system with ancient origins. This idea was apparently reinforced by alignment evidence from archaeological sites. However, a broadly based critique has been offered, and the evidence convincingly refuted.

In a fresh attempt to resolve the technical difficulties, the present author subjected the existing account of Borana time-reckoning to close astronomical scrutiny. The results were published in 1987. At the same time, and unknown to this author, ethnoastronomical fieldwork was being undertaken amongst the Borana, in an attempt to resolve the problem of the calendar. The first report on this fieldwork was published in 1988. A comparison of these two accounts gives the opportunity to assess, retrospectively, the success of a disciplined archaeoastronomical approach in reconstructing the Borana calendar from the evidence then available.

A broad-based ethnographic approach

During the 1960s, Asmarom Legesse undertook extensive fieldwork with the Borana of southern Ethiopia as part of his study of the Gada system, the central institution of the Galla people which forms the basis of their cognitive world view. His account included a detailed account of Borana time-reckoning, since

> Borana have an unusually deep awareness of time and history...they have the same degree of involvement with time as we find in the Western world. They schedule their lives, their rituals, their ceremonies, their political and economic activities to a very high degree.
>
> (Legesse 1973: 179).

Legesse's description is based on oral accounts by Borana experts on sky observation (*ayantu*). According to this, the Borana calendar is based upon lunar synodic cycles. There are twelve named months, a new month being marked by the appearance of the new crescent moon. However, there are also twenty-seven named days. The cycles of days and months run independently and continuously, so that the starting day of a given month is two or three days later than that of its predecessor, and in any given month certain days appear twice, once at the beginning and once at the end. The permutation of the two cycles is completed in twelve months.

Ayantu, we learn from Legesse's description, can tell the day and month by examining the relative position of the moon and stars in order to determine the month astronomically (observations of the sun are not used). For half of the year the new moon is identified relative to one of seven stars or star groups—Triangulum, the Pleiades, Aldebaran, Bellatrix, Orion's belt and sword, Saiph, and Sirius. During the other half *ayantu* identify the moon at different phases in relation to a particular star group, Triangulum.

However, the astronomical parts of Legesse's account are at times very garbled. The main reason is that oral accounts are notoriously difficult to translate, a problem compounded in this case by the investigator not himself being familiar with the astronomical phenomena being described.

A particular difficulty with Legesse's description is the following. He refers to the reference star groups which identify successive months appearing 'in conjunction with' and 'side by side with' the new moon. At the latitude of Borana country (approx. 5°N), this seems to imply that the relevant star group is roughly at the same right ascension as the moon when the observations are made. However, the monthly advance in RA of the moon at a given phase (i.e. at a given time in the synodic month), $1^h 56^m$ on average, is such that it will take only something like two and half months, rather than the six required, for the moon at a given phase to advance past the seven reference star groups. Here, as elsewhere in Legesse's description, there is no clear way to resolve the inconsistency.

An archaeoastronomical approach

Doyle (1986) has offered the suggestion that Legesse, when he talks of 'rising in conjunction', might actually mean rising at the same horizon position, i.e. at the same declination. On the face of it this interpretation is unable to reconcile Legesse's account and the astronomical facts, since (for one thing) the declinations of β Tri and γ Tri are well to the north of the northerly declination limit of the moon. However, Doyle argues that a reconciliation can be achieved by precessing to a date of around 300 BC. This leads him to suggest that the present-day Borana calendrical system derives from a Cushitic calendar set up around 300 BC.

In support of this idea, Doyle refers to the Namoratung'a II (or Kalokol) archaeological site in north-west Kenya, where standing stones have been discovered that could have been used to mark relevant horizon positions and hence to set up the calendar (Lynch and Robbins 1978; see also Lynch and Robbins 1979; Paul 1979).

Unfortunately, the whole idea of an ancient Cushitic calendar rests upon a succession of ideas, each of which is susceptible to criticism. In a wide-ranging reassessment, reminiscent of those applied to the more speculative claims of British archaeoastronomy in the early 1980s, Soper (1982) has called into question a number of aspects of the evidence from Namoratung'a II. The present author (Ruggles 1987: S36–40) has offered a detailed critique of the astronomical aspects of the hypothesis.

An astronomical approach

The present author has examined a number of problems and ambiguities which arise in Legesse's account when it is examined in detail from an astronomical point of view, of which the anomaly mentioned above is merely one.

First, it was stated by Legesse and assumed by all other commentators that the Borana year was 354 days long, and hence divorced from the seasons. However, this is inconsistent with the interpretation of Legesse's 'in conjunction with' as meaning 'at the same RA as', since the phase of the moon is a consequence of its RA relative to that of the sun (Ruggles 1987: S42–43).[1] If this obvious interpretation of 'conjunction' is to be restored, then the Borana calendar *must* somehow keep in step with the solar calendar.

By a sequence of logical steps, and noting coincidences between certain star-group names and day names, it was concluded that the Borana calendar consists of two independent rounds:

---

[1] This fact does not seem to have been appreciated by Doyle (1986).

- A month-round of 29/30 days corresponding to the synodic month.

- A day-name round corresponding to the *sidereal* month. The current day is fixed by observing the RA of the moon, i.e. the star group with which it rises or sets on a level in the sky.

The two cycles are independent, since RA is not a consideration in the first and phase is not a consideration in the second.

It was suggested that Borana months were identified by the star groups mentioned earlier, but at different phases in different months. It was also suggested that the occasional adjustment of approximately three days needed to keep the day-name round of 27 days in step with the sidereal month of 27·32 days could be achieved by inexactitude and retrospective adjustment, as with the nearby Mursi (Turton and Ruggles 1978).

It is not the purpose of this paper to recount the details of this argument and its conclusions, as they are given elsewhere (Ruggles 1987). The essential point is that, at the time, this gave a possible explanation of Borana calendar, arguably consistent with Legesse's account, which (i) had a basis in simple, practical observations and (ii) did not rest on a complex secondary hypothesis concerning a postulated calendrical system that needed to have evolved over many centuries.[2]

An ethnoastronomical approach

In 1986, and unknown at the time to the present author, new anthropological fieldwork was being undertaken amongst the Borana of northern Kenya, with the specific aim of clarifying issues relating to the Borana calendar. This new fieldwork involved direct observations under the guidance of a respected elder and *ayantu*.

These new data (Bassi 1988) confirm that the Borana calendar is indeed based on a combination of synodic and sidereal lunar months, with day-names corresponding to star groups identifying the RA of the moon. Other major features are also in broad agreement, such as the mechanisms for inserting intercalary months. Indeed, the success of a disciplined deduction, on purely astronomical and logical grounds, from a very incomplete account of Borana astronomy, in predicting the major features of the true Borana calendar is impressive, and a clear vindication of this type of approach where further ethnographic fieldwork is impossible.

There are, however, a number of details which the new fieldwork has clarified. For example, it is now clear that Legesse's remark 'In six

---

[2] We note, however, that the idea of some form of proto-calendar is not ruled out by this conclusion.

out of the twelve lunar months the seven constellations appear successively, in conjunction with the moon' refers to successive nightly observations within a month; an interpretation which, with the benefit of hindsight, now seems obvious. Ironically, the suggestion by Ruggles (1987) that has been brought most strongly into question by the new fieldwork relates to the role of institutionalised disagreement, a suggestion arising from an attempt to add a cultural dimension to the astronomical analysis by drawing analogical evidence from the nearby Mursi (Turton and Ruggles 1978).

Conclusions

On the one hand, the Borana calendar serves to illustrate the dangers of unrestrained archaeoastronomical speculation. On the other hand, and more importantly, it serves as a case study demonstrating the extent to which archaeoastronomical methods, properly applied, can reconstruct important aspects of the astronomy of remote cultures from partial evidence.

The Borana gives us a rare opportunity to evaluate the results of an archaeoastronomical analysis using existing cultural data (Ruggles 1987) against fresh, directly relevant, field data (Bassi 1988). The conclusions of the two papers were gratifyingly similar, demonstrating the extent to which reasonably reliable conclusions can be obtained by archaeoastronomical analysis, provided that attention is paid to strict methodological constraints.

Clearly, first-hand ethnographic fieldwork, where available, is a far more promising resource than any form of archaeoastronomical data. But archaeoastronomical methods, including purely technical astronomical deductions, can, when properly applied, achieve significant results.

References

Bassi, M. (1988). On the Borana calendrical system: a preliminary field report. *Current Anthropology* **29**, 619–24.

Doyle, L.R. (1986). The Borana calendar reinterpreted. *Current anthropology* **27**, 286–87.

Legesse, A. (1973). *Gada: three approaches to the study of African society*. New York NY: Macmillan.

Lynch, B.M. and Robbins, L.H. (1978). Namoratunga: the first archaeoastronomical evidence in sub-Saharan Africa. *Science* **200**, 766–68.

Lynch, B.M. and Robbins, L.H. (1979). Cushitic and Nilotic prehistory: new archaeological evidence from north-west Kenya. *Journal of African History* **20**, 319–28.

Paul, G. (1979). The astronomical dating of a northeast African stone configuration. *The Observatory* **99**, 206–9.

Ruggles, C.L.N. (1987). The Borana calendar: some observations. *Archaeoastronomy* no. 11 (supplement to *Journal for the History of Astronomy* **18**), S35–53.

Soper, R. (1982). Archaeo-astronomical Cushites: some comments. *Azania* **17**, 145–62.

Turton, D.A. and Ruggles, C.L.N. (1978). Agreeing to disagree: the measurement of duration in a southwestern Ethiopian community. *Current Anthropology* **19**, 585–600.

12

Astronomy in the ancient written sources of the Far East

ILDIKÓ ECSEDY AND KATALIN BARLAI

Chinese written records survive from as far back as the second half of the second millennium BC, and provide evidence of a long tradition of observational astronomy. This paper considers the *Chou* 周 period and its antecedents, as far back as the mythic period. Characters and texts from the *Chou* period preserve, in our view, a very early astronomical tradition, originating well before the time when the characters originated or the texts were recorded. There is an abundance of material on this topic, but we restrict ourselves here to three main points.

Ch'en and *Shen* representing two hostile principalities in *Chou* times

Amongst the oldest variants of Chinese characters the constellations *Scorpio* (*Ch'en*, 辰) and *Orion* (*Shen*, 参) are already represented. These first pictograms depicting stars originate from the last centuries of the second millennium BC, at the time of the *Shang-Yin* 商殷 dynasty (17th–11th century BC), that is a few hundred years earlier than the continuous records began (in 841 BC). These two constellations seem to pursue each other through the heavens, the difference in their mean right ascensions being about 180°. The European explanation—the reason for their successive regular appearance and disappearance—is that Scorpio wants to bite the heel of the heavenly hunter Orion. This picture was formed in the Mediterranean region.

In China, however, the same phenomenon, recorded in the mythic agricultural tradition, became the symbol of a fratricidal war (Ecsedy 1981: footnote on pp. 268–69). In short, the story is about two brothers hostile to each other. Their offspring became the peoples of two *Chou* time principalities, *Chin* 晉, the conquered people of *Hsia* 夏, and *Shang* 商. *Shen*—three stars of *Orion* (probably its belt)—has been related to the state *Chin*, while *Ch'en*—the central three stars of Scorpio together with the 'Fire Star' (Antares)—represents the state *Shang*.

It is not at all astonishing that ancient peoples of different cultural backgrounds, but living within the same geographic belt, notice the same heavenly phenomenon, although the mythic explanations are different in China and in Mediterranean Europe.

The 'sidereal year' of the Fire Star

The cave dweller ancestors of the peoples of eastern Asia had used fire for at least half a million years. In ancient China a fire-coloured star surrounded by a triad of stars could have been watched at dawn in springtime. It must have been a remarkable experience for them that the return of the 'Fire Star' took place in mild weather at the beginning of the agricultural season. For numerous thousands of years this vernal return of the star in the early morning had promised sunshine and fertility during the coming season. After its heliacal rise, the Fire Star suddenly vanished amidst the rays of the rising sun. This phenomenon could have been considered as a sign to start working in the fields. The tradition of honouring the Fire Star may have preceded written history by several thousands of years; it certainly survived until the time of calendar makers, mythic and historic rulers.

The characters *ch'en* 辰 and *ju* 辱 were formed in the centuries immediately before and in the first centuries of the *Chou* era, and referred to the fact that the rising of Antares in the early morning at springtime was the sign for the beginning of agricultural activities (Karlgren 1957: 314–15, no. 1223a; Ecsedy 1981: footnote on pp. 268–69). What date does this imply for the beginning of the Chinese agricultural civilisation?

Ecsedy *et al.* (1989) calculate that at the latitude of Peking the heliacal rising of Antares coincided with the vernal equinox as early as about 17,000 BC.[1] However, we are not concerned with the sort of exactitude that contemporary astronomy would require (Schaefer 1987); merely with the early rising of a red star in springtime.

Following heliacal rise at around the spring equinox, the Fire Star rises about four minutes earlier every day. Sunrise, however, also occurs earlier every morning, so the rise of the star and the sun remain close to each other in time, and the brilliant red star remains prominent in the pre-dawn springtime sky for many weeks (Fig. 1). Furthermore, this general

---

[1] They also obtain 2300 BC as the date when its heliacal setting coincides with the autumnal equinox.

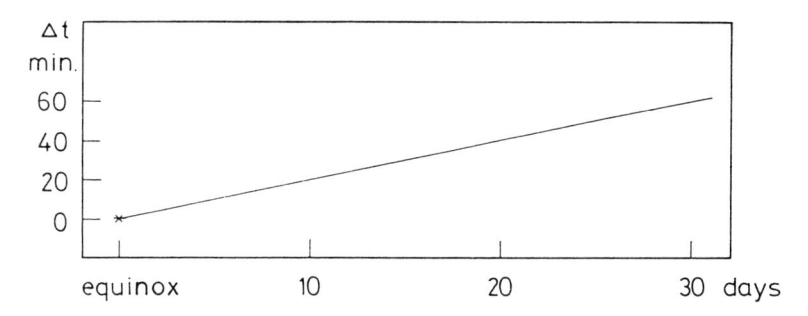

FIG. 1. Schematic representation of the time difference Δt between the rising of a star and the sun, after the star's heliacal rising at spring equinox. It can be clearly seen that the difference in time does not exceed one hour within a month (the data are for latitude $47°·5$).

phenomenon survived for several thousands of years after the date at which the heliacal rise of Antares actually took place at spring equinox.

Nonetheless, observation of the rising of Antares in the early morning at springtime still implies a tradition spreading back some thousands of years before the *Chou* period, and recent radiocarbon dates do not contradict this interpretation, attesting the presence of agricultural civilisation in China as early as the sixth millennium BC (Chen and Xi 1993).

By watching the Fire Star in spring, a sidereal time reckoning was applied, which got out of step with the solar year as the millennia passed. This process was recorded in the Canon of the mythic emperor *Yao* 尧. This document dates from the sixth century BC, and preserves traditions on heavenly phenomena from various earlier historic and prehistoric epochs (*ibid.*). In this document the heliacal rise of the Fire Star is connected to the summer solstice or to the middle of summer.

The interval between two successive heliacal risings of Antares is practically one sidereal year, so the rate of the slow continuous shift between the 'Antares year' and the seasonal, solar year can be estimated. It is roughly one week per millennium.[2]

---

[2] The rate of divergence between the Julian and Gregorian calendars is the same as this, because the Julian year is reckoned sidereally (Barlai and Ecsedy 1990). In the Indian calendar, which is purely sidereal (Bhatnagar 1990), this kind of shift is compensated by continuous correction. One wonders if, for example, the African Mursi tribe is aware of the fact that the correlation between the heliacal setting of four stars in the area of the Southern Cross and the flood of the Omo river (Turton and Ruggles 1978) is very gradually slipping, so that it will eventually suffer the same deviation as the inundation of the Nile and the summer solstice in ancient Egypt.

The agricultural calendar in the Book of Odes (*Shi Ching*)

Village life in the millennium prior to the imperial era has been depicted in an artistic way in the *Book of Odes* (*Shi Ching* 詩經). This collection of songs or odes originates from the *Chou* period. Song no. 154 enumerates the events and works of the peasant's life in a poetically loose order of the months. The first of the eight verses of this song in Karlgren's (1950) rough translation is as follows:

> In the seventh month there is a declining Fire-star;
> in the ninth month we give out the clothes;
> in the days of first, there is a rushing wind;
> in the days of second, it is bitterly cold;
> if we have no robes, no coarse-cloth [garments], wherewith
>   should we finish the year?
> In the days of the third we go to plough;
> in the days of the fourth we lift the heels;
> all our wives and children carry food [to us] in those
>   southern acres;
> the inspector of the fields comes and is pleased.

If the different months are set in order, it becomes clear that no more than ten are mentioned. A considerable part of the Sinological literature deals with the problem of how these numbered months might fit with the solar year. The simpler the explanation, the more probable—in our view. The eight verses of the ode enumerate agricultural activities that take place in the course of the growing season, that is from spring until late autumn. The second month may settle the question. It is about thunderstorms, i.e. the first storms at the end of winter. So it suggests that the year begins at the end of the winter. This remained the time of new year during later, imperial times.

There are two ways of interpreting the first sentence of ode 154. One is to consider a meteor, or 'floating star' (*liu hsing*), in the seventh month (Granet 1919). This attitude seemed so attractive that even a poetic Hungarian translation in rhyme (by S. Weöres) was inspired by it. In his translation he speaks about a meteor shower usual in the autumn, in the seventh month. Nonetheless, Karlgren's (1950) point of view seems more probable to us. His interpretation deals with *huo hsing* (大火), the Fire Star. We accept this approach because of the outstanding importance of Antares in Chinese ethnoastronomy.

Finally, reference should inevitably be made to an astonishing parallel in European antiquity, namely Plutarch's *Lives* (Plutarch 1959), where in the biography of *Numa Pompilius*, the second king of Rome, a completely analogous method of time reckoning is described, in use amongst the agricultural population of the early Roman state in the first millennium BC.

References

Barlai, K. and Ecsedy, I. (1990). Sidereal years—catalogue uses in archaeoastronomy. In *Inertial coordinate system on the sky*, eds. J. Lieske and V. Abalakin, pp. 197–98. Dordrecht: Kluwer Academic Publishers (IAU symposium no. 141).

Bhatnagar, A.K. (1990). Effect of the new equinox definition on the zero-point of longitude of the Indian calendar. In *Inertial coordinate system on the sky*, eds. J. Lieske and V. Abalakin, p. 186. Dordrecht: Kluwer Academic Publishers (IAU symposium no. 141).

Chen Cheng-Yih 程貞一 and Xi Zezong 嘴澤宗 (1993). The *Yáo Diǎn* 尧典 and the origins of astronomy in China. In *Astronomies and cultures*, eds. C.L.N. Ruggles and N.J. Saunders. Niwot CO: University Press of Colorado. In press.

Ecsedy. I. (1981). Far Eastern sources on the history of the Steppe region. *Bulletin de l'École Française d'Extrême-Orient* **69**, 263–76.

Ecsedy, I., Barlai, K., Dvorak, R. and Schult, R. (1989). Antares year in ancient China. In *World Archaeoastronomy*, ed. A.F. Aveni, pp. 183–85. Cambridge: Cambridge University Press.

Granet, M. (1919). *Fêtes et chansons anciennes de la Chine*. Paris: Bibliothèque de l'École des Hautes Études (Sciences Religieuses, tome 34).

Karlgren, B. (1950). *The Book of Odes, Chinese text, transcription and translation*. Stockholm: The Museum of Far Eastern Antiquities.

Karlgren, B. (1957). *Grammata serica recensa*. Stockholm: The Museum of Far Eastern Antiquities (Bulletin, no. 29).

Plutarch (1959). *Lives*. London: Heinemann.

Schaefer, B. (1987). Heliacal rise phenomena. *Archaeoastronomy* no. 11 (Supplement to *Journal for the History of Astronomy* **18**), S19–33.

Turton, D.A. and Ruggles, C.L.N. (1978). Agreeing to disagree: the measurement of duration in a south-west Ethiopian community. *Current Anthropology* **19**, 585–600.

13

Orientations of religious and ceremonial structures in Polynesia

WILLIAM LILLER

Introduction

Serious scientists are in general agreement that the first Polynesians arrived from islands to the west between three and four thousand years ago. From New Guinea and Fiji, they moved initially into the Tongan and Samoan Islands, later spread eastwards into the Cook Islands and French Polynesia, and finally, a few centuries after the beginning of the Christian era, reached the corners of the Polynesian Triangle marked by Easter Island, Hawaii, and New Zealand. When this vast continent of scattered islands first became known to the Europeans, the inhabitants shared a common basic language and worshipped many gods in similar ways.

Architecturally, their temples (*marae*) were dominated by a single raised platform (*ahu*), upon or near to which stood substantial stone statues (Easter Island), upright stone or wooden slabs (Hawaiian Islands), or stone and coral slabs (Society and Cook Islands) representing gods or departed chiefs. In Hawaii, the most important *marae*, or *heiau*, often consisted of a rectangular walled enclosure with various structures inside.

Prior to 1955, virtually no archaeoastronomical studies had been made of any of these temples or other structures in Polynesia. My own research into Easter Island (Liller 1989) followed the pioneering but limited archaeoastronomical work carried out there by William Mulloy and others participating in Heyerdahl's Norwegian Archaeological Expedition in 1955–56. Elsewhere in Polynesia, perhaps the best known result was the discovery in 1967 by H.M. King Taufa'ahau Tupou IV of Tonga that a massive coral trilithon dating from AD 1200 was solstitially oriented and bore markings indicating all three significant solar directions.

Marae, heiau, and *ahu* in Polynesia must number thousands. Although there were some grand monuments that, like cathedrals, served the needs of large groups of people, there were also the village and family *marae*, usually much smaller and simpler in construction and of course much more numerous. A definitive archaeoastronomical study of Polynesia would take the extensive—and expensive—dedicated time and energy of a team of investigators that would go from island to island measuring all possibly significant structures and alignments. And this would have to be accompanied by an in-depth survey of the literature devoted to Polynesian anthropology and ethnology.

In this report, I describe a preliminary survey undertaken at intervals over the past two and a half years. The aim was to become familiar with the basic literature, and to measure—or find measurements for—the orientations of the more important and generally better known structures and monuments of a religious or ceremonial nature. It is hoped that the interpretation of these measurements, still admittedly naive owing to my incomplete knowledge of the ethnology of Polynesia, will become more secure as I become more familiar with the literature.

The Easter Island survey

More than three hundred Easter Island *ahu* are relatively well-preserved. I have concluded (Liller 1989) that about twenty of these were intentionally oriented with solstitial or equinoctial rising/setting azimuths. One relatively modest *ahu*, Huri A Urenga, displays a high degree of astronomical sophistication with the critical solar directions marked in about a dozen different ways to high accuracy (usually of the order of $\pm 1°$).

Previous archaeological work on other Polynesian islands

The pioneering work on several hundred stone structures in French Polynesia was carried out over fifty years ago by Emory (1933; 1947), and since then, Sinoto (1969) has been the leader in archaeological research in these islands. In the Cook Islands, Bellwood (1978) has surveyed over two hundred sites including the intriguing alignments of standing stones on the island of Aitutaki. Unfortunately, only Sinoto has paid close attention to the establishment of precise bearings, and his and my magnetic bearings differ on the average by $4°·2$.

In Samoa and Tonga (and Niue), the 'paved stone *marae* with its raised stone platform is conspicuously absent' (Buck 1938). However, in Tonga there exist impressive stone tombs and the above-mentioned trilithon, while in Samoa, several multi-pointed star-shaped earthen mounds, some measuring dozens of metres across, have been preserved and probably once supported 'god-houses' (Davidson 1979).

In New Zealand, a number of Maori sites remain, and they have been the subject of investigations by numerous people (see Davidson 1984). However, having not yet personally studied this, the last settled corner of Polynesia, I will omit further mention of it here.

By far the best studied islands in Polynesia are those in the Hawaiian chain. Both Emory and Sinoto have devoted much time and effort to the archaeology of these islands, and the list of other investigators is long. Excellent topographical maps exist that show the outlines of some of the larger *heiau*. A good introduction to Hawaiian archaeology and prehistory with numerous references has been written by Kirch (1985).

Many other islands and island groups are to be found in Polynesia, but those listed here include most of the major centres of prehistoric population, and their study should provide an ample database to permit a statistically significant study to be made of the orientations of religious and ceremonial structures.

Orientations of well-measured structures

In order to select those *marae*, *heiau*, and *ahu* that might have been intentionally aligned solstitially or equinoctially, we should first consider a histogram that shows the orientations of these structures. Fig. 1 includes 53 structures on nine islands (Aitutaki and Easter Island excluded) that have been carefully measured by me in the field or whose outlines appear on good-quality topographical maps. The quantity plotted is the azimuth towards which a perpendicular to a long dimension of the platform or other significant structure is directed. The majority of my measurements were made with a hand-held sighting magnetic compass, and the result is usually the average of four readings, two in each of the two opposing directions, direct and reverse. On all islands but Moorea, the magnetic declinations published in 1988 by the US Coast and Geodetic Survey were checked by observing the rising or setting sun or moon. Comparisons of my compass-determined orientations with those found by direct measurements of the sun or moon indicate that my readings are accurate to within $\pm 0°\cdot 5$. A greater uncertainty usually comes from trying to decide exactly what is to be measured. The possibility of the occurrence of significant magnetic anomalies is felt to be small because the structures were for the most part located on uplifted coral reefs.

Examination of Fig. 1 reveals no obsession on the part of the *marae* constructors to orient these structures astronomically. Only three *marae* are oriented at all near to sunrise at the summer solstice—Kapa'akai, Oahu (declination $+26°\cdot 7$), Arahurahu, Tahiti ($+20°\cdot 2$) and Taata, Tahiti ($+21°\cdot 2$); and one near winter solstice—Umarea, Moorea ($-23°\cdot 6$).

Orientations of religious and ceremonial structures in Polynesia 131

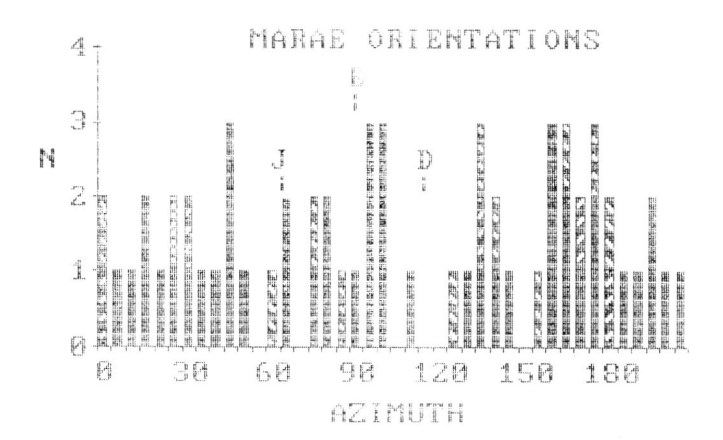

FIG. 1. The orientations of 53 well-measured *marae* in the Society Islands (solid bars), Hawaiian Islands (\-slanting lines), Rarotonga (/-slanting lines), and the Trilithon in Tonga (solid). The values shown are the true azimuths of perpendiculars to the long dimensions of the primary structures. *J*, *E*, and *D* mark the azimuths of the rising solstitial and equinoctial sun.

It is, however, of interest that the *marae* oriented closest to a solstitial rising point, Umarea, is said by Kay (1988: 93) to be the oldest on Moorea (AD 900). It is located at the end of a narrow peninsula, perhaps manmade, on the east coast of the island with a spectacular view of Tahiti a mere 20 km distant. December solstice sunrise would have occurred over Tahiti's mountainous interior (but not over the highest peak). From the two *marae* on Tahiti, June solstice sunset would have occurred over Moorea (from Taata, over the highest peak, Tohiea). Both Tahitian *marae* are roughly a half kilometre inland from the west coast with long axes skewed by about 30° to the relatively smooth coastline. The *heiau* near Kapa'akai Point was much overgrown but seemed not to be well constructed.

We should also consider those five structures oriented within 5° of azimuth 0° and the two near 90°. If we assume that *marae* in the first group were constructed so as to have their long axes *aligned* with the desired direction, then the rising declinations would be +1°·9 (Ofata, Huahine), +0°·2 (Fareura, Bora Bora), +4°·3 (Kainuku, Rarotonga), +0°·9 (Mo'okini, Hawaii) and –1°·3 (Mailekini, Hawaii). In the second group we obtain +2°·8 (Moa, Hawaii) and –0°·7 (Haumaru, Huahine).

The most noteworthy of these temples are Kainuku, perhaps the largest *marae* on Rarotonga and located about 400m inland from the south-east coast, and Mailekini, one of the oldest *heiau* on the island of Hawaii and located next to Mo'okini on the north-west coast.

Special mention must be made of the *marae* Taputapu-atea located on a point of land on the east coast of Raiatea. This greatest of temples, traditionally the first constructed and the most sacred in all of ancient Polynesia, is oriented such that a perpendicular to its long dimension is directed towards a rising declination $-6°·0$. It also lies nearly perpendicular to the nearest shoreline. According to legends, *marae* built afterwards had to incorporate a stone taken from Taputapu-atea 'in order to acquire religious prestige' (Buck 1938). It is interesting to note, then, that the four next most important temples in the leeward (western) Society Islands are all close to the same size as Taputapu-atea (43m long), in three cases strikingly similar in design, and all similarly oriented: Tainuu, on the west coast and the second most important on Raiatea (declination $-8°·3$ rising); Anini on Huahine Iti ($-8°·3$); Matairea-rahi on Huahine Nui, said to be the second most important temple in Polynesia ($-11°·7$); and Marotetini, the most important *marae* on Bora Bora ($-13°·3$).

However, there is no obvious astronomical explanation for this pattern. The most conspicuous celestial objects in this declination range a thousand years ago were the southern half of Orion and Spica, but according to Johnson and Mahelona (1975) neither has particular significance in Polynesian ethnology. None of these directions is very close to the declination of the sun at the time of zenith passage, which is about $-16°·5$ for these islands. This leaves the possibility that importance was attached to the time of year when the sun rose at declination $-6°$, i.e. early March or early October.

The Tongan trilithon, Ha'Amonga-a-Maui

According to Tongan tradition, in AD 1200 the eleventh King of Tonga ordered the construction of a trilithon made of massive coral blocks (Fig. 2). Various reasons have been given for why he had it built, but the two uprights are said to symbolise his two sons, and the lintel resting in slots at the top of the uprights to symbolise either the strong bond between them, or their sister. Approximately 50m to the north is a monolith 2·1m tall and 1·8m × 0·6m in cross-section. One legend says that it served as a sort of stand-up throne from which the King could watch the construction of the trilithon; another claims that the trilithon was a gate through which bearers of food passed as the King waited at his throne.

In 1967 it occurred to the newly-crowned King that the trilithon might have an astronomical connection. The idea came to him while he was pondering the significance of a small marking on top of the lintel (see Fig. 2); his thought was that the arms of the double zig-zag indicated

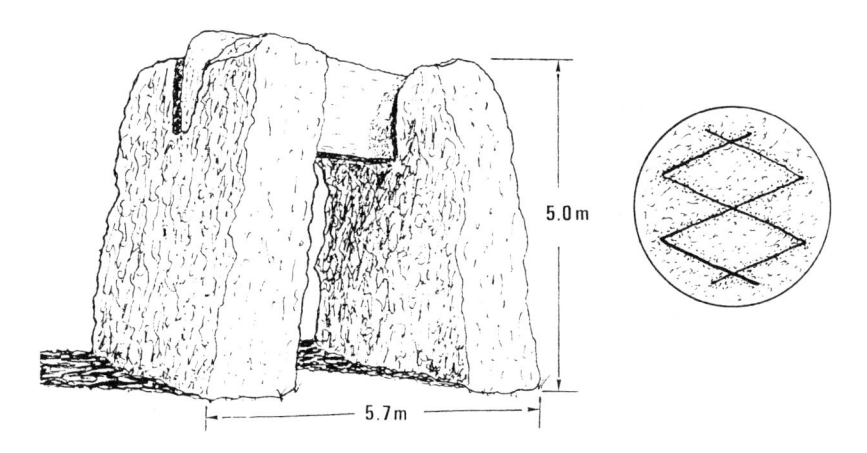

Orientations of religious and ceremonial structures in Polynesia 133

FIG. 2. The Trilithon of Tonga. Cross-section of left pillar: 1·5m × 2·3m. The 6cm-figure shown in the inset at the right is carved into the top of the lintel and has its arms closely aligned with the solstices.

the directions to the solstice rising points. To quote from the official guide book: '...in his Majesty's gracious presence surveyors took accurate sightings of the rising sun on the morning of the 21st June, and lo and behold it was found that the sun did rise according to His Majesty's expectations... The lintel itself is undoubtedly aligned along the southern-most point at which the sun rises marking the Summer Solstice.'

My own measurements fully confirm these statements. Although a small figure cut in rough coral cannot define directions with high accuracy, the smoothly cut lintel 5·7m long can; it is directed towards a rising declination of −24°·7. (In AD 1200 the sun's declination at the summer solstice was −23°·54.) Also well defined is the true azimuth from the centre of the trilithon to the throne stone: 10°·6.

Without further ethnological information, we cannot positively say that the trilithon was intentionally oriented upon the summer solstice. However, the W-shaped mark does indicate the directions to the three significant solar rising points reasonably well (to better than ±6°) and suggests that at some point, someone did want to record these directions. The direction towards the throne stone would seem to have no astronomical significance.

The trilithon is unique in Polynesia. Moreover, in Tonga there are about fifteen stone-terraced tombs of impressive size and construction dating from the thirteenth century. The orientations of these should be measured; they may prove to have archaeoastronomical significance.

The standing stones of Aitutaki

In the *marae* of Polynesia it is usual to find—scattered about, often more or less at random—a half dozen or so vertical stone or coral slabs within a walled area. On the Cook Island of Aitutaki the *marae* borders are unmarked, and the upright stones, all of basalt and many over a metre tall, frequently appear in straight rows. In Marae Paengariki, there were once some sixty such stones arranged in six roughly parallel rows, the largest being 40m long. The azimuthal orientation of the best defined row is 168°·8. This corresponds to a rising declination of −68°·1, and any putative astronomical correlations are novel (e.g., five hundred years ago the Large Magellanic Cloud was at −69°). On the other hand Marae Tapere, with approximately fifty uprights (the tallest measures 2·8m), consists of three roughly parallel rows, the longest (65m) oriented with azimuth 98°·5 (dec. −8°·0, rising). Marae Rangi O Karo again has three rows, oriented with azimuths of 180°·8, 179°·3, and 178°·8. All three rows, again made up of a total of about fifty stones, are approximately 85m long.

Bellwood (1978), who made his archaeological survey in 1969–70, lists ten other *marae* having stone alignments, although some of those had already been badly destroyed or heavily damaged. According to my local guide, the three listed above are the only ones that have been preserved and that have not become overgrown by jungle. Two others for which Bellwood shows plans are rather closely aligned north-south: 43m-long Kaionu at azimuth 13° and 24m-long Maramanui at 5°. However, as noted earlier, Bellwood's direction indications are only approximate (±11°). Perhaps the most cogent observation is that all five alignments are relatively close to (say within 13° of) a cardinal direction.

So far, I have been unable to find any ethnological reference to these *marae*, but it should be remembered that the Polynesians were expert navigators. Aitutaki is located within 350 km of eight other islands, none visible from Aitutaki; travel to and from these islands must have been frequent, and Aitutaki, despite its small size (1 km × 4 km), obviously supported a sizeable population. These alignments, never more than a few hundred metres from the shore, could have been seen from canoes offshore and would have served a practical navigational purpose—and almost certainly a ritual one, too.

Conclusions

The large number of impressive religious and ceremonial sites in Polynesia begs further archaeoastronomical investigations. The majority of religious structures are, like those on Easter Island, oriented parallel to the nearby shoreline, but again, as in Easter Island, there are some that strongly suggest astronomical usage and deserve more study. Especially noteworthy are the alignments on Aitutaki and the coral monuments on Tonga.

References

Bellwood, P.S. (1978). *Archaeological research in the Cook Islands, no. 27: Pacific anthropological records.* Honolulu HI: B.P. Bishop Museum Press.

Buck, P.H. (1938). Vikings of the Pacific. Chicago IL: University of Chicago Press.

Davidson, J.M. (1979). Samoa and Tonga. In *The prehistory of Polynesia*, ed. J.D. Jennings. Cambridge MA: Harvard University Press.

Davidson, J.M. (1984). *The prehistory of New Zealand.* Auckland: Longman Paul.

Emory, K.P. (1933). *Stone remains in the Society Islands.* Honolulu HI: B.P. Bishop Museum Press (B.P. Bishop Museum Bulletin, 116).

Emory, K.P. (1947). *Tuamotuan religious structures and ceremonies.* Honolulu HI: B.P. Bishop Museum Press (B.P. Bishop Museum Bulletin, 191).

Johnson, R.K. and Mahelona, J.K. (1975). *Na Inoa Hoku: a catalogue of Hawaiian and Pacific star names*, 1st edn. Honolulu HI: Topgallant Publishing Company.

Kay, R.F. (1988). *Tahiti and French Polynesia.* South Yarra, Victoria, Australia: Lonely Planet Publications.

Kirch, P.V. (1985). *Feathered gods and fishhooks.* Honolulu HI: University of Hawaii Press.

Liller, W. (1989). The megalithic astronomy of Easter Island: orientations of *ahu* and *moai*. *Archaeoastronomy* no. 13 (supplement to *Journal for the History of Astronomy* **20**), S21–48.

Sinoto, Y. (1969). Restauration de marae aux Iles de la Société. *Bulletin de la Société des Études Océanistes* **14**(7–8), 236–44.

14

Aboriginal sky-mapping?
Possible astronomical interpretation of Australian Aboriginal ethnographic and archaeological material

HUGH CAIRNS

1 Introduction

It is known that Aboriginal people within historic times intently watched celestial phenomena (Bell 1983), depicting many in story (Berndt and Berndt 1989) and in engraved and painted art (Allen 1976; Sutton 1989). It has also been suggested that Aboriginal people represented empirical phenomena in rock art (Cairns 1983; 1986). Recently, Cairns and Branagan (1992) have hypothesised that some of the carefully depicted dot and line symbols found in rock, bone and wood engravings and in paint represent major celestial movements and events. They go on to suggest a theory of 'sky-mapping' according to which, at certain places and times, Aboriginal people used such markings in relation to day-to-day living, survival techniques, story-culture and ceremonial life.

This paper attempts, following suggestions from Alexander Marshack and W. Breen Murray, to integrate ethnographic and archaeological data in support of this 'sky-mapping' theory, juxtaposing Australia-wide material with material from the environs of Sydney, and archaeological sites with human artefact and story. The paper proceeds to an interpretation of these data within a cognitive mapping framework, and finishes by presenting sky-mapping as an experimental hypothesis. The ultimate aim is some understanding of the dot and line symbols which are so common in Australia, and which are also found in many other areas of the world.

2 Ethnographic and archaeological data

Traditional Aboriginal activities and interests were (and are) expressed in story, song, art, and ceremony, along with language and mythology (Edwards 1987). Empirical knowledge of the natural world was extensive and detailed. However, although many of the 650 Aboriginal languages and cultures have been recorded, artefactual records often do not adequately match particular ethnographic and archaeological data. Even archaeological sites in historic use normally lack records of continuity of tradition, and traditional interpretation of ancient sites (and paintings) cannot always be taken at face value (Morphy 1991). Accounts concerning ethnographic acquisitions (e.g. Roth 1897; Howitt 1904) are normally inadequate. Those museum artefacts on which dot or line symbols occur—message-sticks, phallocrypts, cylcons, bull-roarers, boomerangs, and ceremonial boards—seldom have more information attached to them than the date of acquisition. Most, however, will have meaning associated with corroboree-meetings for particular ceremonies, which are managed in traditional Aboriginal societies by ritual and totemic-ancestral rules (Dixon *et al.* 1990; Morphy 1991).

Some dot and line patterns are more exact and identifiable than others. In general, regular spacing of series and patterns points to human ordering and meaning. Our speculations on this meaning must depend upon the prominence of a celestial phenomenon or event (e.g. the magnitude of a star), desirable accuracy, received Aboriginal lore's idiosyncratic highlighting of particular stars, and the position, cultural responsibility, design-skill and age (relating to the normally-excellent eyesight of Aboriginal people) of the engravers. In Aboriginal cultures, actual star-groups are totemic entities with mythological reality, operating dynamically in relation to Dreamtime Ancestors in song, story and ceremonial practice. The zenith, falling and appearing stars have spatial and/or directional significance for the timing and practice of some ceremonies, as well as for celestial-mythological presence (Mountford 1976).

Descriptions of Aboriginal message-sticks earlier this century included detailed accounts of messenger being taught the message, the contents of the message and the way it was then recounted in practice (Roth 1897), for instance which people were asked to attend a particular ceremony, when it was occurring, and why (Howitt 1904), and also where, and how much *pituri* should be brought (Horne and Aiston 1924). On the other hand, a recent study of message-sticks and shields (Edgar 1986) found that no 'universal language' seemed to exist on message-sticks: their markings do not correlate territorial, totemic or kinship meanings.

Such conclusions are based on inadequate data. The handful of artefacts found in museums represent the uncertain selection processes of museum donation and eccentric acquisition (Mulvaney 1989) in the context of the near-destruction of a culture (Broome 1982). There are inadequacies of language, linguistic analysis and productive questioning in the pioneering ethnography (Horton 1991). Furthermore, the Aboriginal culture is one where much ceremonial is secret (Berndt and Berndt 1964; Elkin 1977), as is much knowledge, whether sacred (Morphy 1991), women's (Bell 1983) or astronomical (Mountford 1976): so answers to questions are unreliable.

Moreover, even where Aboriginal artefacts have a secure cultural context, explanation of specific markings may not exist because a dot or a line can symbolise anything (or nothing). Ornamentation, often present in contemporary Aboriginal paintings (Sutton 1989), must be presumed present on artefacts (Roth 1897) and in the rock art: so the problem is, what is ornament, and what is symbol and representation? For archaeologists, historic carved tree-trunks, cylcons left strewn on bora-grounds, and even flat stones 'inscribed with seeming purpose' (McCarthy 1967) pose complex questions. The problems are greatest where no local story remains from informants, although analogy from other areas and groups can sometimes suggest a satisfactory explanation (Berndt and Berndt 1989).

Concerning sky-knowledge, some artefacts do have symbols relating by analogy to natural celestial phenomena: for example, rock-engravings where the shape (Fig. 1), described in the literature as 'boomerang', matches the moon's crescent: elsewhere, the same word was being used for

FIG. 1. Rock engravings near Dampier, Western Australia. This relatively recent, unpatinated engraving from the Pilbara shows mythological figure(s), 'boomerangs' and a 'comb' capable of lunar interpretation. (Photograph by the author.)

FIG. 2. Bark painting of moon, sun and yam. This shows crescent and full moon phases in association with the yam. It may represent a particular story, but the thirteen-month yam cycle links directly with the thirteen lunar-month approximation to the solar year. (Reproduced from Berndt (1964: pl. 31) by permission of the Aboriginal Arts Management Association of the Aboriginal Arts Board and Ure Smith, Sydney.)

both (Dixon *et al.* 1991). Historic paintings exist where moon and sun are represented in relation both to each other and to a natural plant (Fig. 2). Such works of art have symbols correlating reasonably through known associations (in this case astronomy and zoology). Moreover, Morning and Evening Star ceremonies include paintings and ritual equipment depicting (and explained by informants as *being*) the planet(s): one particular Morning Star ceremonial centre focuses on a large hole in the rock, which represents and *is* the Morning Star (Thomson 1926: file no. 133, photo. 2246). Gillen (1968 [1902]) records a stone column at a totemic centre which represents a particular star which is stationed to guard the sacred black stone. Tindale (1935) records the Aboriginal 'remarkable interest in the movement of the planets': Mars with the two women, Venus and Jupiter. It is clear, then, that a strong general case can be put forward for Aboriginal sky-mapping despite the dearth of specific local data.

Aboriginal stories are full of celestial content, marrying mythological personages to empirical celestial phenomena, which become or are totemic actors in the drama (Berndt and Berndt 1989). The songlines of the cult centres (popularised by Chatwin 1986), marking the 'historic' travels of the totemic Ancestors, relate consistently to the empirical realities such as waterhole, whirlwind or heavenly body; and this is communicated in the art, music and story within the pervasive symbolism and associations of mythology (Berndt and Berndt 1964).

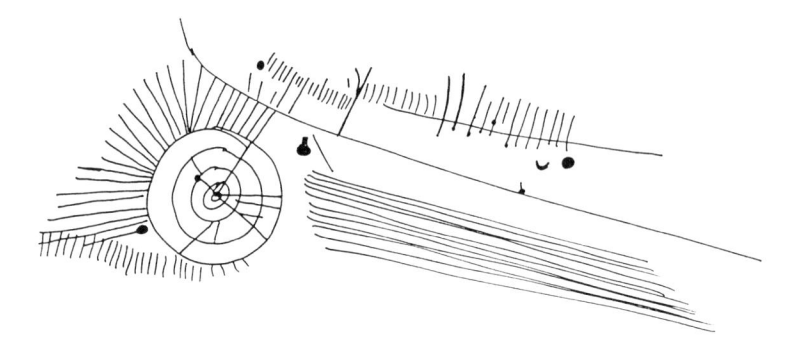

FIG. 3. Selection of motifs at a set of Victoria River sites at Innesvale, Northern Territory, described by the local traditional Aboriginal woman owner as 'Moon Dreaming'. The set of three rings, some 40 cm in diameter, has 28 grooved rays started (or completed) with a cup-and-groove, but only occupying the left two-fifths. Juxtaposed outside the circles is a series of 28 grooves with an engraved crescent and a cup; then 14, then 11 (with a cup), returning to the first set. A series of over 24 less prominent grooves carry along under the circles, making a possible lunar count of four months in all. (Sketch by the author from photograph by Howard McNickle, reproduced by permission. See also Walsh 1988: 97.)

Both men and women had stories, made art, worked wood and stone, and had ceremonial sites (sometimes unknown to each other). It is said that women kept menstrual sticks for checking the gestation period (Thompson 1981): if so, message-stick 8394 (Macleay Museum, University of Sydney), which has 29 and 30 notches along its edges, totalling precisely the number of days in two lunar synodic months, might denote just this. The 'Moon Dreaming' sites in Western Australia and the Northern Territory, owned at present by women, include Ancestral figures with lunar-number-like details (see Fig. 3). If there is lunar information present for intimate and personal reasons, it is now subsumed under mythological guise and used for activities in the wider group. Information at mother's knee might become gestation-notation for girls who, after all, have the crescent moon or 'rainbow' painted below the breasts at menstruation ceremony (Berndt and Berndt 1989); but young men will recognise lunar timing in the ceremonial realities they enter at initiation, the months planned ahead being reckoned in 'moons'.

The numbers between 27 and 31 appear widely, and, if of lunar significance (Marshack 1972), seem therefore of special importance to men as well as to women. Examples are ceremonial board C342 (Macleay Museum, University of Sydney) with 28 marks along the edge; ceremonial boomerang QE342 (Queensland Museum), with 27 special-ochred twine-coils; phallocrypt A2044 (Macleay Museum), with two careful and distinct sets of dots, 28 + 7; message stick 13379 (Roth Collection, Australian Museum) with 27 grooves; and bull-roarer QE5892 (Queensland Museum) with 28 edge-marks.

Statistically, the percentage of museum artefacts with series or sets of possible astronomical significance is not low: out of some hundreds of artefacts studied by the present author, 97 appeared interesting (by a subjective assessment) and 45 had symbols and/or numbers which *could* have an astronomical relation. But the value of such statistics is uncertain, for percentages are non-locality-specific, artefact selection is necessarily biased, researchers may be insufficiently open-minded, and symbols may have many meanings. More importantly, not every artefact or part of a rock site or ceremonial complex will necessarily need to embody (let alone be impregnated with) astronomical information, even knowledge from mother's knee; and 'gestation' information would only be expected at some 'women's places'. But information relating to specific ceremony, specific timing and Dreamtime-celestial figures may reasonably be expected in some guise on men's ceremonial artefacts and at rock art sites; not necessarily in abundance, but at some point in the art; just as it normally appears 'at some point' in the cultural concept-cluster and symbol-system in the stories, songs and, indeed, in the Aboriginal languages themselves.

Rock-art sites demonstrate some permanence and a traditional special position in Aboriginal culture (Walsh 1988). Though particular sites (like particular artefacts) may provide no obvious symbol, pattern or series, many do: recently, interpreters have judged batches of grooves and cups to be 'entranced routine' or phosphene/entoptic replay (Lewis-Williams and Dowson 1988, Steinbring and Lanteign 1990), emus' eggs or idle/children's scratchings, or tally-counts with 'precise meaning', ideas that are unsustainable in the absence of informants (Flood 1983; Stanbury and Clegg 1990). While specific interpretation is precarious, most rock-art site areas in Australia provide elements that can be interpreted generally in terms of sky-mapping. Furthermore, there can be some interpretative feedback between this evidence and the artefacts through special examples such as the natural but enlarged hole in the rock face which represents and 'is' the Morning Star and the body-painted menstruation 'crescent', both already mentioned.

Within suggestive cultural environments where such to-and-fro analogy is possible, it is possible that some dot-and-line incisions on wooden artefacts, and cup-and-groove engravings on rock, may reasonably be interpreted as representing and symbolising celestial phenomena. Identical dot marks are used to symbolise many different pieces of information in Aboriginal art (Roth 1897; Munn 1973; Reser 1979; Sutton 1989), but one meaning is astronomical (e.g. Fig. 4). By inference, the

FIG. 4. Bark painting of Orion and the Pleiades. The inside of a 'humpy' could be painted with myriad mythological themes, the night-sky relationships being deeply studied, incorporated into stories, and mirrored in art. These post-contact paintings depict Orion and the Pleiades with recognisable isomorphic (Harris 1986) representation. The canoe, women and hunters of the stories are displayed by dot symbols, but two of the four 'incorrect' dots depict campfires and food (Sutton 1989). The circle/dot symbol in Aboriginal iconography can represent around seventeen different items from (e.g.) camp to sacred waterhole to Dreamtime Ancestor (Munn 1973; Reser 1979). This multi-level symbolism emphasises that the art, already permeated with totemic symbol, has layer upon layer of concepts, the multiple associations of which can include empirical celestial realities such as individual stars (Morphy 1991). (Painting by Minimini of Groot Eylandt. Reproduced from Isaacs (1980: 152) by permission of the Aboriginal Arts Management Association of the Aboriginal Arts Board and Weldon Pty. Ltd., Sydney.)

sky-information evinced by ethnographers Dawson (1881), Roth (1897), Howitt (1904), Tindale (1963), Strehlow (1971), Mountford (1976), Elkin (1977), and Berndt and Berndt (1964; 1989), amongst many others, will sometimes be present in Aboriginal ceremonial artefact and rock art, particularly in the dot and line symbols.

3 Astronomical knowledge as part of cognitive mapping

The antiquity of the dot/groove symbolism may be considerable, in the form of series of incisions. The 'series' may be the medium for the 'message': ancient ways of mind, when noting some particular empirical phenomenon, often expressed it by means of what appears to us as artistic sensibility. Early series may as likely have been probing and reflecting astronomical movements as any other tally. It is possible that the Bilzinsleben artefact of Holstein Interglacial's Homo Erectus dating to 350,000–250,000 BP (Bednarik 1992), the most ancient work of human art yet discovered, may transmit lunar phase information: if Davidson's reconstruction is correct, it had 28 beautiful groove-incisions (Mania and Mania 1988; Davidson 1990).

In Australia, the 28 incisions on the tooth of a (now-extinct) diprotodon (Vanderwal and Fullagar 1989; Fig. 5) dated to 19,800±390 BP may convey similar information. These two examples suggest that independent note of the lunar phase cycle, reflected in some form of tally marks, may have been a very ancient human activity. Recent cation dating suggests that in Australia the cup-and-groove tradition extends back beyond 30,000 BP (Sutton 1989; Nobbs and Dorn 1988).

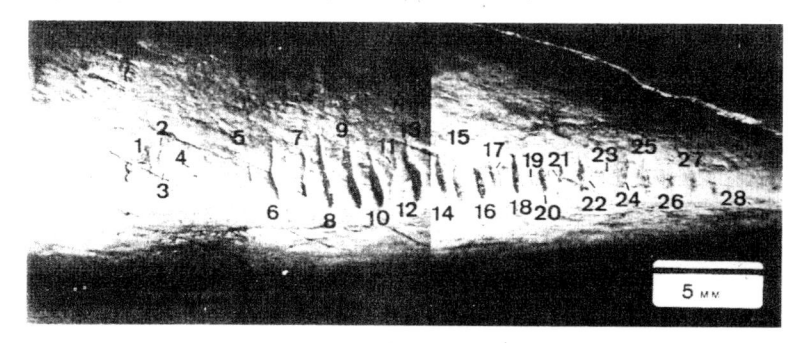

FIG. 5. Markings on an extinct mammal's upper incisor, 19800±390 BP. The 28 marks on this South Australian diprotodon tooth suggest the nights of viewed moon. The mean distance between grooves is 1·2mm. (Museum of Victoria, MNV 177,945. Photograph by Richard Fullagar. Reproduced by permission.)

FIG. 6. Marks in a tessellated pavement at Elvina Track, New South Wales. Patterns on this pavement closely resemble particular star-groups in the Milky Way from the Southern Cross and Scorpio to Orion and the Pleiades. (Photograph by the author.)

The marks on the tessellated pavement at Elvina Track, in Ku-ring-gai Chase National Park near Sydney (Fig. 6), were probably constructed between 6000 and 4500 years ago (Stanbury and Clegg 1990). At the site are groups of drilled cups, transverse grooves and enhanced natural indentations,[1] found in close proximity to prominent engraved representations of the ancestral sky-heroes Daramulan and Biame, the former specifically linked to Ritual Law and Initiation (Cairns and Branagan 1992: figs. 1–3). There are no informants for Ku-ring-gai Chase, although language records star-names. It is, however, almost inconceivable that the milky river cascading from Scorpio to Orion and the Pleiades across the crystal-clear firmament, moving by night and by month, and so prominent in Aboriginal story and ritual, would not be expressed or alluded to in some form.

---

[1] Although some of the site's multiple pockmarks are subject to geological dispute (Bednarik 1990; Cairns and Branagan 1992) the regular spacing, diameter and depth of the drilled cups, in series with artificial form and pattern, keep alive the question of purpose.

Fig. 6 shows one small section of a pavement filled with polygons and providing more than a hundred worked and enhanced dot-patterns (Cairns and Branagan 1992). If the sky-map speculation is correct, many combinations can be suggested which appear in Aboriginal lore. For example, the four-sided polygons may represent star groups in Argo, Sagittarius, Musca and (at zenith) Corvus, as well as the always-visible Crux, standing out particularly with its linking groove. The central pattern in Fig. 6 is identical to that on the 'Calendar Stone' 30m away (Cairns and Branagan 1992: pl. 1), and may represent the scorpion's sting, or Orion's belt, so important as the Two Guardians at zenith during the Aboriginal New Year Initiation Season (Mountford 1976). However, it also resembles the Canis Major pattern which is found close by within its own stone 'frame', near other New Year sky shapes which resemble parts of Leo, Musca, and the Southern Triangle.

Organising tribal meetings and ceremonies needed sensible knowledge of the lunar cycle and stellar movements. According to ethnographic information, the sun was valuable only for determining the time of day: longer periods were reckoned by 'moons', specific ceremonies being fixed for specific lunar phases, and ritual served to relate myths to particular stars (Roth 1897; Howitt 1904; Mountford 1976; see also Haynes et al. 1993). This basic cultural and ceremonial information could be reflected in message-stick (and other artefactual) markings (cf. Fig. 7), and permanently recorded in the form of rock engravings. Thus star-group markings might be expected elsewhere in Australia (cf. Fig. 8).

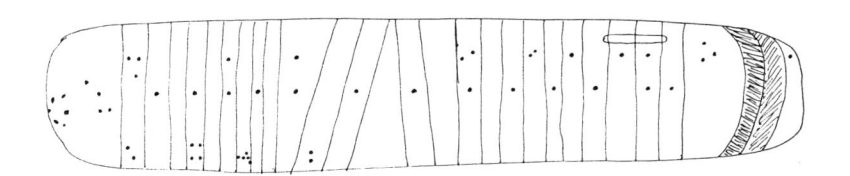

FIG. 7. Wooden message-stick X873 (Museum of Victoria). Length c. 75cm, width c. 2·5cm. This message-stick was collected in 1890 from the Whajah and Ballarding tribes of Western Australia, and described by the original (anonymous) informant as being a 'corroboree stick' with 'inscribed rules', but nothing is known of the ceremony involved. In this sketch (by the author), the cylindrical shape of the original has been opened out to form a flat surface. The burn-gouged dots, in regular series and particular groupings, do not seem to the writer to be ornamentation, recordings of a game, representations of artefacts or (except for the non-uniform group at the very left) of people. Speculatively, the dot-symbols on the stick to the left and right of the central line of thirteen dots may represent star groups, all of which figure expansively in Aboriginal lore, within the Western constellations of Scorpio, Crux, Argo and Orion.

FIG. 8. Series of cups at a ceremonial site, Victoria River, Northern Territory. This series of large cups numbers 30. It continues along the rock with 44 very small cups. Vertical grooves enter the cups at positions (from the right) 7, 15, 17–22, and 31 and occur after cups 11, 15 and 28. If the out-of-line and crescent-shaped cup (no. 4) is taken as First Crescent, the major series of cups-with-grooves occurs at full moon, two 'tandem' cups mark the beginning and end of season, and the deep groove accompanying the six tandem-limited cups underline the season's ceremonial importance. (Photograph by the author.)

For all the examples illustrated so far in the figures (Fig. 5 excepted), there is some direct evidence relating to cultural context. However, the overwhelming majority of collected artefacts and sites exhibit no such evidence. The complexity and sophistication of Aboriginal languages, and normal mammalian mapping of spatial environment, provide some solace. Linguistic analyses indicate that Aboriginal lore may have its roots as long ago as 80,000 BP (Stanbury and Clegg 1991), and the language is said to be as highly developed as, say, French or English in describing experience (Ridley 1875; Capell 1970; Blake 1991). Aboriginal stories carry a great deal of astronomical information within vocabularies and concepts as well as in the content of tales learned at the knee (Berndt and Berndt 1989), in ceremonial meetings (Thomson 1926), in female education (Berndt and Berndt 1964) and at high-point male initiation (Mountford 1976). Different tribes and generations may have 'lost' and 'rediscovered' what is now 'sacred-secret' astronomical information since the days when it was knowledge needed for simple survival (perhaps linked to original navigation used in crossing continents and the ocean from Asia).

The mapping of social reality in Aboriginal kinship structure (Tomlinson 1974), the mapping of geography in its art (Sutton 1989), the totemic and conceptual mapping of flora-and-fauna (J. Mathews, unpublished manuscript 1988) and the stars overhead (Dawson 1881; Turner 1989), suggest to the present writer that the mythological relationships of the empirical nocturnal moving points of light as totemic persons, and the sun as compass, are prime examples of 'cognitive mapping' of permanent spatial points. Noting the regularities and anomalies of the experienced and observed total environment is as basic to normal human survival as it is to the modern scientific quest (Giere 1990): thus naming, and recording, natural phenomena in story may be a parallel activity to the production of cups (series or group-patterns) and grooves (lines, combs or ladders) on the same sites where mythological sky-story and sky-ritual took place. The overall process involved is the 'cognitive mapping' that is now agreed to be a normal factor in every mammal's perceptual apparatus (McGaugh 1990).

General and specific astronomical knowledge in the (schematised) 'cognitive map' of Aboriginal people is evidenced by abundant ethnographic material. The mapping of astronomical phenomena runs from (the common-sense statement that) 'North, South, East and West are *like this in my head*' (my italics, to denote that the informant is waving to indicate directions with his hands), to bark paintings specifically depicting dynamic celestial phenomena such as the moon, sun, Pleiades, Scorpio and Orion; to abundant stories of individual planets, stars and star-groups, sometimes specifically related to ceremonial activities or to navigation, and at other times bringing together different types of 'coincidental' phenomena, for instance the falling Pleiades linked to first frosts and first dingo pups (Mountford 1935; Tindale 1963); to markings on stones, rock and wood that show lunar symbols together with number-series that suggest the lunar phase sequence, or groups of dots that represent particular stars.

It is on this basis of the concept of cognitive mapping, as well as on the evidence of astronomical interests within language, other ethnographic layers, and analogous correlations/associations on artefacts in wood and stone and at archaeological sites, that the sky-mapping theory can now be tested in relation to 'cold' (context-free) data.

4 Sky-mapping as an experimental hypothesis

When standing stones have a specific orientation amidst sky symbols, or numbers associated with lunar cycles and other series or patterns appear in close proximity to mythological sky heroes or abstract symbols, these then become interpolations within a growing theory. The basic evidence for a fundamental interest in celestial bodies rests in ritual use expressed in myth and cult action. Examples include the holes in the rock-face central to the Morning Star ceremony in Arnhem Land, and effigies of the Morning Star crucial to funerary rites (Thomson 1926: file no. 133, photo. 2246); the ritual objects 'Kulpidji' present in the heavens in the Southern Cross (Mountford 1976); the moon being of special interest at particular sites where it is represented with its own paintings and abstract symbols (Fig. 3); special stones polished in crescent shape (McCarthy 1967); and the elder pointing to Antares (α Sco), Shaula (λ Sco) and the Boy-and-Girl pair of stars (μ Sco), relating them directly to Dreamtime Ancestors, within the crucial initiation rites for young men (Mountford 1976). All this is in the ethnographic material which shows that the Aboriginal people have an encyclopaedic knowledge of and value for the celestial bodies and their movements, placed centrally within their all-embracing religious culture and ceremonial life.

The question is whether there was, as well as cognitive intellectual mapping, abstract physical mapping of lunar phase complemented by isomorphic representation of star patterns, both underpinning the ceremonial world with empirical celestial realities, and gathered together in symbolic painting and engraving as well as in the art of song and story.

If 'yes', the archaeological data from prehistoric rock art in Australia would reflect specific aspects of Aboriginal perception in relation to astronomical realities, specific accurate knowledge of them, and certain ceremonial and real-world orientations relating to them. It would formally express imaginative intellect.

The pre-mathematical realities of arithmetical-behavioural notation (Marshack 1972) were present in indigenous Australia, if the lunar interpretation of the diprotodon bone's series (Fig. 5) is accepted. Historic evidence does not seem to include actual monthly calendar days, but Aboriginal people at contact could accurately gauge number in relation to quantities and time/space, had words denoting specific numbers up to 31 (Roth 1897; Howitt 1904). This might be expected if the lunar phase cycle was observed and months counted, and the year was geared to relationships between biological, meteorological and astronomical events, with major ceremonies tied to the positions of particular celestial bodies in the heavens. The presence of geometrical as well as numerical perception is shown crucially in Aboriginal art (Sutton 1989): the abstract patterning of sets of empirical celestial realities may also present itself as empirical notation in forms such as those noted in this paper.

5 Conclusion

In summary, the sky-mapping theory perceives empirical pattern as layered symbol, and Aboriginal mind-process as cognitive mapping. It proceeds with open mind from ethnographic data to anthropological and archaeological speculation, and back again. There is no shortage of suitable 'cold' archaeological data in the form of engraved and painted cups/dots and grooves/lines, often in association with abstract and representational figures, from around Australia. In series, these can resemble lunar phase number; in groups, star-patterns. In some instances, details within associated rock art and ceremony also relate to sky phenomena. The suggested sky-mapping meanings are subject to legitimate criticism in the normal manner, and we can expect critics to provide reasonable alternatives or further-developed interpretations which enlarge the discussion.

What is principally evidenced in these pre-literate and environmentally successful people is sensible cognitive mapping of the total environment, with the starry night as much part of empirical reality as the practical day, the social kinship system, and ceremonial life. The totemic religion in which history vied with empathy and mythic reconstruction, empirical reality with myth-morality, and individual creativity with communal survival-necessity was firmly based in the movements and patterns of the celestial bodies. The stars were the spirit-persons and campfires of the specific ancestors, the Milky Way a tracing-board of mythological story, the moon a major way of measuring time.

Assuming that the concept of 'cognitive mapping' is a fair description of scientific endeavour (Giere 1990), and given that Aboriginal people were competent explorers, both physically and geographically, who stored discovery and information in language, art, story and song, it is reasonable to suggest that what seem to us today to be simple or eroded scratches on stone may be clues to methods of thinking as well as map-making. The resulting marks and patterns elucidate the mentality of the maker because astronomical knowledge juxtaposes with other meaning. To counter positivist claims to the contrary, ethnographic evidence shows that knowledge and perception, throughout Australian Aboriginal cultures, is expressed in the form of art. It would be extraordinary indeed if important meaning, including astronomical realities, were not in evidence in the rock engravings and paintings of indigenous Australia.

Acknowledgements

Material studied for this ethnographic background was found in the Macleay Museum, University of Sydney; the Art Gallery of the University of Melbourne; the University of Queensland Museum; the Australian Museum, Sydney; the Museums of Queensland, South Australia and Victoria; and the State Libraries of New South Wales and South Australia. My thanks are due to the curators and staff who made this possible.

John Clegg, David Branagan and James McGaugh have played a specially important part in the formation of this paper. However, the particular sky-mapping speculations and the interpretations of the cultural material, as well as any shortcomings, remain my own.

References

Allen, L.A. (1976). *Time before morning*. Sydney: Rigby.

Bednarik, R.G. (1990). Comment. *Australian Rock Art Research (AURA) Newsletter* 7(1), 3.

Bednarik, R.G. (1992). Palaeoart and archaeological myths. *Cambridge Archaeological Journal* 2(1), 27–57.

Bell, D. (1983). *Daughters of the Dreaming*. Melbourne: McPhee Gribble.

Berndt, R.M., ed. (1964). *Australian Aboriginal art*. Sydney: Ure Smith.

Berndt, R.M. and Berndt, C.H. (1964). *The first Australians*. Sydney: Ure Smith.

Berndt, R.M. and Berndt, C.H. (1989). *The Speaking land*. Sydney: Penguin.

Blake, B.J. (1991). *Australian Aboriginal languages*. Brisbane: University of Queensland Press.

Broome, R. (1982). *Aboriginal Australians*. Sydney: Allen and Unwin.

Cairns, H.C. (1983). Intellectual expressions of pre-historic man in relation to Aboriginal people in Australia. In *Prehistoric art and religion*, ed. E. Anati, pp. 63–67. Milan: Jaca Book Spa.

Cairns, H.C. (1986). The searching intellect of man: map, model, chart. *Rock Art Research* 3(1), 20–25.

Cairns, H.C. and Branagan, D.F. (1992). Artificial patterns on rock surfaces in the Sydney region, New South Wales. Evidence for Aboriginal time charts and sky maps. In *State of the Art: regional art studies in Australia and Melanesia*, ed. J. Macdonald and I. Haskovic, pp. 24–31. Melbourne: Australian Rock Art Research Association (Occasional AURA publication no. 6).

Capell, A. (1970). Aboriginal languages in the central south coast of New South Wales. *Oceania* 41(1): 20–27.

Chatwin, B. (1986). *The songlines*. London: Pan Picador.

Davidson, I. (1990). Bilzingsleben and early marking. *Rock Art Research* **7(1)**, 52–56.
Dawson, J. (1881). *Australian Aborigines: the languages and customs of various tribes of Aborigines in the western deserts of Victoria.* Sydney: Government Printer.
Dixon, R.M.W., Ramson, W.S. and Thomas, M. (1990). Australian Aboriginal words in English. London: Oxford University Press.
Edgar, R.J. (1986). *Stylish messages.* Unpublished BA Thesis, University of Sydney.
Edwards, W.H., ed. (1987). *Traditional Aboriginal society.* Melbourne: Macmillan.
Elkin, A.P. (1977). *Aboriginal men of high degree.* Brisbane: University of Queensland Press.
Flood, J. (1983). *Archaeology of the dreamtime.* London: Collins.
Giere, R. (1990). *Explaining science.* Chicago IL: University of Chicago Press.
Gillen, F.J. (1968 [1902]). *Gillen's diary. The camp jottings of F.J. Gillen on the Spencer and Gillen expedition across Australia 1901–2.* Adelaide: South Australian Libraries Board.
Harris, R. (1986). *The origin of writing.* London: Duckworth.
Haynes, R.F., Haynes, R.D., McGee, R.X. and Malin, M.D. (1993). *Opening the Southern skies: two hundred years of Australian astronomy.* Cambridge: Cambridge University Press.
Horne, G.A. and Aiston, G. (1924). *Savage life in central Australia.* London: Macmillan.
Horton, D. (1991). *Recovering the tracks.* Canberra: Aboriginal Studies Press.
Howitt, R. (1904). *Native tribes of south east Australia.* London: Macmillan.
Isaacs, J. (1980). *Australian heritage.* Sydney: Lansdowne Press.
Lewis-Williams, J. and Dowson, T. (1988). *Images of power.* Johannesburg: Southern Books.
McCarthy, F.D. (1967). *Australian Aboriginal stone implements.* Sydney: The Australian Museum.
McGaugh, J.L., ed. (1990). *Brain organization and memory.* London: Oxford University Press.
Mania, D. and Mania, U. (1988). Deliberate engravings on bone artefacts of Homo Erectus. *Rock Art Research* **5(2)**, 91–107.
Marshack, A. (1972). *The roots of civilization.* London: Weidenfeld and Nicholson.

Morphy, H. (1991). *Ancestral connections*. Chicago: University of Chicago Press.

Mountford, C.P. (1935). *Mountford collection: vol. 6*. Unpublished manuscript, State Library of South Australia, Adelaide.

Mountford, C.P. (1976). *Nomads of the Australian desert*. Sydney: Rigby.

Mulvaney, J. (1989). *Encounters in place: outsiders and Aboriginal Australians 1606–1985*. St Lucia: University of Queensland Press.

Munn, N.D. (1973). *Walbiri iconography*. London: Cornell University Press.

Nobbs, M.F. and Dorn, R.I. (1988). Age determinations for rock varnish formation within petroglyphs. *Rock Art Research* **5**(2), 108–46.

Reser, J. (1979). Values in bark. *Hemisphere* **22**, 27–36.

Ridley, W. (1875). *Kamilaroi and other Australian languages*. Sydney: New South Wales Government Printer.

Roth, W.E. (1897). *Ethnological studies among the north-west Central Queensland Aborigines*. Brisbane: Government Printer.

Stanbury, P. and Clegg, J. (1990). *Aboriginal rock engravings*. Sydney: Sydney University Press.

Steinbring, J. and Lanteign, M. (1990). Comment. *Rock Art Research* **7**(1), 57–62.

Strehlow, T.G.H. (1971). *Songs of central Australia*. Sydney: Angus and Robertson.

Sutton, P., ed. (1989). *Dreamings*. London: Penguin.

Thompson, W.I. (1981). *The time falling bodies take to light*. New York NY: Saint Martin's Press.

Thomson, D.F. (1926). *Arnhemland field-trip notes*. Thomson Collection, Museum of Victoria, Melbourne.

Tindale, N.B. (1935). The Story of the man Waijungari who became the planet Mars. Records of the Museum of South Australia, Adelaide.

Tindale, N.B. (1963). *Aboriginal Australians*. Brisbane: Jacaranda Press.

Tomlinson, R. (1974). *The Mardudjara Aborigines*. Orlando FL: Holt, Rhinehart and Winston.

Turner, D. (1989). *Return to Eden*. New York NY: Peter Lang.

Vanderwal, R. and Fullagar, R. (1989). Engraved *diprotodon* tooth from the Spring Creek locality, Victoria. *Archaeologia Oceania* **24**, 13–16.

Walsh, G.L. (1988). *Australia's greatest rock art*. Bathurst, NSW: E.J. Brill / Robert Brown and Associates.

III

NEW TECHNIQUES, METHODS AND APPROACHES

15

Basic research in astronomy and its applications to archaeoastronomy

BRADLEY E. SCHAEFER

1 Introduction

Before the invention of the telescope, all astronomy was done with the human eye. Thus, to understand many old observations, we need to know the eye's capabilities. In modern times, enough is known about astronomy, meteorology, and physiology to allow for precise modelling of the eye's capabilities as applied to practical situations. Unfortunately, the professional astronomers are not interested in visual observations and other researchers do not have the background or inclination to pursue the required complex calculations.

For the past decade, I have been gathering the necessary mathematical models for all relevant phenomena, so that I can now calculate from first principles what should and what should not be visible to the eye. Just as important, I have also developed a formalism for establishing the uncertainties in my model predictions.

Whenever a scientist advances a new theoretical model, the model must be tested against reality. Therefore, for every phenomenon that I have modelled, I have collected data to test my model. Frequently, these tests consist of myself or my friends making the relevant naked eye observations. So, for example, I have timed the first and last appearances of hundreds of stars as they rise and set, I have timed many sunsets over ocean horizons, I have looked for many hundreds of stars in the dusk and dawn sky, I have looked for stars near the moon over a wide range of phases, I have monitored the sun for naked-eye sunspots daily since 1988, and I have measured shadow penumbras both on small gnomons and on the Washington Monument. Frequently, these tests involve my soliciting data from the public at large. So, for example, I have published appeals for observations of lunar occultation visibility and the limiting magnitude of visual telescopic observations, and with LeRoy Doggett of the US Naval Observatory have organised several moonwatches involving roughly two thousand observers across North America. Frequently, these tests consist of locating appropriate data from the published literature. In this way several naked eye sunspot surveys have been found, as were many observations of lunar crescent visibility, and A. Thom has published a summary of his terrestrial refraction measurements in Scotland.

If the theoretical models are verified observationally, then the predictions can be confidently applied to many areas involving archaeoastronomy, the history of astronomy, and amateur astronomy. The historical applications discussed in this paper are taken from all over the world and from many cultures.

2 Tools

2.1 Source brightness

A fundamental factor in determining the visibility of any astronomical object is its apparent brightness above the atmosphere. The brightness and colour of stars and planets have been tabulated in many books, of which the *Astronomical Almanac* is one definitive source. The integrated brightness of the moon can be found in section 66 of Allen (1976), while the spectral reflectivities are given in McCord and Johnson (1970). The lunar surface brightness is given by the complex Hapke equations (Hapke 1984) and the parameters given by Helfenstein and Veverka (1987). The brightness and spectral energy distribution of the sun are given in sections 75 and 82 of Allen (1976), while sunspot information is presented in section 88.

2.2 Sky brightness

The brightness of the daytime sky is given by Weaver (1947), Koomen *et al.* (1952), and Tousey and Hulburt (1947). The night sky brightness is given observationally by Krisciunas *et al.* (1987), Pilachowski *et al.* (1989) and Garstang (1989), while Schaefer (1990a) gives general formulae and methods for estimation. Krisciunas and Schaefer (1991) present observation, theory and simple formulae for calculating the brightness of moonlight and sunlight. The sky brightness during twilight has been observed by Barteneva and Boyarova (1960) and Koomen *et al.* (1952), while Schaefer (1987a) uses additional observations to develop a general formula for twilight. The effects of city light pollution have been observationally and theoretically addressed by Walker (1973), Garstang (1989) and Lockwood *et al.* (1990).

2.3 Refraction

When light travels through the atmosphere, its path is bent by refraction. For altitudes well above the horizon, the classic treatment of Smart (1979: 58–73) is valid. For altitudes near and even below the horizon, the complex algorithm of Garfinkel (1967) is valid if the atmospheric thermal structure is similar to the US Standard Atmosphere. However, Schaefer and Liller (1990) have shown that significant thermal inversions are ubiquitous and will drastically change the refraction near the horizon. Schaefer (1989) presents a program for calculating refraction (and air mass) for arbitrary thermal structure in the atmosphere of any planet.

2.4 Extinction

The Rayleigh scattering component of the extinction is reliably calculated as a function of wavelength, altitude, and barometric pressure by Hayes and Latham (1975). The ozone component of the extinction can be calculated from the cross-sections in Allen (1976) and the ozone depth from Bower and Ward (1982) as formulated as a function of latitude and time of year by Schaefer (1990e). The aerosol extinction coefficient is tabulated as a function of the time of year for 253 sites around the world (*ibid*.), so that most sites are close to a location with known extinction. Schaefer (*ibid*.) has also formulated the aerosol component as a function of altitude, latitude, longitude, relative humidity, year, and time of year. Unfortunately, the aerosol content has significant variation around the calculated value, and this uncertainty is always the largest source of uncertainty in any calculation involving astronomical sources near to the horizon.

2.5 Air mass

The air mass is the optical path length of a ray of light in units of the optical path length of a vertical ray. For altitudes well above the horizon, the classical secant formula is valid. Near to the horizon (but not below it), the formula of Rozenberg (1966) is convenient and fairly good for reasonable elevations and aerosol density. However, for an accurate calculation, the various extinction components must be handled separately (Schaefer 1990e). The air mass for the Rayleigh and aerosol component can be calculated for the appropriate scale heights (*ibid*.) with the program of Schaefer (1989) for arbitrary atmospheric conditions and viewing directions. The ozone air mass can be calculated in a closed form from simple trigonometry (Schaefer 1990e).

2.6 Optics

For questions of visibility in modern times, the effects of telescope optics may be needed. The optics of telescopes have well-known effects, but the researcher must beware the many subtle phenomena that can change the result. The best discussions from a visibility point of view are in Tousey and Hulburt (1948) and Schaefer (1990a).

A pinhole camera represents a simple form of optics that needs no lens. The resolution of pinhole cameras is discussed in many textbooks (e.g. Wood 1934), but no theoretical work has been reported on the visibility of sources. Schaefer (1992a) presents calculations of sizes, brightnesses, resolutions, and thresholds for use in examining a pinhole image of the sun.

2.7 Visibility

All the preceding tools can be used to calculate the apparent brightness of the astronomical source and the surface brightness of the background. The question of visibility then depends on the physiological question of whether such a source can be detected by the human eye. (Note that sensitivity variations from person to person are usually smaller than other uncertainties.) For point sources of light, the thresholds given by Hecht (1947) are convenient and accurate. For uniform extended sources, the vast work of Blackwell (1946) and Siedentopf (1941) are useful. Note that many corrections (e.g. colour temperature, binocular, pupil size, age, observing time, and detection probability) are required before these data can be applied even to a naked-eye sighting. For non-uniform extended sources, the visibility must be calculated from a detailed probability calculation for each visual element.

2.8 Glare

A bright source of light can mask a nearby dim source because of scattering of light either in the atmosphere, in the eye, or in the telescope. The excess background from atmospheric scattering is quantified by King (1971). The scattering in the eye is quantified by Holladay (1926). Schaefer (1991a) presents a complete theoretical treatment, an array of observational tests, and discussions of many applications for glare in astronomical settings.

2.9 Shadows

The sun does not cast a shadow that has a perfectly sharp edge, since the sun subtends an angle of half a degree in diameter. I have derived an analytic expression for the brightness of the shadow in the penumbral zone. There are several ways that an observer could try do define the edge of the shadow so that its positional uncertainty is smaller than the size of the penumbra. The accuracy of these various methods has been assessed on the basis of the thresholds for distinguishing shades of grey.

3 Results

3.1 Extinction angle

The extinction angle is the altitude above the horizon at which a star first becomes visible in a dark sky. The extinction angle used by Thom (1967) and by many other archaeoastronomers is based on the prescription of Neugebauer (1929) which was deduced from Babylonian data. Unfortunately, Thom's test of Neugebauer's rule is a circular argument since Thom's sample of stars was chosen because of a close fit to Neugebauer's rule. Schaefer (1986) derives the extinction angle as a function of the extinction coefficient and stellar magnitude with the use of the mathematical tools discussed above in Sections 2.1–2.5 and 2.7. The uncertainty in the extinction angle and the azimuth of first sighting is also derived. Schaefer (1987b) gives a program for calculating the extinction angle. Schaefer (1986) presents the only collection of modern extinction angle observations, and these are well fitted by the theoretical model.

3.2 Heliacal rise

The heliacal rising of a star or planet happens on the first day that the object is visible in the dawn light after the conjunction with the sun. The previous theoretical work on this topic dates back to Ptolemy (see Aveni 1972), while the last published set of observations are Babylonian. Schaefer (1987a) presents a model for predicting the dates and azimuths of heliacal risings and settings based upon the various mathematical tools discussed in Section 2. Estimates of the uncertainty in these quantities caused by normal variations of the moonlight and weather as well as uncertainties in the observing conditions are presented. In addition, 556 observations of heliacal rise are reported, virtually all of which confirm the theoretical calculations, with the few exceptions being directly caused by clouds near to the horizon. Schaefer (1985) gives a short computer program that calculates the date of heliacal rising.

The achronal rising of a star is when the star first appears in the eastern evening sky just after sunset. Unfortunately, the phenomenon is poorly defined. Did the ancient cultures mean that the star actually rose at sunset? If so, then the event is unobservable. Did the ancient cultures adopt the end of twilight as the time when the star must be visible? If so, then what was their definition of the end of twilight? The possibilities include civil, nautical, or astronomical twilight. These varying definitions can change the date of achronal rising by up to a month while the azimuth will vary by many degrees for temperate latitudes. Thus any archaeoastronomer who tries to invoke achronal risings must either concede that they are unobservable or highly uncertain.

3.3 Visibility of the lunar crescent

Lunar calendars usually define the first day of the month by when the thin crescent is first visible after new moon. The prediction of crescent visibility is a very difficult problem, in which astronomers since Babylonian times have invested great effort (see Ilyas 1984). In modern times, the utility of a calendar is greatly reduced if dates cannot be predicted. Given that more than 20% of the world's population has the Islamic faith, I would claim that crescent visibility is the 'non-trivial' astronomical problem that affects the greatest number of humans.

Until recently, all prediction algorithms were based on empirical data sets from one site (in modern times, all algorithms used Schmidt's observations from Athens in the late 1800s). A basic problem with this approach is that different workers have interpreted the same data and reached very different conclusions (Maunder 1911). This approach also has the severe difficulty that the algorithm does not account for local observing conditions, so that the same criterion is applied to the jungles of the Yucatan as to the antiplano of Peru. Bruin (1977) proposed a new approach, which was to model the various physical phenomena involved, to allow for varying conditions and to avoid dependence on a particular dataset. Unfortunately, Bruin made a variety of errors in his mathematical modelling, the results of which were wrong by many orders of magnitude (Schaefer 1988a). Despite these difficulties, Bruin's method is the same as that which I have used throughout this paper, although I discovered his paper only after my own work was completed. My algorithm for predicting crescent visibility is presented in Schaefer (1988a; 1988b) and incorporated into a computer program, *LunarCal* (Schaefer 1990c).

Extensive efforts have been made to collect observations as a test of the historical criteria as well as my new algorithm. Schaefer, Ahmad and Doggett (1993) have identified seven widely publicised reports that are provably erroneous, some owing to simple poor reporting and others on the evidence of independent weather reports showing that it was raining at the site at the time of the reported observation. Schaefer (1988a) has collected 201 observations from the astronomical literature and finds that the ancient algorithms have a zone of uncertainty larger than the earth's circumference and that the new algorithm is 2·3 times better than all other algorithms. Doggett *et al.* (1988) and Doggett and Schaefer (1989; 1990) have recruited thousands of observers across the North American continent to participate in five 'moonwatches'. The results (Doggett and Schaefer 1989; 1992) show that all algorithms except the new one are grossly wrong for at least some lunations. Another moonwatch on 1 June 1992 run by R.C. Victor closely fitted the *LunarCal* predictions, while the predictions of Ilyas (1984) were almost 5000 km in error. The success of the new theoretical algorithm arises primarily because no other algorithm even tries to account for variations in the observing conditions.

3.4 Length of the lunar crescent

If the moon were an ideal sphere, then the crescent should appear to extend from pole to pole, for a total angle of 180° subtended around the centre of the lunar disc. Danjon (1932; 1936) shows that the actual angle is considerably less than 180° when the moon is close to the sun. McNally (1983) proposed an explanation based on the scintillation of the atmosphere, although this idea has since been disproved on the basis of data collected during the moonwatches (Doggett and Schaefer 1992). Schaefer (1990a) explains the shortening as a natural consequence of the crescent's rapid decrease in brightness towards the cusps, such that the threshold for visibility to the eye is reached at angles less than 180°. The model is shown to match closely all of Danjon's data as well as those of the moonwatches.

3.5 Lunar eclipse visibility

When an eclipsed moon is low on the horizon, the sun must also be near to the horizon so that the high sky brightness and the large extinction may combine to render the moon invisible. Schaefer (1990b) presents the first-ever theoretical or observational discussion on the visibility of low-altitude lunar eclipses. Relatively few observations have been made to test the theoretical calculations, and it appears that the theory may be optimistic by perhaps a degree of altitude. Schaefer (1991a) presents the theory for visibility of the partial phases. Schaefer (1991c) calculates the visibility of lunar surface features during a total solar eclipse.

3.6 Lunar appulse visibility

Whenever a star or planet is close to the moon, it may be hidden by the glare. I have analysed this question using the mathematical tools from Sections 2.1–2.4 and 2.7–2.8. The result is a predicted threshold for the visibility of a point source as a function of the source's magnitude, the distance from the moon, and the phase of the moon (Schaefer 1991a). I have also collected roughly two hundred observations from various observers which confirm the model predictions in detail.

A related question is the visibility of lunar occultation. This is theoretically modelled by Schaefer, Bulder and Bourgeois (1992). The model is confirmed by observations of 1739 events.

3.7 Sunspot visibility

Large sunspots are easily visible to the human eye (with proper protection) and were extensively recorded by the ancient Chinese (Yau and Stephenson 1988). Until recently, the best rule of thumb was that a sunspot must be larger than one arc-minute in order to be visible (Eddy *et al.* 1989). However, this rule has now been strongly refuted by observation (Mossman 1989; Keller 1980; 1986; Schaefer 1991d). Schaefer (1991d) has developed a detailed model that predicts the visibility of sunspots as a function of the observer's Snellen rating. Schaefer (1992a) has slightly improved this model and has extended it to include observations with telescopes and pinhole cameras. My theoretical work shows that, for equally motivated observers, only resolution can effect sunspot visibility. The filter technique, the observer's age, their diet, and the history of close-up work such as reading will not change the sunspot visibility (*pace* Keedy 1989). Observationally, my wife and I have been monitoring the sun daily since 1988, and I have access to the similar surveys by Stephen J. O'Meara (unpublished), Mossman (1989), and Keller (1986). These datasets amply confirm the threshold for visibility, the quantitative dependence on acuity, and the relation between the telescopic and unaided sunspot counts.

The sunspot count is the premier record of long-term solar activity. This record covers the last two millennia, most of the time with unaided vision, yet even in modern times the counts are based on what the human eye sees. Schaefer (1992a) has used the model of sunspot visibility to suggest improvements of technique so as to stabilise the K correction factor. Also, it was discovered that the K factor is accurately given as $14 \cdot 0/(10+\rho)$, with ρ being the average number of spots per group. Since this formula does not depend upon other observers, it can be applied without long-term biases. Hence, the relative heights of all previous sunspot cycles must be re-evaluated.

3.8 Refraction near to the horizon

In any calculation of the apparent altitude, and hence the apparent azimuth, of a celestial phenomenon at a given declination, a refraction correction is needed. All archaeoastronomers have adopted refraction corrections similar to those of Garfinkel (1967) which assumes a standard atmosphere with no inversions. Those archaeoastronomers who worried about the day-to-day variation in the refraction (e.g. Thom 1974; Patrick 1979; Heggie 1981: 134–35) all concluded that the changes were typically one arc-minute in magnitude. The fallacy in their argument is that they did not account for the ubiquitous thermal inversion layers that change on fast timescales. Schaefer and Liller (1990) use the program of Schaefer

(1989) and the synoptic temperature data of Lettau and Davidson (1957: 396–566) to show that the variation is roughly an order of magnitude larger than previously suspected. Schaefer and Liller (1990) also report more than 144 measurements of refraction, taken using four different experimental methods from eight sites, which show that at the 95% probability level the range in the refraction correction at the horizon is $0°\cdot64$.

4 Applications

4.1 Archaeoastronomy

The archaeoastronomy paradigm is the contention that some ancient monuments are aligned with the rising and setting of celestial sources. To establish whether a source will appear at the azimuth indicated by the monument, a calculation is made relating the object's declination to the observed azimuth. This declination-to-azimuth conversion has mandatory corrections for refraction, parallax, and altitude of first visibility.

The variation of the refraction correction near to the horizon is large (see Section 3.8 above). This translates into a large uncertainty in the azimuth of first appearance for temperate latitudes. At the 95% probability level for a latitude of 45°, the conversion will have an error that ranges over two thirds of a degree or more. With such a large day-to-day variation, it is impossible to achieve an accuracy very much better than this. In fact, this provides a fundamental limit on the accuracy of the archaeoastronomy paradigm. Most claimed alignments have an accuracy that is at best comparable to the limit, and so are not affected by the refraction variations.

The altitude of first visibility for stars will be the extinction angle. However, Schaefer (1986) has shown that random variations in sky conditions will cause a large variance in the extinction angle on a case-to-case basis. Thus, there will be a large intrinsic scatter in the observed azimuth for any stellar alignment which depends upon the site's latitude. For temperate latitudes, Schaefer (*ibid.*) proves that even under the most optimistic conditions, no stellar alignment can ever be statistically significant, and hence convincingly demonstrated to be deliberate, in the absence of independent corroborating evidence (e.g. ethnographic information).

4.2 Megalithic archaeoastronomy

One of the few proposed types of alignment toward a horizon event that require high accuracy are the famous megalithic lunar observatories of Thom (1971). He claims that the accuracy of these orientations is sufficiently good that Neolithic people detected the lunar nutation, whose amplitude in declination is a mere nine arc-minutes. However, the day-to-day variation in refraction renders this technically quite impossible, even if data over many millennia are averaged together (Schaefer and Liller 1990). Hence, one of the most influential results in archaeoastronomy can be easily refuted by a methodology similar to that of Thom himself (cf. Thom 1958). Thom's claimed accurate alignments have already been rejected by a variety of conclusive arguments (Ruggles 1989), although the refraction has the virtue of being the simplest refutation.

Thom (1967) proposes 67 stellar sightlines for megalithic monuments in Scotland. Unfortunately, he used the extinction angle law of Neugebauer which is optimistic by several degrees. For the high latitudes of Scotland and the declinations relevant for lunar standstills, an error of several degrees in the extinction angle translates into an average error of 3·6 degrees in azimuth (Schaefer 1986). In other words, the majority of Thom's stellar alignments have systematic errors of fourteen lunar radii.

Robinson's (1983) claimed alignment on Regulus at Stonehenge is similarly in error by 2·9 degrees for a dark sky (Schaefer 1986). For achronal rising, the azimuth would be further in error, by an amount depending upon the definition of twilight adopted. The rising of the partially eclipsed moon over the Heel Stone (*ibid.*: fig. 2) would hardly be 'an awesome omen', since the extinction and bright twilight would hide the moon until it was several degrees above the horizon, and even then it would be inconspicuous (cf. Schaefer 1990b). This high altitude implies that the moon will not be aligned with the Heel Stone when first sighted. Finally, series of winter solstitial lunar eclipses occur during 170 out of every 400 years, so it is not surprising that such a series could be found for any specific phase of building, the date of which is uncertain within several centuries.

4.3 The Middle East

The date of the Crucifixion of Jesus Christ is important as much of the chronology of the New Testament is tied to it. Unlike the birth of Christ, there may be enough information in the New Testament (along with historical evidence) to specify an unambiguous date. The problem is usually stated as looking for a Friday that was either the fourteenth or fifteenth day of the Jewish (lunar) month *Nisan*, most likely in the years AD 29–33 (Finegan 1964). The primary non-systematic (Doggett 1976) uncertainty in this calculation is when the lunar months started. With my new algorithm for predicting visibility, I have re-examined this old question and find that the best of the earlier calculations (Humphreys and Waddington 1983) yields correct results in critical cases (Schaefer 1990b). This program yields two candidate dates: Apr 7, AD 30 and Apr 3, AD 33. It has generally been thought that the eclipse of the moon on the later of the candidate dates (Link 1969), mentioned in a prophecy by Joel that was repeated by Peter soon after the crucifixion ('...and the moon shall turn to blood...'), provides a tie-break between the two candidates (Humphreys and Waddington 1983). However, it turns out that the eclipse would not have been visible from Jerusalem and that it would not have changed the colour of the moon (Schaefer 1990b). So we now are back in the situation where the crucifixion is likely to have happened on one of two dates, and we do not know which.

The Great Pyramid of Cheops in Giza, Egypt, is cardinally aligned with the awesome accuracy of three arc-minutes (Petrie 1940). The task of establishing a meridian line must be based solely on astronomical observations. Many methods have been proposed over the years (Zaba 1953; Isler 1989), but there is no reliable textual or archaeological evidence that enables us to choose between them. Several proposed methods are clearly inadequate, including Petrie's trench method and the method of looking for a near-pole star. The methods that use low horizon observations (such as that of Haack 1984) are all too poor to achieve the necessary accuracy because of refraction variations (Schaefer and Liller 1990). I calculate that the various methods involving the sun's shadow (e.g. Neugebauer 1980) will be too inaccurate unless some luck is invoked. However, simple variants on the shadow ideas (e.g. that of Isler 1989 or if the shadow edge were defined by the eye spotting the position with a small gleam of sun around a large gnomon) can be good enough. In addition, various stellar sighting schemes (like that of Edwards 1961) can provide the necessary accuracy. In summary, although several proposals can be rejected, there remains more than one valid method by which the ancient Egyptians could have oriented the Great Pyramid.

The ancient Egyptians made observations of the heliacal rising of Sirius as the basis for a sidereal calendar (Lockyer 1894). Schaefer (1987a) finds that the uncertainty in the heliacal rise date for a bright star is so large that the regulation of the Sothic year could only be accurately accomplished on a long term basis. I calculate that the heliacal rising of Sirius would have happened at the solstice in the year 3000 BC.

4.4 The star and crescent

In modern times, the star and crescent is usually identified as a symbol of the Islamic religion and is one of the most widespread astronomical symbols in the world. Its origins are currently unknown, although I have counted eleven mutually exclusive hypotheses ranging from 640 BC to AD 1922. One edition of the Encyclopaedia Britannica even repeats five theories in various articles. I have divided my study into two separate approaches (see Schaefer 1991b for details).

First, I have tested all the theories that claim a specific astronomical event for the symbol's origin.

- The 'Betsy Ross' myth taught to Turkish youngsters is that the flag commemorates the Battle of Sakarya in 1922 when the young republic ejected the Greek army from Anatolia. The flag depicted a close conjunction of the moon and Venus as reflected in a river of Greek blood. Unfortunately, the conjunction never occurred during the battle.

- A whole set of theories associates the symbol with the Fall of Constantinople. For example, I learned in high school that a close conjunction of the moon and Venus on the night before the Fall was commemorated by the Turks as a good omen. Alternatively, Runciman (1965) points out that a lunar eclipse four days before the Fall might have been seen as a crescent shape. Runciman also repeats the claim that the crescent was adopted as the symbol of the Turks because the Sultan Mehmet triumphantly entered the sacked city with the crescent high in the sky. All these claims can be shown to be wrong on astronomical grounds. First, the moon was gibbous at the time of the Fall. Second, the night of the attack was chosen because it was dark and stormy so as to cover up the troop movements. Third, the eclipsed moon (actually six days before the Fall) left the umbra when it was at an altitude of 10° so that the visibility would have been unimpressive because of the extinction and bright twilight (Schaefer 1990b); thus even at best the moon had only a very small portion taken out of the full circle and hence was not crescent shaped. Fourth, the (gibbous) moon had already long set when Mehmet made his entry.

- The Night of Power is the night on which the archangel Gabriel first brought the revelations of the Koran to the prophet Muhammad. Tradition places the Night of Power on an odd-numbered night in the last ten days of Ramadan. Hawkins (1978) has interpreted various passages of the Koran to place this important date on Jul 23, AD 610 when a moon/Venus conjunction occurred. Ahmad (1989) advanced the idea that this conjunction inspired the symbol either directly or through later astrological calculations. Unfortunately, my improved crescent visibility algorithm places Hawkins' conjunction on an even-numbered night in Ramadan. Furthermore, Ahmad has found a hadith that places the Night of Power on a Monday, whereas Hawkins' conjunction is on a Thursday.

- Byzantine historians quote a tradition that relates the origin of the star and crescent to a failed siege of Byzantium by Phillip of Macedon in 339 BC. The best source is Hesychius of Miletus, whose account has been rendered in three different versions in secondary references. The most popular version says that Phillip's men managed to tunnel under the wall at night, and that the moon suddenly appeared which started dogs barking which in turn roused the populace who beat back the attack. I examined the original sources and have had them translated and find that the night was explicitly stated to be moonless and raining heavily. The light source is called a sheet of fire. The event cannot be associated with anything astronomical as not even an aurora can be seen through thick rain clouds.

With all the specific astronomical events excluded, I started to search for historical or archaeological examples of the symbol. An excellent article in the *Encyclopaedia of Islam* (Ettinghausen 1971) traces the usage back to the times of Muhammad. The modern star and crescent has now been adopted as a religious symbol, but it was first used by many Islamic countries because of their historical connection with the Ottoman Empire. The Turks had a long secular tradition of the star and crescent both before and after the Fall of Constantinople. I have been able to push back the first known usage in an Islamic context to AH 18 on a coin (Walker 1941). However, this coin (with seven star-and-crescent pairs) is virtually an identical copy of a Sassanian coin form that predates the Hegira by over sixty years. Coins represent a convenient source of symbols that can be well dated and placed, so I have searched for earlier coins. I find many examples from all over the Middle East, ranging from modern Romania to Italy, Yemen, and Iran. The symbol was used on coins by Romans, Greeks, Jews, and Persians with my earliest example dated to 470 BC (Gardner 1974). Before this time, cylinder seals and stone monuments provide an abundant source of iconography with reliable provenance. The star and crescent was used by many cultures centred around Turkey, with the earliest example dating from 2500 BC. Apparently, the star and crescent was an ubiquitous symbol throughout the region at an early date. So what is its origin? It must have arisen from some tradition, perhaps based on a specific astronomical event, that pre-dates 2500 BC. Given the lack of any earlier evidence and the low likelihood that new evidence will turn up, I conclude that the origin is ancient and is now lost in the mists of time.

4.5 Islam

The Islamic calendar is lunar, in that the first day of the month is defined by the visibility of the crescent. With my improved algorithm, important Islamic dates can be predicted for modern times or 'postdicted' for historical applications. The algorithm has been published in a convenient format for predicting the Islamic months in the twentieth century (Schaefer 1990c). For historical applications, I agree with all modern scholars that the Islamic epoch of 1 Muharram, AH 1 corresponds to the western date of Jul 15/16, AD 622. Another historical application regarding the date of the Night of Power has been discussed in the previous section.

Every Islamic person should pray five times a day, with the prayer times defined in astronomical terms. The time when the sun is setting is a time when the prayers should not be offered. With variations of refraction, the time of sunset may range over four minutes centred around the best prediction. Thus, it might be prudent to provide a larger than expected buffer when starting prayers or breaking the Ramadan fast. The traditional time for the end of twilight is when the sun is 18° below the horizon, and my observations agree closely with this number for ideal conditions. However, normal variations in moonlight, city light, atmospheric clarity, and cloud cover can greatly change the real time of twilight depending on the definition used.

4.6 Asian lunar lodge systems

Throughout Asia, in ancient times, the astronomies of many great civilisations prominently featured a system of lunar lodges. A lunar lodge is a section of sky whose boundaries are delineated with the help of 28 asterisms spread out in a band which circles the sky. These lunar lodge systems are intriguing not only because of their frequent occurrence in Asian scientific thought but also because of their great antiquity. Historical and archaeological evidence demonstrates that the Chinese and Indian lodge systems are over three thousand years old. However, the major problems concerning the date and the place of origin have not been solved. I have approached the problem by using astronomical evidence concerning a variety of correlations involving the lunar lodges. The quality of these correlations for each lodge system depends on the epoch, since star positions precess with time. It is found that the Indian system is derived from the Chinese system and that the Arab system is based on either the Indian or Chinese lodges. The best estimate of the date of formation of the Chinese lodge system is 3250 BC with a statistical uncertainty of roughly a millennium. Similarly, the preferred dates for the formation of the Hindu and Arabic systems are 1750 BC and 200 BC, respectively; again with a statistical uncertainty of roughly a millennium.

4.7 China

The ancient records of sunspot observations from the Far East (Yau and Stephenson 1988) are invaluable as the only direct measure of solar activity for the two millennia before the discovery of the telescope. To better understand these reports, various investigators including myself have been watching daily for naked eye sunspots (Keller 1980; 1986; Schaefer 1991d; 1992a). Keedy (1989) has questioned whether such surveys are applicable to ancient observers, but my results show strongly that such worries are unfounded, the possible errors concerned being negligible. My studies have shown that the spots are visible down to a much smaller size than was earlier realised. In addition, the naked-eye sunspot count is directly proportional to the Zurich relative sunspot number.

Michel (1950) and D. de Solla Price have both proposed that certain ancient Chinese ritual jade discs were used to locate the north celestial pole. This hypothesis has been severely criticised on purely textual and archaeological grounds (Cullen and Farrer 1983). Schaefer (1983) reached the same conclusion on purely astronomical grounds, including the fact the jades could not possibly work as described.

Liu (1988a; 1988b) and Hilton et al. (1989) have tried to constrain the acceleration of the Earth's rotation period by analysis of ancient Chinese records of lunar occultations of planets. A serious inconsistency with their method is that a significant fraction of their events are only appulses for all reasonable accelerations, so that the validity of the observations themselves must be questioned. Their explanation is that the planet would have been lost in the moon's glare and so appeared as an occultation. But my theoretical work shows that all planets are bright enough to be easily spotted up to the very edge of the moon's disc. As, in addition, their results change greatly with their weighting scheme, I conclude that the ancient Chinese observations are not reliable for measuring the Earth's acceleration.

Liu (1984) tabulates the heliacal rise dates for planets as tabulated in Chinese calendars and concludes that Jupiter is brightening at a rate of 0·003 magnitudes per millennium. Schaefer (1987a) shows this result to be derived from an incorrect analysis procedure and that the data suffer from several fatal systematic errors.

4.8 Miscellaneous historical applications

Brecher (1984) has related the possible lunar impact event reported by Gervase of Canterbury (Hartung 1976) to the material in the orbit of Encke's comet (Clube and Napier 1983) which he has termed the Canterbury Swarm. Meeus (1990) points out that the moon could not possibly have been visible on the recorded date, and I agree (Schaefer 1990d). I have also found that Gervase has a history of reporting wrong dates and violent flames in the sky. Any attempt to associate Gervase's event with the Canterbury Swarm by invalidating the quoted date must destroy the only evidence of the association.

The southern wall at Pueblo Bonito in Chaco Canyon is cardinally oriented to an accuracy of eight arc-minutes, while the Great North Road out of Chaco Canyon has a 16km stretch that is within half a degree of true north (Malville and Putnam 1989; Sofaer et al. 1989). The accuracy requirement for the alignments is weaker than that of the Great Pyramid, so there are several methods by which the meridian could have been laid out.

Eddy (1974) and Robinson (1983) have advanced various claims for alignments of Amerindian Medicine Wheels with the heliacal rising of bright stars as well as Fomalhaut. Schaefer (1987a) shows that the actual heliacal rising of these stars will appear at an altitude more than four degrees higher than assumed, so that the claimed alignments systematically point five degrees too far north.

The periods of visibility and invisibility of Venus as viewed from the Yucatan are significantly different from those adopted by the Mayans (Schaefer 1987a), as was already known by Lounsbury (1978).

I have used the crescent visibility algorithm to predict the observed lengths of lunar months with and without cloudy weather (Schaefer 1992b). I find that if clouds are ignored, then the lunar month is always 29 or 30 days long. With clouds, the months will vary widely in duration. My statistical results may be of interest to historians of Islamic calendrics or to researchers interested in calendar sticks.

4.9 Amateur astronomy

I have a wide variety of applications that are relevant to amateur astronomers and may also be of interest to historians: I have generated a complete model and a convenient computer program for predicting the magnitude of the faintest star visible through a telescope (Schaefer 1990a); I have written a computer program that calculates the colours and durations of the green flash for an arbitrary atmosphere, planet, and sun; I have developed a model for predicting the visibility of a star being occulted by the moon; I have written a computer program that calculates the shading inside a mountain shadow; I have quantified the frequency for short optical flashes in the sky; and I have quantified the limiting magnitude for detecting stars in the daytime and as viewed through chimneys.

4.10 Future applications

I am currently working on a variety of projects, including the collection of more observations regarding the time that twilight ends and observations regarding the accuracy with which a naked-eye observer can measure the times of full and quarter moon (cf. Schaefer 1992b).

I plan to pursue several topics in the future: I will derive from first principles the visibility of double stars, galaxies, comets, and comets near the sun; I hope to be able to derive the first reliable aperture correction for visual brightness estimates of comets; and I am looking for a collaboration with some researcher knowledgeable in the history of science for study of questions relating to the origins of the Asian lunar lodge systems.

5 Summary

The work involving the capabilities of the human eye for detecting astronomical sources has thus grown into a separate discipline in its own right. I would like to propose the name of 'celestial visibility' as a concise and descriptive title for this new discipline. I would also like to prescribe the following steps for any investigation of celestial visibility.

(1) Collect the necessary mathematical tools to model the physics of the light and its detection.

(2) Unify these tools into one model that predicts the visibility of the source.

(3) Use the same tools to estimate the uncertainty in the prediction.

(4) Collect actual observations to test the validity of the predictions.

(5) Apply the model prediction to historical or observational questions of interest.

In essence, this prescription merely says that visibility should be treated as a science.

I know of only two articles (Bruin 1977; 1979), that have adopted even part of this same approach for any topic. However, Bruin (1977) made several errors by orders of magnitude, did not estimate the uncertainties, did not test against any real data base, and did not apply his results to any outside questions. Therefore, there has been virtually no competition in this new discipline of celestial visibility. This has the advantage that any work I do in the topic is ground-breaking and all future work can only be built upon the building blocks presented above. It also means that the implications are previously unknown, so that many of my results provide refutations of earlier work and many of my results are surprising.

In conclusion, I would urge all readers to examine the tools, results, and applications in this article to see how it might affect their own area of research.

References

Ahmad, I.A. (1989). The dawn sky on lailat-ul-qadr (the Night Power). *Bulletin of the American Astronomical Society* **21**, 1217–8.

Allen, C.W. (1976). *Astrophysical quantities*. London: Athlone.

Aveni, A.F. (1972). Astronomical tables intended for use in astro-archaeological studies. *American Antiquity* **37**, 531–40.

Barteneva, O.D. and Boyarova, A.N. (1960). *Trudy Glavnoj Geofiz. Observatorii im. A.I. Voeikova* (Leningrad) **100**, 133.

Blackwell, H.R. (1946). Contrast thresholds of the human eye. *Journal of the Optical Society of America* **36**, 624–43.

Bower, F.A. and Ward, R.B. (1982). *Stratospheric ozone and man*. Boca Raton FL: CRC Press Inc.

Brecher, K. (1984). The Canterbury swarm: ancient and modern observations of a new feature of the solar system. *Bulletin of the American Astronomical Society* **16**, 476.

Bruin, F. (1977). The first visibility of the lunar crescent. *Vistas in Astronomy* **21**, 331–58.

Bruin, F. (1979). The heliacal setting of the stars and planets. *Proceedings of the Koninklijke Nederlandse Akademie van Wetenschappen* **B82**, 385–410.

Clube, S.V.M. and Napier, W.M. (1983). The microstructure of terrestrial catastrophism. *Monthly Notices of the Royal Astronomical Society* **211**, 953–68.

Cullen, C. and Farrer, A.S.L. (1983). On the Ferm Hsüan Chi and the flanged trilobate jade discs. *Bulletin of the School of Oriental and African Studies* **46**, 52–76.

Danjon, A. (1932). Jeunes et vieilles lunes. *L'Astronomie* **46**, 57–66.

Danjon, A. (1936). Le croissant lunaire. *L'Astronomie* **50**, 57–65.

Doggett, L.E. (1976). *The date of the crucifixion*. U.S. Naval Observatory pamphlet DP 3–76.

Doggett, L.E. and Schaefer, B.E. (1989). Results of the July moonwatch. *Sky and Telescope* **77**, 373–5.

Doggett, L.E. and Schaefer, B.E. (1990). A challenging lunar eclipse moonwatch, Aug. 21st. *Sky and Telescope* **80**, 174.

Doggett, L.E. and Schaefer, B.E. (1992). Lunar crescent visibility. *Astronomical Journal*, submitted.

Doggett, L.E., Seidelmann, P.K. and Schaefer, B.E. (1988). Moonwatch—July 14, 1988. *Sky and Telescope* **76**, 34–35.

Eddy, J.A. (1974). Astronomical alignment of the Big Horn Medicine Wheel. *Science* **184**, 1035–43.

Eddy, J.A., Stephenson, F.R. and Yau, K.K.C. (1989). On pretelescopic sunspot records. *Quarterly Journal of the Royal Astronomical Society* **30**, 65–73.

Edwards, I.E.S. (1961). *Wisdom of the Egyptians.* New Orleans LA: Pelican.

Ettinghausen, R., ed. (1971). *The encyclopedia of Islam*, vol. 3, pp. 379–85. London: Luzac.

Finegan, J. (1964). *Handbook of Biblical chronology.* Princeton NJ: Princeton University Press.

Gardner, P. (1974). *History of ancient coinage.* Chicago IL: Ares.

Garfinkel, B. (1967). Astronomical refraction in a polytropic atmosphere. *Astronomical Journal* **72**, 235–54.

Garstang, R.H. (1989). Night sky brightness at observatories and sites. *Publications of the Astronomical Society of the Pacific* **101**, 306–29.

Haack, S.C. (1984). The astronomical orientation of the Egyptian pyramids. *Archaeoastronomy* no. 7 (supplement to *Journal for the History of Astronomy* **150**), S119–25.

Hapke, B. (1984). Bidirectional reflectance spectroscopy. *Icarus* **59**, 41–59.

Hartung, J.B. (1976). Was the formation of a 20km-diameter impact crater on the moon observed on June 18, 1178? *Meteoritics* **11**, 187–94.

Hawkins, G.S. (1978). The sky when Islam began. *Archaeoastronomy* (Center for Archaeoastronomy), **1**(3), 6–7.

Hayes, D.S. and Latham, D.W. (1975). A rediscussion of the atmospheric extinction and the absolute spectral energy distribution of Vega. *Astrophysical Journal* **197**, 593–601.

Hecht, S. (1947). Visual thresholds of steady point sources of light in fields of brightness from dark to daylight. *Journal of the Optical Society of America* **37**, 59.

Heggie, D.C. (1981). *Megalithic science.* London: Thames and Hudson.

Helfenstein, P. and Veverka, J. (1987). Photometric properties of lunar terrains derived from Hapke's equation. *Icarus* **72**, 342–57.

Hilton, J.L., Seidelmann, P.K. and Liu, C. (1989). Chinese records of lunar occultations of the planets. *Bulletin of the American Astronomical Society* **21**, 1153.

Holladay, L.L. (1926). The fundamentals of glare and visibility. *Journal of the Optical Society of America* **12**, 271–320.

Humphreys, C.J. and Waddington, W.G. (1983). Dating the crucifixion. *Nature* **306**, 743–46.

Ilyas, M. (1984). *Islamic calendar, times and Qibla.* Kuala Lumpur: Berita.

Isler, M. (1989). An ancient method of finding and extending direction. *Journal of the American Research Center in Egypt* **26**, 191–206.

Keedy, D.R. (1989). Non-telescopic search for sunspots. *Quarterly Journal of the Royal Astronomical Society* **30**, 361.

Keller, H.U. (1980). «A» Sonnenfleckenbeobachtungen von Blossem Auge. *Orion* **181**, 180–84.

Keller, H.U. (1986). Der Sonnenfleckenzyklus Nr. 21 von Blossem Auge registriert. *Orion* **216**, 154–56.

King, I. (1971). The profile of a star image. *Publications of the Astronomical Society of the Pacific* **83**, 199–201.

Koomen, M.J., Lock, C., Packer, D.M., Scolnik, R., Tousey, R. and Hulburt, E.O. (1952). Measurements of the brightness of the twilight sky. *Journal of the Optical Society of America* **42**, 353–56.

Krisciunas, K., Impey, C., Christian, C., Sinton, W., Tholen, D., Tokunaga, A., Golisch, W., Griep, D. and Kaminski, C. (1987). Atmospheric extinction and night-sky brightness at Mauna Kea. *Publications of the Astronomical Society of the Pacific* **99**, 887–94.

Krisciunas, K. and Schaefer, B.E. (1991). A model of the brightness of moonlight. *Publications of the Astronomical Society of the Pacific* **103**, 1033–39.

Lettau, H.H. and Davidson, B. (1957). *Exploring the Earth's first mile*, vol. 2. London: Pergamon Press.

Link, F. (1969). *Eclipse phenomena in astronomy*. Berlin: Springer-Verlag.

Liu, C. (1988a). Secular acceleration of Earth's rotation from Chinese records of lunar occultations of stars before 600 AD. *Chinese Astronomy and Astrophysics* **12**, 169–78.

Liu, C. (1988b). The secular variation of the Earth's rotation obtained from 171 Chinese records of lunar occultations before AD 600. *Chinese Astronomy and Astrophysics* **12**, 260–67.

Liu, J.-Y. (1984). Is Jupiter getting brighter? *Chinese Astronomy and Astrophysics* **8**, 354–56.

Lockwood, G.W., Floyd, R.D. and Thompson, D.T. (1990). Sky glow and outdoor lighting trends since 1976 at Lowell Observatory. *Publications of the Astronomical Society of the Pacific* **102**, 481–91.

Lockyer, J.N. (1894). *The dawn of astronomy*. London: Macmillan.

Lounsbury, F.G. (1978). Mayan numeration, computation, and calendric astronomy. In *Dictionary of scientific biography*, ed. C.C. Gillispie, vol. 15 (supplement 1), pp. 759–818. New York NY: Charles Scribner's Sons.

McCord, T.B. and Johnson, T.V. (1970). Lunar spectral reflectivity (0.30 to 2.50 microns) and implications for remote mineralogical analysis. *Science* **169**, 855–58.

McNally, D. (1983). The length of the lunar crescent. *Quarterly Journal of the Royal Astronomical Society* **24**, 417–29.

Malville, J.M. and Putnam, C. (1989). Prehistoric astronomy in the Southwest. Boulder CO: Johnson.

Maunder, E.W. (1911). On the smallest visible phase of the moon. *Journal of the British Astronomical Association* **21**, 362–65.

Meeus, J. (1990). The 'lunar event' of AD 1178. *Journal of the British Astronomical Association* **100**, 59.

Michel, H. (1950). Chinese astronomical jades. *Popular Astronomy* **58**, 222–30.

Mossman, J.E. (1989). A comprehensive search for sunspots without the aid of a telescope, 1981–82. *Quarterly Journal of the Royal Astronomical Society* **30**, 59–64.

Neugebauer, P.V. (1929). *Tafeln zur astronomischen Chronologie*. Leipzig: J.C. Hinrichssche Buchhandlung.

Neugebauer, O. (1980). On the orientation of pyramids. *Centaurus* **24**, 1–3.

Patrick, J.D. (1979). A reassessment of the lunar observatory hypothesis for the Kilmartin stones. *Archaeoastronomy* no. 1 (supplement to *Journal for the History of Astronomy* **10**), S78–85.

Petrie, F. (1940). *Wisdom of the Egyptians*. London: Quaritch.

Pilachowski, C.A., Africano, J.L., Goodrich, B.D. and Binkert, W.S. (1989). Sky brightness at the Kitt Peak National Observatory. *Publications of the Astronomical Society of the Pacific* **101**, 707–12.

Robinson, J.H. (1983). The solstice eclipses of Stonehenge II. *Archaeoastronomy* (Center for Archaeoastronomy) **6**, 124–31.

Rozenberg, G.V. (1966). *Twilight: a study of atmospheric optics*. New York NY: Plenum.

Ruggles, C.L.N. (1989). Recent developments in megalithic astronomy. In *World archaeoastronomy*, ed. A.F. Aveni, pp. 13–26. Cambridge: Cambridge University Press.

Runciman, S. (1965). *The Fall of Constantinople*. Cambridge: Cambridge University Press.

Schaefer, B.E. (1983). Chinese 'astronomical' jade disks: the pi. *Archaeoastronomy* (Center for Archaeoastronomy) **6**, 99–101.

Schaefer, B.E. (1985). Astronomical computing: predicting heliacal risings and settings. *Sky and Telescope* **70**, 261–63.

Schaefer, B.E. (1986). Atmospheric extinction effects on stellar alignments. *Archaeoastronomy* no. 10 (supplement to *Journal for the History of Astronomy* **17**), S32–42.

Schaefer, B.E. (1987a). Heliacal rise phenomena. *Archaeoastronomy* no. 11 (supplement to *Journal for the History of Astronomy* **18**), S19–33.

Schaefer, B.E. (1987b). Astronomical computing: extinction angles and megaliths. *Sky and Telescope* **73**, 426.

Schaefer, B.E. (1988a). Visibility of the lunar crescent. *Quarterly Journal of the Royal Astronomical Society* **29**, 511–23.

Schaefer, B.E. (1988b). An algorithm for predicting the visibility of the lunar crescent. In *Proceedings of the lunar calendar conference*, ed. I.A. Ahmad, pp. 11-1 – 11-12. Herndon: International Institute of Islamic Thought.

Schaefer, B.E. (1989). Astronomical computing: refraction by Earth's atmosphere. *Sky and Telescope* **77**, 311–13.

Schaefer, B.E. (1990a). Telescopic limiting magnitudes. *Publications of the Astronomical Society of the Pacific* **102**, 212–29.

Schaefer, B.E. (1990b). Lunar visibility and the crucifixion. *Quarterly Journal of the Royal Astronomical Society* **31**, 53–67.

Schaefer, B.E. (1990c). *LunarCal*. Available from Western Research Company Inc., 2127 E. Speedway, Suite 209, Tucson AZ 85719, USA.

Schaefer, B.E. (1990d). The 'lunar event' of AD 1178: a Canterbury tale? *Journal of the British Astronomical Association* **100**, 211.

Schaefer, B.E. (1990e). *Estimating extinction coefficients*. Unpublished manuscript.

Schaefer, B.E. (1991a). Glare and celestial visibility. *Publications of the Astronomical Society of the Pacific* **103**, 645–60.

Schaefer, B.E. (1991b). Heavenly signs. *New Scientist* **132**, 48–51.

Schaefer, B.E. (1991c). Eclipse earthshine. *Publications of the Astronomical Society of the Pacific* **103**, 315–16.

Schaefer, B.E. (1991d). Sunspot visibility. *Quarterly Journal of the Royal Astronomical Society* **32**, 35–44.

Schaefer, B.E. (1992a). Visibility of sunspots. *Astrophysical Journal*, submitted.

Schaefer, B.E. (1992b). The length of the lunar month. *Archaeoastronomy* no. 17 (supplement to *Journal for the History of Astronomy* **23**), S32–42.

Schaefer, B.E., Ahmad, I.A. and Doggett, L.E. (1993). Records for young moon sightings. *Quarterly Journal of the Royal Astronomical Society* **34**(1), 53–56.

Schaefer, B.E., Bulder, H.J.J. and Bourgeois, J. (1992). Lunar occultation visibility. *Icarus*, in press.

Schaefer, B.E. and Liller, W. (1990). Refraction near the horizon. *Publications of the Astronomical Society of the Pacific* **102**, 796–805.

Siedentopf, H. (1941). Neue Messungen der visuellen Kontrastschwelle. *Astronomische Nachrichten* **271**, 193–203.

Smart, W.M. (1979). *Spherical astronomy*. Cambridge: Cambridge University Press.

Sofaer, A., Marshall, M.P. and Sinclair, R.M. (1989). The Great North Road: a cosmographic expression of the Chaco culture of New Mexico. In *World archaeoastronomy*, ed. A.F. Aveni, pp. 365–76. Cambridge, Cambridge University Press.

Thom, A. (1958). An empirical investigation of atmospheric refraction. *Empire Survey Review* **14**, 248–62.

Thom, A. (1967). *Megalithic sites in Britain*. Oxford: Oxford University Press.

Thom, A. (1971). *Megalithic lunar observatories*. Oxford: Oxford University Press.

Thom, A. (1974). A megalithic lunar observatory in Islay. *Journal for the History of Astronomy* **5**, 50–51.

Tousey, R. and Hulburt, E.O. (1947). Brightness of the daytime sky. *Journal of the Optical Society of America* **37**, 78–92.

Tousey, R. and Hulburt, E.O. (1948). The visibility of stars in the daylight sky. *Journal of the Optical Society of America* **38**, 886–96.

Walker, J. (1941). *Catalogue of the Arab-Sassanian coins*. London: British Museum.

Walker, M.F. (1973). Light pollution in California and Arizona. *Publications of the Astronomical Society of the Pacific* **85**, 508–19.

Weaver, H.F. (1947). The visibility of stars without optical aid. *Publications of the Astronomical Society of the Pacific* **59**, 232–43.

Wood, R. W. (1934). *Physical optics*. New York NY: Macmillan.

Yau, K.K.C. and Stephenson, F.R. (1988). A revised catalogue of Far Eastern observations of sunspots (165 BC to AD 1918). *Quarterly Journal of the Royal Astronomical Society* **29**, 175–97.

Zaba, Z. (1953). *L'orientation astronomique dans l'ancienne Egypte et la precession de l'axe du monde*. Prague: L'Academie Tchecoslovaque des Sciences.

16

A method for determining limits on the accuracy of naked-eye locations of astronomical events

ROLF M. SINCLAIR AND ANNA SOFAER

1 Introduction

We describe briefly a method of calculating the rising or setting points of astronomical bodies, and the variations to be expected in repeated observations of these points. These variations set limits on the accuracy with which particular rising or setting points can be located with the naked eye. This accuracy in turn represents one limit on how precisely or reproducibly artificial constructs can be aligned to these points.

The rising and setting points are linked closely to the apparent motion of the sun and moon, and of the stars and planets. Each star or asterism is associated with points on the horizon where it is first or last visible. The rising and setting points of the sun and moon oscillate regularly on the horizon, reflecting the various cycles of their motions. Although in principle any rising or setting of the sun or moon could be of interest, we will focus here on those events that reproducibly signal the limits and mid-points of the long-term periodic solar and lunar cycles. These are 'universals' that have been recognised by many cultures for calendrical and ceremonial purposes. We consider:

(i) the limits in rising and setting of the sun determined by its annual motion, and the midpoint of that motion, i.e. the solstices and equinoxes;

(ii) the limits of the rising and setting of the moon determined by the regression of the lunar nodes (the 'standstill cycle' of 18·61 years), i.e. the major and minor standstills which in turn bound the monthly extremes; and

(iii) the daily motion of stars and asterisms (neglecting the long-term shift due to the precession of the equinoxes).

We do not consider planets, since their motion does not lead to simple periodic extremes in rising or setting, although the method outlined here can be used to treat specific cases. We will describe this work in more detail in a subsequent publication, where we will discuss the implications for archaeoastronomical studies in general.

From the spherical triangle formed by the zenith, the celestial north pole, and the body in question, we obtain the relationship

$$\sin \delta = \sin \phi \sin a + \cos \phi \cos a \cos A , \qquad \qquad (1)$$

where δ is the declination of the body, ϕ is the latitude of the site, and a is the geocentric altitude of the body at the time of observed rising or setting at azimuth A. We define a nominal value

$$A_0 (a = 0) = \cos^{-1}\left[\frac{\sin \delta}{\cos \phi}\right],$$

and find the dependence of A on each independent variable in (1) via the first term in the expansion

$$A \approx A_0 + \Delta A \approx A_0 + (x - x_0)\left(\frac{\partial A}{\partial x}\right)_{x = x_0} , \qquad \qquad (2)$$

where x is one of the variables $\{a, \phi, \delta\}$ and x_0 corresponds to A_0. We will consider each of these variables in turn. We assume that the values of x at repeated times of observation will have a normal distribution characterised by a standard deviation σ_x, which in turn leads to a normal distribution of values of A characterised by

$$\sigma_A \approx \sigma_x \left(\frac{\partial A}{\partial x}\right)_{x = x_0} . \qquad \qquad (3)$$

In the absence of any specific knowledge about the actual techniques used by a culture, we must consider a reasonable range of variation in these variables at the actual moment taken to be 'rising' or 'setting' by that culture, which leads to a variation in repeated observations of the azimuth of an event and an uncertainty in any one observation.

2 The effect of altitude

The observed altitude a' is related to the geocentric altitude a by

$$a = a' - R(a') + P(a') , \qquad \qquad (4)$$

where R and P are corrections due to atmospheric refraction and parallax. The value of A depends on the value of a', and hence of a, at the moment taken to be 'rising' or 'setting'. In the absence of specific knowledge of the value of a' at that moment, we must take the full range of possibilities, from sighting on the first (last) flash to sighting on the full disc of the sun or moon tangent to the horizon (Hawkins 1963). There is an equivalent range due to the finite size of an asterism. This leads to a mean value $A_0 + \Delta A_D$, which from (2) is

$$A(a) \approx A_0 + a\left(\frac{\partial A}{\partial a}\right)_{A_0} \approx A_0 + a\frac{\tan \phi}{\sin A_0},$$

and a spread in values characterised by a standard deviation

$$\sigma_D \approx \sigma_a \frac{\tan \phi}{\sin A_0}.$$

Variations in refraction (Schaefer and Liller 1990) and parallax do not introduce any further shift, but do introduce further spreading (σ_R and σ_P). Planets and stars have a further shift (ΔA_E) and spreading (σ_E) in the azimuth of rising or setting due to atmospheric extinction (Hawkins 1973: 274; Schaefer 1986): $\Delta A_E = 1°·8$ and $\sigma_E = 0°·8$ for the Pleiades, for the conditions of Table 1. There is also a possible correction for the height of the local horizon at a particular site.

3 The effect of declination and latitude

If observations were made of a sunrise/set or moonrise/set at a declination shifted by $\Delta\delta$ from an extremum, there is a corresponding shift away from the limiting azimuth of

$$\Delta A_\delta \approx \Delta\delta\left(\frac{\partial A}{\partial \delta}\right)_{A_0} \approx -\Delta\delta\frac{\cos \delta}{\cos \phi \sin A_0}.$$

We take half the shift in azimuth due to missing the limiting event by a day (solstice or monthly lunar extreme) or a year (lunar standstill extreme) as a typical value of this shift (ΔA_d or ΔA_Y). This corresponds to missing the actual event by 0·7 days or 0·7 years. We then take half of this shift as an estimate of the corresponding standard deviation of repeated measurements (σ_d or σ_Y), so that ~95% of the observations will fall within a day (or a year) of the extreme.

The effect of a change in latitude on the azimuth of a particular event is given from (2) by

$$\Delta A_\phi \approx \Delta\phi\left(\frac{\partial A}{\partial \phi}\right)_{A_0} \approx -\Delta\phi\frac{\tan \phi}{\tan A_0}.$$

From this we can calculate how far apart in latitude two sites must be before we expect a significant difference in this azimuth.

Event	δ	A_0	ΔA_D	σ_D	σ_R	σ_P	σ_H	ΔA_d	σ_d	ΔA_Y	σ_Y	$\tilde{A}$	$\tilde{\sigma}$
Major standstill N	+28·72	53·53	0·44	0·33	0·12	0·03	0·13	0·51	0·25	0·20	0·10	54·68	0·46
Summer solstice	+23·57	60·36	−0·39	0·30	0·11	0·00	0·13	0·002	0·001	—	—	59·97	0·34
Pleiades	+20·51	64·32	0	0·53	0·05	0·00	0·25	—	—	—	—	66·1	1·0
Minor standstill N	+18·42	66·99	0·38	0·28	0·10	0·02	0·13	0·33	0·17	−0·19	0·09	67·51	0·38
Minor standstill S	−18·42	113·01	0·38	0·28	0·10	0·02	0·13	−0·33	0·17	0·19	0·09	113·25	0·38
Winter solstice	−23·57	119·64	−0·39	0·30	0·11	0·00	0·13	−0·002	0·001	—	—	119·25	0·34
Major standstill S	−28·72	126·47	0·44	0·33	0·12	0·03	0·13	−0·51	0·25	−0·20	0·10	126·20	0·46

TABLE 1. Azimuths and spreads of various risings observed at Chaco Canyon (lat. +36°·05, apparent horizon ht. 0°, AD 1000) expressed in degrees.

4 Other effects

There will be a further uncertainty in sighting on an extended body because of its horizontal size. This introduces no net shift but does introduce a further spread σ_H which we take to be half the horizontal semi-diameter. This is not unreasonable, considering that one has only a minute or so to sight on a moving target which can present further observational difficulties (in the case of the sun, its brightness; in that of the moon, its varying phase; and in that of an asterism, its gradual appearance or disappearance owing to extinction).

Equinox presents a different problem. Since the azimuth of the rising/setting sun simply shifts at a constant rate (~0°·5 per day) along the horizon near to the equinoxes, this rising/setting cannot be used by itself to determine accurately either the event of equinox or the direct east/west line. Either east/west or the time of equinox must be separately determined in order to find the other. There are several ways to estimate the time or azimuth of equinox rise or set from horizon observations of the solstices; these lead to a range of values for the azimuth of equinox centred on east/west with a σ_e of ~1°. Since a number of cardinally-aligned building features (for example at Chaco Canyon, New Mexico) lie closer to east/west than this would indicate, we conclude they were not laid out with reference to the equinox azimuths but that some other method (probably using the symmetry of the sun's daily motion via the motion of a shadow) was employed. We estimate that the cardinal directions could be located with no net shift and a standard deviation $\leq 0°·25$ from non-horizon methods.

5 Conclusions

We present in Table 1 the shifts and standard deviations associated with the above effects, for the latitude (+36°·05) and altitude (1900 m) of Chaco Canyon for the era AD 1000. We show calculations for the most prominent phenomena of the sun and moon and (for comparison) for one asterism, the Pleiades, known to be particularly regarded by the Pre-columbian cultures of North and Central America. We show the results for the rising of the indicated bodies; the results for the settings are symmetric about north ($A = 0°$). The risings are not symmetric about $A = \pm 90°$, however, with the consequence that risings in the NE (SE) quadrant are not diametrically opposed to settings in the SW (NW) quadrant, although the corresponding distributions are close enough that they do overlap to a certain extent *modulo* 180°. Since we cannot resolve these opposing events we can combine their distributions, and then have a still broader range with which artificial features could be correlated.

Event	$4\tilde{\sigma}$	$4\tilde{\sigma}'$
Solstice rise/set	1°·4	0°·6
Major standstill rise/set	1°·8	1°·2
Minor standstill rise/set	1°·5	0°·9
Pleiades rise/set	4°·0	3°·2

TABLE 2. Values of $4\tilde{\sigma}$ and $4\tilde{\sigma}'$ at Chaco Canyon.

Having assumed that each of the independent variables determining A is normally distributed, and that each in turn produces a normal distribution of values of A, we now assume that the variations in each variable are independent of the others (e.g. that variations in refraction are not correlated with variations in declination at successive attempts to 'catch' a lunar extreme). We can thus add the various shifts linearly, and add the various standard deviations in quadrature. We term the results $\tilde{A}$ and $\tilde{\sigma}$. These are also shown in Table 1.

If we focus on the various standard deviations in Table 1, we see that the dominant effect is due to the uncertainty in the definition of rise actually utilised. We would expect ~95% of the determinations of the azimuth of a rising or setting to fall within an interval of width $4\tilde{\sigma}$. Values of this at Chaco Canyon are given in Table 2.

If we know from other sources (or from the internal consistency of sightlines) that a particular culture used certain techniques that reduced σ_D (and perhaps σ_H) significantly, we are then still limited by the remaining uncertainties to a new value $4\tilde{\sigma}'$, also calculated for Chaco Canyon in Table 2.

We see that these spreads in reproducibility set inherent limits on the accuracy of naked-eye determinations of these azimuths. We thus feel that some claims of prehistoric precision in determining limiting azimuths (e.g. Atkinson 1966; Thom 1971) are overly optimistic and that many prehistoric alignments were of necessity more symbolic than accurate (as measured by present-day standards).

These results can help in determining the correlation between sightlines or architectural features and astronomical events, by making reasonable assumptions about the observing techniques of the culture involved.

References

Atkinson, R.J.C. (1966). Moonshine on Stonehenge. *Antiquity* **40**, 212–16.

Hawkins, G.S. (1963). Stonehenge decoded. *Nature* **200**, 306–8.

Hawkins, G.S. (1973). *Beyond Stonehenge.* New York NY: Harper and Row.

Schaefer, B.E. and Liller, W. (1990). Refraction near the horizon. *Publications of the Astronomical Society of the Pacific* **102**, 796–805.

Schaefer, B.E. (1986). Atmospheric extinction effects on stellar alignments. *Archaeoastronomy* no. 10 (supplement to *Journal for the History of Astronomy* **17**), S32–42.

Thom, A. (1971). *Megalithic lunar observatories.* Oxford: Oxford University Press.

17

An integrated approach to the investigation of astronomical evidence in the prehistoric record: the North Mull project

CLIVE RUGGLES AND ROGER MARTLEW

Introduction

The work of the North Mull project epitomises the shift in emphasis within prehistoric British archaeoastronomy away from the mere consideration of astronomical alignments towards more integrated approaches, and illustrates the problems in trying to integrate statistical and cultural evidence in a context where the latter is very thin on the ground.

The project arose as a result of statistical surveys of megalithic alignments in western Scotland, which demonstrated that the orientations of the short (three-, four- or five-) stone rows found in western Argyll exhibit a pattern that is related to the rising and setting positions of the moon in the south (Ruggles 1984a; 1985; 1988).

The aim of the North Mull Project was to examine this result in its fuller archaeological context through an integrated program of excavation, horizon survey, locational analysis and statistical investigation in a particular area. In northern Mull there are the remains of seven short stone rows. Five (Quinish, Maol Mor, Dervaig N, Dervaig S and Balliscate) occur in isolation; the remaining two (Ardnacross) are found together, in close association with three burial cairns. In addition to these seven rows there is a triangular setting of three standing stones at Glengorm. Since two of these stones were known to have been re-erected in recent times, and this type of setting was unknown elsewhere locally, it was suspected that the site had originally been a three-stone row similar to the others in the group.

The North Mull project is sponsored by Earthwatch and the Center for Field Research, Boston, MA. Its detailed objectives, together with its overall research design and some initial results, have been described in the first of a series of detailed reports (Ruggles and Martlew 1989). This paper gives a brief summary of progress in the project.

Excavations and related work

Excavations and related environmental and survey work have taken place at two sites. The principal aim of work at Glengorm in 1987 and 1988 was to establish whether the stones there did indeed originally form a linear setting, and if so, to establish whether their orientation fitted the general pattern of southerly lunar orientation observed at the other linear settings. The work demonstrated that the Glengorm stones had indeed once formed a row, but that its orientation was farther to the east of south than had been assumed from surface indications alone (Ruggles and Martlew 1989).

The Glengorm site was generally disappointing stratigraphically, since only a thin layer of topsoil separated the bases of the stones from bedrock, although the presence of quartz fragments and pieces of burnt flint, and particularly the presence of a small post hole carved into the bedrock in the original alignment to the south of the stones, were of considerable interest in themselves (*ibid.*).

Ardnacross, where two stone rows and three cairns were sited (Fig. 1), has proved stratigraphically much richer, with excavations since 1989 revealing a long sequence of activity. Ploughing pre-dated the erection of the two stone rows (Fig. 2): the stone holes cut into a sub-soil that is extensively scarred by ard-marks. At some time after their erection, the end stones in each row were pushed over into large pits and partially buried, perhaps in an attempt to obliterate the original symbolism of the monument. It has unfortunately proved impossible to date this event precisely, and the best that can be hoped for is to establish it within the general sequence of activity on the site. Later, 'lazy beds' were dug around the stones. These are banks of soil and manure heaped up between roughly parallel ditches, representing a system of cultivation for poor soils that survived in the Outer Hebrides until the nineteenth century. There may be an important stratigraphical link between the different elements of the site at Ardnacross—the two stone rows and the three kerb cairns—represented by a layer of dark soil containing many small fragments of charcoal. This has been found widely across the site, and may represent repeated clearances of the area by burning off the vegetation. It parallels evidence found at Glengorm, and may indicate similar ritual activity taking place at both sites. The creation of the ard-

FIG. 1. The location of the stone row sites of northern Mull.

marks at Ardnacross pre-dates the appearance of this layer, while the standing stones are set in it. It is assumed that the kerb cairns were constructed after the lifetime of the stone rows, and this should be demonstrable stratigraphically. Any dating evidence obtained from the cairns will therefore give a *terminus ante quem* for the rows if no direct dating evidence is found in the stone-holes.

Horizon surveys

The original orientation of the stone alignment at Glengorm, as determined by excavation, yields an azimuth of approximately 156° (Ruggles and Martlew 1989). An interesting fact is that the isolated peak of Ben More, the highest mountain on Mull, lies prominently on the horizon on this alignment. It is not now visible from the stones themselves, owing to trees on an intervening ridge some 2 km away, but it can be clearly seen either from a few metres to the west of the stones or from higher ground to the north.

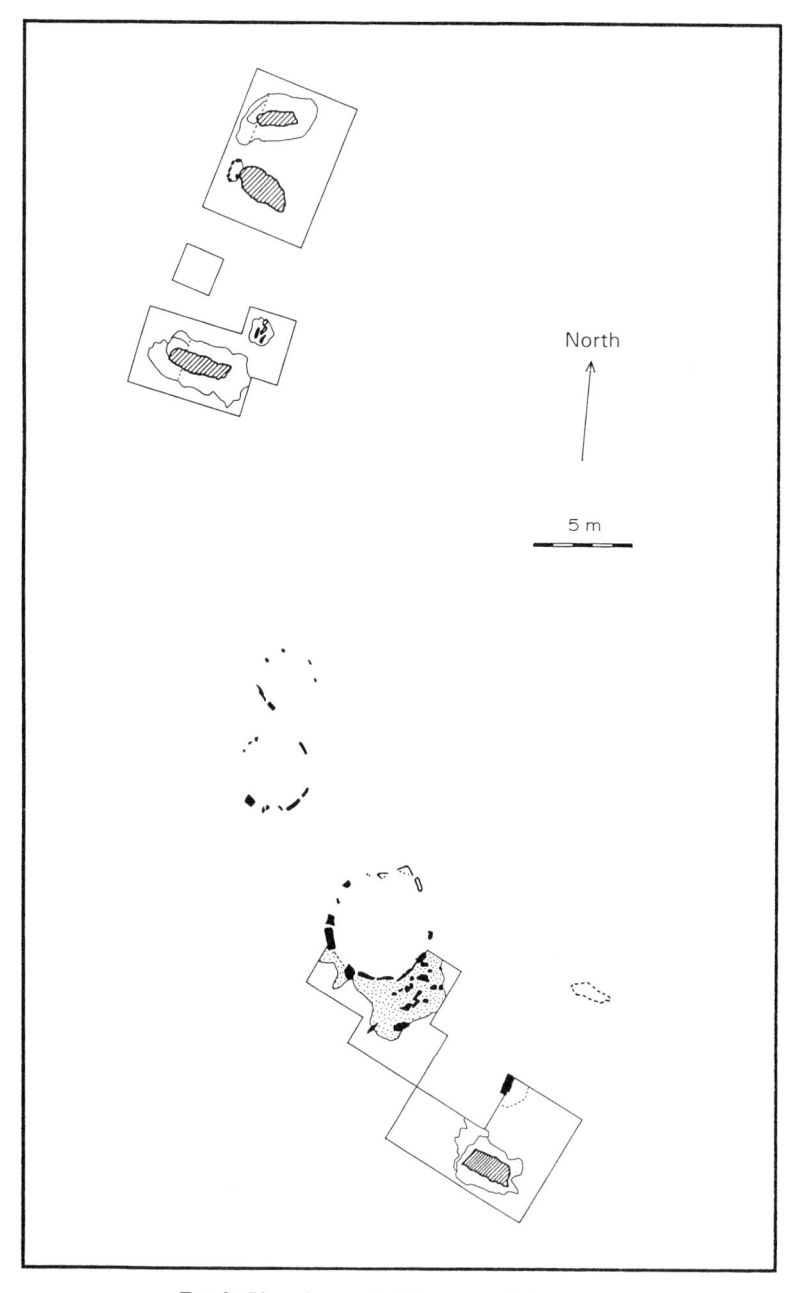

FIG. 2. Plan of excavated features at Ardnacross.

Five of the North Mull sites—Glengorm, Quinish, Maol Mor, Dervaig N and Dervaig S—form a relatively concentrated group towards the northern tip of the island. A hitherto unsuspected phenomenon, that only emerged during survey fieldwork in 1990, is that each of the sites seems to be placed on the very limits of visibility of the peak behind more local ground, so that the mountain is clearly visible from some points within a few metres of the site and obscured from others. This phenomenon had been noted at Quinish in the first season of fieldwork but had been unsuspected at the other sites owing to factors such as poor weather and local tree cover. The peak yields a declination between $-28°·5$ and $-26°·5$ in each case.

The row at Dervaig North is aligned to within a couple of degrees of Ben More, but the other three rows are oriented a few degrees to the right, upon declinations between about $-31°$ and $-29°$. The largest difference between the two directions, and the lowest declination, both occur at Quinish, where only a single stone stands and the original orientation of the row is in greatest doubt.

Table 1 summarises the available information. All the data for Glengorm are taken from recent work. At the other sites, the figures for the stone row orientations are taken from Ruggles (1988: Table 9.1) and those for Ben More are obtained by calculation.

There is, however, no possibility of seeing Ben More from the northeastern strip of the island where the two remaining North Mull stone rows, Balliscate and Ardnacross, are located. The rows at both sites are aligned upon a southern horizon which is relatively close, high and featureless. At Ardnacross it was possible to reconstruct the original orientations of the rows as a result of the excavation. These are significantly different: the northern row, in which all the stones are now prostrate, yields a declination of approximately $-26°$, slightly lower than was previously estimated from the surface evidence alone (see Ruggles 1988:

Site name	Orientation	Ben More:		
		Azimuth	Altitude	Declination
Glengorm	156°	156°·0	2°·0	−28°·6
Quinish	168°	149°·6	2°·0	−26°·7
Maol Mor	162°	152°·3	2°·1	−27°·5
Dervaig N	150°	151°·9	2°·2	−27°·2
Dervaig S	157°	151°·3	2°·3	−26°·9

TABLE 1. Ben More and the northernmost stone rows on Mull (after Ruggles and Martlew 1992: Table 9).

Table 9.1), while the southern row yields a declination of approximately −22°·5. The Balliscate row yields a declination of −28°·5. Both rows are oriented to the west of south. Intriguingly, the most southerly prominent peak visible at Ardnacross, Beinn Talaidh, yields a declination of −26° to the east of south—in other words, a celestial body seen to rise behind Beinn Talaidh would set behind the more local horizon at a point in line with the northern row—and the only more distant peak showing behind the nearby horizon at Balliscate, Speinne Mór, appears in a similar way to the east of south yielding a declination of −28°·9.

Statistical investigations

Once the possibility is raised that prominent horizon peaks, not necessarily in the direction of alignment of a site, may have been of significance (astronomical or otherwise) to the builders, then a great many possibilities are raised of correlations between site location and orientation, the visibility of certain prominent peaks and the rising and setting positions of celestial bodies of possible interest. It is clearly imperative that a serious attempt be made to give an objective assessment of whether observed correlations (such as those mentioned above) really were of significance to the builders, or may instead be easily explained away as chance occurrences.

In the North Mull project, two factors have been chosen for statistical analysis. The first is the variation of horizon distance with azimuth, both absolute and relative to site orientation. If it was important to be able to view certain (specific or classes of) horizon astronomical phenomena, then this would require at least moderately distant horizons in the relevant directions. The second factor is the visibility of prominent hills and their relationship to astronomical phenomena.

Three ways have been explored of obtaining objective data samples from the archaeological record. The first is simply to compare data from the sites themselves, as was done in an earlier analysis of the eastern Scottish Recumbent Stone Circles (RSCs) (Ruggles 1984b; Ruggles and Burl 1985). The second and third methods involve comparing data from the sites with data from a set of 'control' locations chosen by a suitable selection procedure. In one case, the control group are selected using a procedure that generates spatially pseudo-random points within the area under consideration. In the other, an attempt is made to identify locational factors other than astronomical potential that appear to link the locations chosen by the builders of the stone rows. Having done so, an attempt is made to identify alternative locations in the vicinity of each site that appear to satisfy the locational prerequisites just as well as the locations actually chosen. This permits a 'local' analysis at each site, where the

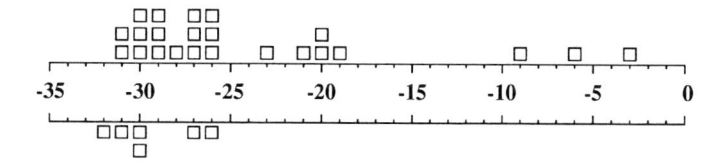

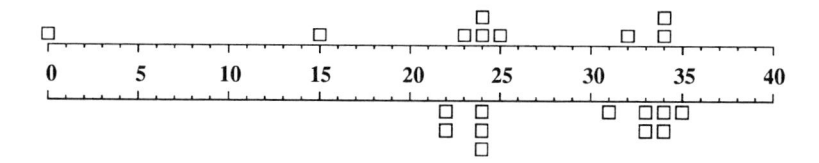

FIG. 3. Declinations of prominent hill summits at the north Mull sites. Easterly (rising) lines are plotted upwards; westerly (setting) ones downwards. (After Ruggles and Martlew 1992: fig. 13.)

astronomical potential of the location chosen is compared with a set of alternatives that appear equally plausible on other grounds.

Some results of these investigations have been reported elsewhere (Ruggles *et al.* 1991; Ruggles and Martlew 1992). On the azimuthal distribution of horizon distance, there is strong evidence that the North Mull stone rows were deliberately placed so as to have non-local horizons between SSE and WSW, against the general trend of horizon visibility distribution. A comparative study of the stone rows in mid-Argyll (Ruggles *et al.* 1991: Appendix) shows a similar trend at the sites towards non-local horizons in the south and south-west, but this is not against the general trend at the control sites. The conclusion is that there was a definite interest in the horizon quadrant centred on SSW both in Mull and mid-Argyll, but in Mull this interest was at odds with the horizon distance profiles most often occurring naturally, and site placement was presumably a trickier business. The interest in the horizon quadrant centred on SSW mirrors what is found at the RSCs in eastern Scotland (Ruggles 1984b: fig. 1).

The declinations of prominent hill summits visible at the north Mull sites cluster into four groups (Fig. 3). The first consists of values around –30° and up to –25° and contains at least one peak from each site. Some of the sites in the northern group contribute several peaks, Ben More being most prominent amongst them in each case. Balliscate and Ardnacross, however, each contribute only one: Speinne Mór and Beinn Talaidh respectively, the peaks already mentioned in the section above. Each of the these peaks marks a celestial rising point to the east of south. The second cluster contains peaks yielding declinations around –20°, from various sites, and again east of south in each case. The third cluster contains nine rising and setting lines with declinations close to +24°, and the last contains another nine with declinations close to +34°.

An interpretation: sacred mountains and the moon

Earlier work (Ruggles 1985) had uncovered a specific pattern of lunar orientation amongst the concentrations of short stone rows in north Mull and mid-Argyll. Wherever one of these stone rows is found in isolation, it marks a southern horizon declination within a couple of degrees of −30°, the lunar major standstill. However, a number of the rows are found in adjacent or nearby pairs, and where this occurs one of them indicates a declination between about −24° and −21°, which corresponds to the position of the moon towards the southern monthly limit but not necessarily close to the maximum of the 18·6-year cycle. (It also, of course, corresponds to other astronomical bodies such as the midwinter sun.) This pattern of 'primary' and 'secondary' orientation appeared to apply to every site in the two concentrations, with the exception of Ardnacross, where both orientations seemed to be secondary. Other stone rows in Argyll and the Inner Hebrides fit a more general pattern of southern orientation within the range −30° to −19°.

The new evidence described above provides an extremely interesting and plausible picture which somewhat modifies but greatly amplifies the earlier conclusions. At each of the seven north Mull sites there is a prominent peak in the south-east marking the rising point of the moon close to, but not precisely at, the major standstill. In the case of the five sites of the northern group this is Ben More, yielding declinations between about −28°·5 and −26°·5 (see Table 1). At Balliscate and Ardnacross the relevant peaks are Speinne Mór (−29°) and Beinn Talaidh (−25·5°) respectively. This is consistent with a primary interest in a prominent horizon peak beyond which the moon rises relatively rarely. Since the phase of the moon at its southern monthly limit is close to full near to the summer solstice, the most plausible time of observation is near to full moon and near to the summer solstice, when moonrise would occur around dusk and be clearly visible. The complexity of the moon's motions and the large variations in the rising position from night to night, month to month and year to year perhaps accounts for the lack of peaks any closer to the theoretical major standstill limit of −30°. At Ardnacross, where the declination of the relevant peak is furthest from this limit, the altitude of the southern horizon is so high that the moon with declinations much lower than this value would not rise at all.

The orientations of the rows themselves fall between the rising positions of the moon behind these peaks and the position where it would set again in the west. The excavation at Glengorm has shown that the row there was aligned upon Ben More; this is also the case at Dervaig N. The rows at Quinish, Maol Mor and Dervaig S are oriented between about 6° and 18° to the right of Ben More (see Table 1), upon declinations between about −29° and −31°. Finally, the row at Balliscate and the northern row at Ardnacross are oriented upon the position where the moon would set to the west of south having risen over the prominent peak to the east.

Three of the sites also possess further peaks yielding declinations between about −23° and −20°; these could be marking the position of midsummer full moonrise in years well before or after major standstill, or equally well moonrise a couple of nights before or after its appearance behind the primary peak. A group of three such 'secondary' peaks is visible across the Sound of Mull from Ardnacross, and interestingly the rightmost of these, Sgurr Dearg, yields exactly the same declination (−22°·5) as the second, southern, alignment at the site. Thus at Ardnacross, when the moon, close to its visibility limit in the south, rises behind Beinn Talaidh, it sets in line with the northern row; but when, a couple of days earlier or later, it rises in the vicinity of Sgurr Dearg, it sets in line with the southern row.

A further secondary peak is found at Dervaig N, but here the horizon between Ben More and the peak in question (Cruachan Druim na Croise) is occupied by a high local hill. The last example is at Maol Mor, where the secondary peak is Beinn Talaidh (declination −20°·3). Could it be that at the construction of this site on an unusually high ridge was undertaken because the builders could incorporate the two most prominent peaks in southern Mull—Beinn Talaidh and Ben More—to mark roughly the opposite limits of the rising position of the midsummer full moon?

Finally, it is interesting to note the cluster of prominent peaks around the declination of the summer solstice. Quinish is remarkably placed to mark the setting solstitial sun behind the three distant peaks of Beinn Mhor, Ben Corodale and Hecla in South Uist and the simultaneously rising full moon behind Ben More. This is an association that is well attested in ethnographic examples such as the Sun Dance of North American Plains Indians (e.g. Voget 1984).

It seems most likely, then, that it was of paramount importance to place these sites where the full moon around midsummer would rise behind a prominent horizon peak when it was somewhere near the southerly limit of its motions around the year of major standstill. Where possible, related positions, such as the position of moonrise a couple of days earlier or later (or of midsummer full moonrise a few years earlier or later), of moonset at these times, or of the rising or setting position of the midsummer sun, might also be marked by the orientation of the alignment or other prominent peaks.

Looking ahead

A fourth aspect to the North Mull project has been a locational analysis in which predictive site modelling is undertaken by considering factors such as topographic position, local geology, altitude, proximity to water and so on. Preliminary analyses in the Glengorm area have been undertaken using the IDRISI™ raster-based Geographical Information System (GIS). The data are currently being transferred to the more powerful system GRASS™ to enable the analysis to be done in full.

GISs are also a powerful tool for tackling the question: what are the chances that the observed relationships between sites, prominent peaks and astronomical events could have arisen fortuitously, that is through the action of factors quite unrelated to astronomy? In particular we need to consider the 'negative data'. For instance, we must ask for each prominent peak of apparent significance and in particular for Ben More: from how many parts of northern Mull is it visible? What is its declination? Do areas where it is visible and has a certain declination tend to coincide with areas where sites might have been located for other reasons anyway, or do their positions appear to have been carefully chosen within the available parameters?

A first attempt to answer such questions for Ben More led, during the third season of fieldwork, to the production by manual means of a 'declination contour map' (Ruggles and Martlew 1992: fig. 14). This exercise raises some interesting questions, particularly with regard to the other six standing stone sites on northern Mull listed by Ruggles (1984a): namely the four-stone setting at Tenga (ML13), a prostrate stone pair at Calgary (ML8), and single standing stones at Lag (ML6), Cillchriosd (ML7), Tostarie (ML14) and Killichronan (ML15).

Some or all of these sites may, of course, be quite unrelated to the stone rows: Tenga is an enigmatic site in itself, consisting of four stones spaced out in blanket peat that may represent the remains of a site of a number of different forms; single standing stones are notoriously difficult to date and may have been erected in almost any period, including the relatively modern, and represent a variety of motivations and uses. However, including these sites on the declination contour map produces an interesting result. Three of the sites, Tenga, Tostarie and Killichronan, are within sight of Ben More. Of these, Tenga and Killichronan yield declinations for Ben More of −29° and −28° respectively, continuing the trend found at the five stone rows in the north. Yet Tenga is situated in the centre of northern Mull, overlooking Loch Frisa, and Killichronan is in the south, near to the narrow neck joining the two halves of the island. Only Tostarie, on the west coast, yields a completely different declination, around $-18°·5$. It is, of course, notable that while the other values fall between one and three degrees from the extreme declination of the major standstill moon, this value is within one degree of the minor standstill.

It is also worthy of note that the prostrate three-stone row at Uluvalt (ML25) in the south of the island, situated only 4 km from Ben More to the SE, is also aligned upon it. Thus there is strong evidence for the dominance of Ben More in relation to lunar symbolism as a critical factor in the placement of free-standing megalithic sites in northern Mull. It is even tempting to speculate that Ben More was of importance to the builders of Balliscate and Ardnacross, who attempted to align their rows upon it despite the fact that it was invisible behind local horizons to the south and south-west. The azimuth of Ben More from Balliscate is some 15° to the left of the row orientation, and at Ardnacross some 12° to the left of that of the northern row. Even the slab at Cillchriosd is oriented to within 10° of Ben More.

On the other hand, there is ample evidence to suggest that mountains other than Ben More may have been of importance at various sites. For this reason, work is now under way to investigate visibility and astronomical potential using viewshed functionality within GIS. The techniques being developed are described elsewhere (Ruggles *et al.* 1993). The end result will be the ability to overlay information on favourable sites from the point of view of the visibility of certain mountains and their astronomical potential with maps showing other factors that might have influenced site locations, such as the elevation, slope and local terrain form (Kvamme 1989; 1992), land use potential and surface geology.

Discussion

In the absence of the many other, and often far richer, types of cultural data available to those studying astronomy in protohistoric, historic and modern contexts, patterns discernible in the archaeological record are of considerable importance. The work of the North Mull Project illustrates the difficulties encountered in trying to interpret the symbolism of ritual monuments from the material record alone, and the emphasis placed on methodology as a consequence.

In general terms, the results of the North Mull Project suggest that the tendency within prehistoric British archaeoastronomy to focus attention upon horizon astronomy alone may have been simplistic. At the present stage of the analysis it seems that distant but prominent peaks may have played a major part in the symbolism associated with the position of a sacred site within the physical, social and ritual landscape.

Most aspects of this position may be difficult to recover archaeologically, but preferential location with respect to topography, either to enhance the visibility of the site from nearby areas, or to ensure the visibility of a significant natural feature or astronomical phenomenon, is still available for analysis. The extent to which this partial view permits the reconstruction of ephemeral beliefs and rituals is a matter for theoretical debate beyond the limits of this paper. The North Mull Project, however, has shown that patterns of ritual behaviour involving a complex interplay of different factors can be discerned in the archaeological record. This is only the first step in attempting to interpret such patterns.

References

Kvamme, K.L. (1989). Geographical Information Systems in regional archaeological research and data management. In *Archaeological method and theory*, vol. 1, ed. M.B. Schiffer, pp. 139–203. Tucson AZ: University of Arizona Press.

Kvamme, K.L. (1992). Terrain form analysis of archaeological location through Geographic Information Systems. In *Computer Applications and Quantitative Methods in Archaeology 1991*, eds. G. Lock and J. Moffett, pp. 127–36. Oxford: Tempus Reparatum (BAR International Series 577).

Ruggles, C.L.N. (1984a). *Megalithic astronomy: a new archaeological and statistical study of 300 western Scottish sites*. Oxford: British Archaeological Reports (BAR British Series 123).

Ruggles, C.L.N. (1984b). A new study of the Scottish Recumbent Stone Circles, 1: Site data. *Archaeoastronomy* no. 6 (supplement to *Journal for the History of Astronomy* **15**), S55–79.

Ruggles, C.L.N. (1985). The linear settings of Argyll and Mull. *Archaeoastronomy* no. 9 (supplement to *Journal for the History of Astronomy* **16**), S105–32.

Ruggles, C.L.N. (1988). The stone alignments of Argyll and Mull: a perspective on the statistical approach in archaeoastronomy. In *Records in stone: papers in memory of Alexander Thom*, ed. C.L.N. Ruggles, pp. 232–50. Cambridge: Cambridge University Press.

Ruggles, C.L.N. and Burl, H.A.W. (1985). A new study of the Scottish Recumbent Stone Circles, 2: Interpretation. *Archaeoastronomy* no. 8 (supplement to *Journal for the History of Astronomy* **16**), S25–60.

Ruggles, C.L.N. and Martlew, R.D. (1989). The North Mull project, 1: Excavations at Glengorm 1987-88. *Archaeoastronomy* no. 14 (supplement to *Journal for the History of Astronomy* **20**), S137–49.

Ruggles, C.L.N. and Martlew, R.D. (1992). The North Mull project, 3: Prominent hill summits and their astronomical potential. *Archaeoastronomy* no. 17 (supplement to *Journal for the History of Astronomy* **23**), S1–13.

Ruggles, C.L.N., Martlew, R.D. and Hinge, P.D. (1991). The North Mull project, 2: The wider astronomical potential of the sites. *Archaeoastronomy* no. 16 (supplement to *Journal for the History of Astronomy* **22**), S51–75.

Ruggles, C.L.N., Medyckyj-Scott, D.J. and Gruffydd, A. (1993). Multiple viewshed analysis using GIS and its archaeological application: a case study in northern Mull. In *Computer Applications and Quantitative Methods in Archaeology 1992*, eds. J. Andresen, T. Madsen and I. Scollar. Oxford: Tempus Reparatum, in press.

Voget, F. (1984). *The Shoshoni-Crow Sun Dance*. Norman OK: University of Oklahoma Press.

18

The astronomy and geometry of Irish passage grave cemeteries: a systematic approach

JON D. PATRICK

Object of the research

A preliminary investigation of some of the prehistoric monuments in the Boyne Valley, Co. Meath made in 1972 came to the tentative conclusion that they were positioned according to a deliberate design (Patrick 1971). It was contended that the builders oriented the group of monuments on a single line that defined sunrise at the summer solstice and sunset at the winter solstice. One of the monuments on this line is quite obvious while the other monument was predicted by a geometrical design proposed as associated with the orientation. A field investigation revealed what appeared to be a badly destroyed site. The investigation also proposed that the builders of the monuments embodied the maximum altitude the sun attains on each of the days of the solstices and the equinoxes in the groundplan of the sites. It was argued that the evidence that this was done was that all six angles (excluding right angles) from these triangles could be found at one monument formed by the horizontal lines radiating out to the other monuments.

The purpose of the current research was to conduct a systematic test of the hypothesis that the Boyne Valley passage graves were set out to the plan as advocated in the original report. Two independent methods were used to test this theory. First, the probability level was determined as which parts of the proposed design plan could occur by chance. Second, two other locations with collections of comparable archaeological sites were investigated for similar design plans and probability levels for parts of these designs determined.

Site descriptions and their archaeological associations

The three sites covered by this study are the Boyne Valley and Loughcrew in Co. Meath and Carrowkeel in Co. Sligo, each known as a passage grave cemetery. 'Passage grave' has been a common archaeological term since Montelius introduced a classification of Scandinavian megalithic tombs in the early 1900s. Powell (1938) listed and grouped Irish Passage graves and thereby essentially defined their characteristics. He divided the Irish passage graves into four groups, of which the Boyne Valley, Loughcrew and Carrowkeel cemeteries constitute the Boyne Group. This classification is still retained to this day. A comprehensive definition of a passage grave was not advanced until the work of De Valera and Ó Nualláin (1961).

Investigation criteria for prehistoric astronomy and geometry

Hawkins (1968) defined criteria as guidelines for conducting the overall investigation of any site but they do not effectively come to grips with the surveying and accuracy problems. Ruggles (1984) has dealt with many criteria issues in great detail. This paper presents a number of new criteria to control the methodological approach so that it remains consistent and rigorous. These criteria form the basis on which alignments are defined and are accepted or rejected in the computations of probability levels. A detailed justification of these criteria can found in Patrick (1991).

1. *Non-intervisible monuments.* Non-intervisible monuments should be excluded as alignments. In a group of monuments some are frequently not intervisible owing to local topography or other obstructions. Such alignments should be excluded despite our modern-day ability to determine the azimuth of the alignment.

2. *Comparison of measured and theoretical declinations.* A measured declination is accepted as defining a theoretical declination if it is in agreement by better than 1°. The solar and lunar declinations used in this study are $\pm 24°\cdot 0$ and $\pm 0°\cdot 5$ for the sun and $\pm 29°\cdot 2$ and $\pm 18°\cdot 9$ for the moon, as Newgrange has been radiocarbon dated to c. 3300 BC (O'Kelly 1971).

3. *Accuracy of azimuth and altitude.* If the azimuth and altitude of an alignment cannot be established accurately enough to give a declination to better than 1° then it is excluded from the analysis. This is a corollary to criterion 2.

4. *Collinear monuments.* Three or more monuments form one alignment if the directions from one monument to the others are spread less than the accuracy to which the least accurate direction from that monument can be defined. This criterion is best illustrated by a numerical example. If from *A* the direction to *B* is 54° 20′ ± 30′ and to *C* 54° 05′ ± 5′, then since 54° 20′ − 54° 05′ < 30′, *B* and *C* are accepted as being on the same alignment. In any calculations for declination the mean azimuth is used.

5. *Clustered monuments.* If the monuments are clustered into groups, either the most prominent monument or the centroid of the centres of the monuments is used to define alignments. If all the monuments could be ranked as equally important then the centroid is used. Any alignment with this centroid would have an uncertainty equivalent to the separation between the two most extreme monuments of the cluster. In many cases the uncertainty may be so large as to rule out the alignment under criterion 3.

In a cluster of monuments there is often one which is a dominating structure and seems to be much more important than its companions. Under these circumstances it would seem permissible to use the centre of this monument rather than the centroid of all the monuments.

6. *Repeatability of astronomical and geometric designs on similar sites.* Any comprehensive astronomical and/or geometrical theory assigned to a site or monument should be repeatable on a structure with similar morphological or archaeological characteristics. However, this criterion may be tempered when only a small part of the evidence that would explain the functions of any site is available. This criterion is discussed by Heggie (1981) and was developed for a study of two sites in Scotland (Patrick 1974; 1979).

Some methodological considerations in the application of the above criteria

Accuracy of determining gnomon angles. A unique aspect of this work is the investigation of the frequency of occurrence of angles in a geometric figure. The preliminary study showed that the inaccuracies of preserving angles by round off to simple ratios of the three sides of a triangle was of the order of 20′. This information suggested that the field angles should agree with the hypothetical angles if the discrepancy is not greater than 2°.

Accuracy of the monuments' centres. A second consideration that arose in the course of the work was the accuracy to which the centre of a monument can be determined. The centroid of a single standing stone can be determined to within 10 or 20 cm. However, with a passage grave three 'centres' can be defined: (a) the 'centre' of the chamber, (b) the 'centre' of the kerb, and (c) the 'centre' of the cairn. In this study, whenever the kerb could be readily defined its 'centre' was used for calculations. The chamber 'centre' was only used when neither of the other two centres could be determined to a reliable accuracy. Where the kerb is not visible the centre of an obviously symmetrical cairn provides a reasonable alternative 'centre'.

No monument in this work gave a good appearance of being perfectly circular even when it had a defined kerb. Therefore the uncertainty in determining any 'centre' must be at least of the order of 0·5m and on average about 1·0m. The very large monuments have even greater uncertainties.

The Boyne Valley study

In the Boyne Valley (Grid Ref. O 0173) are three large mounds known as Newgrange, Knowth and Dowth (Fig. 1). Newgrange has been one of Ireland's best known monuments since it was first opened in 1699. Dowth was opened in 1847 and at Knowth excavations have just recently been completed. There are no other monuments between Knowth and Newgrange apart from small 'satellite' tombs clustered around their bases. Between Newgrange and Dowth are the mounds A, B, E, F and H. Site U appears to be a badly destroyed passage grave while C and D are standing stones. The sites I and J which would normally be included with the previous mounds are excluded from this analysis, because the heavy tree cover around them made it impossible to establish the conditions of intervisibility in any direction. Monument X was deduced from geometric designs derived in the preliminary study to this work and independently observed on aerial photographs by Mr. L. Swan (priv. comm.). Monument Y was derived from the earlier study (Patrick 1971) and in the field is marked by a very small cairn.

Line admissibility. Table 1 gives the accuracy and admissibility classification of each line between monuments on the basis of the criteria already set down. It should be noted that although the lines XH and XF are listed as non-intervisible this is almost certainly due to the existing tree cover and not local topography.

Astronomical orientations. Some alignments may be considered less reliable because the distant horizon is at a substantially higher elevation than the monument acting as a foresight. These alignments are Newgrange to U, X and D with elevation differences of 3° 37′, 3° 16′ and 3° 30′ respectively, X to Y of 5° 22′ and Dowth to C of 2° 33′. There are 91 admissible lines out of a total of 156 possible lines and from these there are eight unique solar and eight unique lunar alignments (see Table 1).

The Binomial theorem has been used by Hawkins (1968) and Barber (1973) to justify claims of convincing evidence in support of astronomical orientations. Following their approach for solar alignments, a declination variation of 1° represents an azimuth variation of approximately 2°·0 at the solstices and 1°·7 at the equinox. Thus the total target area for an alignment is $4 \times 4·0 + 2 \times 1·7 = 19°·4$. By the binomial theorem the probability that eight hits could occur by chance is 6%. Hence there is some likelihood that the solar orientations of the alignments were set out deliberately.

The target area for lunar alignments is larger. The lunar positions of $(\pm\varepsilon+i)$ and $(\pm\varepsilon-i)$ give an azimuth range of 2°·4 and 2°·0 respectively. Thus the total target area is $4 \times 4.8 + 4 \times 4.0 = 35°·2$ and with eight hits yields a probability of 13·7%. Thus there seems little likelihood that the lunar orientations were deliberate. If both lunar and solar alignments are grouped together the resultant probability level is 9·9%.

The χ^2 test. An alternative to the use of the Binomial Theorem is to test the variation of the distribution of the computed declinations against a rectangular distribution. A distribution was constructed exploiting the direct relationship of azimuth with declination. This curve was evaluated for 1° intervals assuming constant altitude and yielded a ten-cell table of declination frequencies expected under uniform azimuth distribution. The 91 alignment declinations were compared to this table using a χ^2 test and showed no significant result (χ^2 ($n=10$) = 7.43; $p > 0·05$). A discussion of the merits and weaknesses of both tests is presented in Patrick (1974).

The frequency of monument X in the data. Whilst the probability tests on the astronomical alignments weigh against the deliberate orientation of lines, there are other factors that should not be overlooked. Point X has nine admissible lines radiating from it and therefore another nine lines running into it. Six of these lines define solar alignments and two define lunar alignments. After the rejection of redundant lines as previously discussed, there are five solar and one lunar alignments. It seems disproportionate for any one point to dominate the results with five out of eight solar alignments.

Gnomon angles. The Boyne Valley is at a latitude where a vertical gnomon would cast a shadow, at local noon on midsummer's day, to form a triangle very close to the 3-4-5 Pythagorean triple. The triangle angles for the summer solstice, equinox and winter solstice are <29° 42′, 60° 18′>, <53° 12′, 36° 48′> and <77° 42′, 12° 18′> respectively. All angles in the Boyne Valley geometry were calculated and matched against the gnomon angles for an agreement of better than 1° 00′. An angle was classified as admissible if one of its lines was categorised as admissible (see Table 1). Of 858 angles there are 452

The astronomy and geometry of Irish passage grave cemeteries 203

Monument	Accuracy	Kn	Ng	A	B	C	D	E	F	H	Y	U	X	Dt
Kn	1·5m		−19·1R	NV	NV	NV	NV	NV	NV	NV	NV	NV	NV	19·1R
Ng	0·5											0·4R	24·4R	
A	1·0	NV							NV	NV				
B	1·5	NV	24·1S						NV	NV	23·2R			
C	0·5	NV					29·9R		NV	NV				
D	0·5	NV							NV	NV		29·5S		
E	1·0	NV												23·1R
F	2·0	NV		NV	NV	NV	NV			TC	NV	−19·6S	NV	
H	2·0	NV		NV	NV	NV	NV	29·5S	TC		NV	−19·6S	NV	
Y	1·0	NV		−30·0S					NV				28·8S	
U	1·5	NV		−23·3S				29·5S	NV	NV	−24·3R		1·1R	NV
X	2·5	NV	1·2S								NV	1·2S		
Dt	2·0	NV				−29·9S								

TABLE 1. Boyne Valley line accuracy, admissibility (blank squares) and solar and lunar declinations. NV–Not visible, TC–Too close.

Monument	Accuracy	A	B	C–D	E	F	G	H	K	M	N	O	P	S
A	3·0m		NV	NV	NV	NV	NV	NV	NV	NV	NV	NV	NV	NV
B	1·0	NV		28·7R	−29·9R					NV	NV	NV	NV	NV
C–D	2·0	NV				NV	23·0R							
E	0·5	NV		NV			NV	NV		NV	NV	NV	NV	NV
F	0·5	NV						TC		NV	NV	NV	NV	NV
G	0·5	NV			−19·6S	NV	TC			NV	NV			
H	0·5	NV				−23·4S								
K	0·5	NV	NV			1·4S	NV	NV		−28·8R				
M	0·5	NV	NV				NV	NV	NV			NV		
N	0·5	NV	NV			NV	NV					NV	NV	
O	1·0	NV	NV							NV			NV	NV
P	1·0	NV	NV					NV				29·0S		NV
S	0·2	NV	NV				NV					NV	NV	

TABLE 3. Carrowkeel line accuracy, admissibility (blank squares) and solar and lunar declinations. NV–Not visible, TC–Too close.

admissible angles whilst there are thirty admissible hits on gnomon angles. No one point has a significantly greater number of gnomon angle hits, either admissible or inadmissible. The task of laying out thirteen points using angles solely requires the selection of 22 angles. An inspection of the histograms of angles at each point shows that points on the periphery of the area have a large amount of smaller angles (<40°) whereas more centrally located points have a roughly equal distribution from 0° to 180°. Given the size of gnomon angles, the number of hits appears to be consistent with the success rate from points randomly distributed in space.

Right angles. The accuracy of the right-angle was relaxed slightly because if two of the angles were within 1° of a theoretically ideal angle the third would have a larger discrepancy. It was thought that the theoretical angles in this case could be either gnomon angles or the angles from Pythagorean triples. The right-angled triangles are *BAE, XANg, UDF* (inadmissible), *EDA, UDDt* (winter solstice), *NgEC, RUA, RXH* (summer solstice, inadmissible) and *AXE*. None of the triangles represents to any reasonable order of accuracy the six smallest Pythagorean triples or the very close approximation 8-9-12.

Line lengths. There are thirteen independent groups of lines of the same length (i.e. within five metres of each other). The groups, with lines in order of ascending length (metres) are: *UX*(241), *CX, EU*; *DU*(317), *CU, EX*; *AB*(471), *BD, DE*; *BY*(500), *HD*; *AU*(527), *AC*; *FDt*(574), *ANg*; *BX*(646), *UNg*; *AX*(687), *XF*; *EH*(817), *DF*; *CNg*(890), *XNg, BE*; *DH*(898), *YH, CF*; *EDt*(1227), *YNg*; *ADt*(1936), *NgDt*. The lines that have mutual dependence on the same point are illustrated in Fig. 1. The figure that stands out immediately is the parallelogram *EXCU* with the diagonal *UX* nearly the same length as sides *EU* and *XC*. This geometric figure is confirmed by the bearings of the sides as listed here: *CX* 6° 19'; *UX* 137° 50'; *UE* 7° 24'; *EX* 139° 02'.

Note that *AU* and *AC* form an isosceles triangle off the base of the parallelogram using side *UC*. Also, triangle *UXC* is an identical construction. None of the other similar lengths could be integrated with these figures although *ED, DB* and *BA* form a peculiarly interlocking pair of isosceles triangles.

Parallel lines and collinear points. All lines from 0° to 180° of similar azimuths (within 1°) are listed in Table 2. Any line with a bearing greater than 180° is defined by its reverse bearing. Groups of collinear lines according to criterion 4 are: 1 – *Ng, E, F, H*; 2 – *Ng, U, X*; 3 – *B, C, X*; 4 – *Ng, C, Y*; and 5 – *C, D, H*. This appears to be a large proportion of coincidences for a random distribution of points. Indeed, the inclusion of two near misses as straight lines of *AXH* and *EXY* produces a very striking shape as illustrated by Fig. 2.

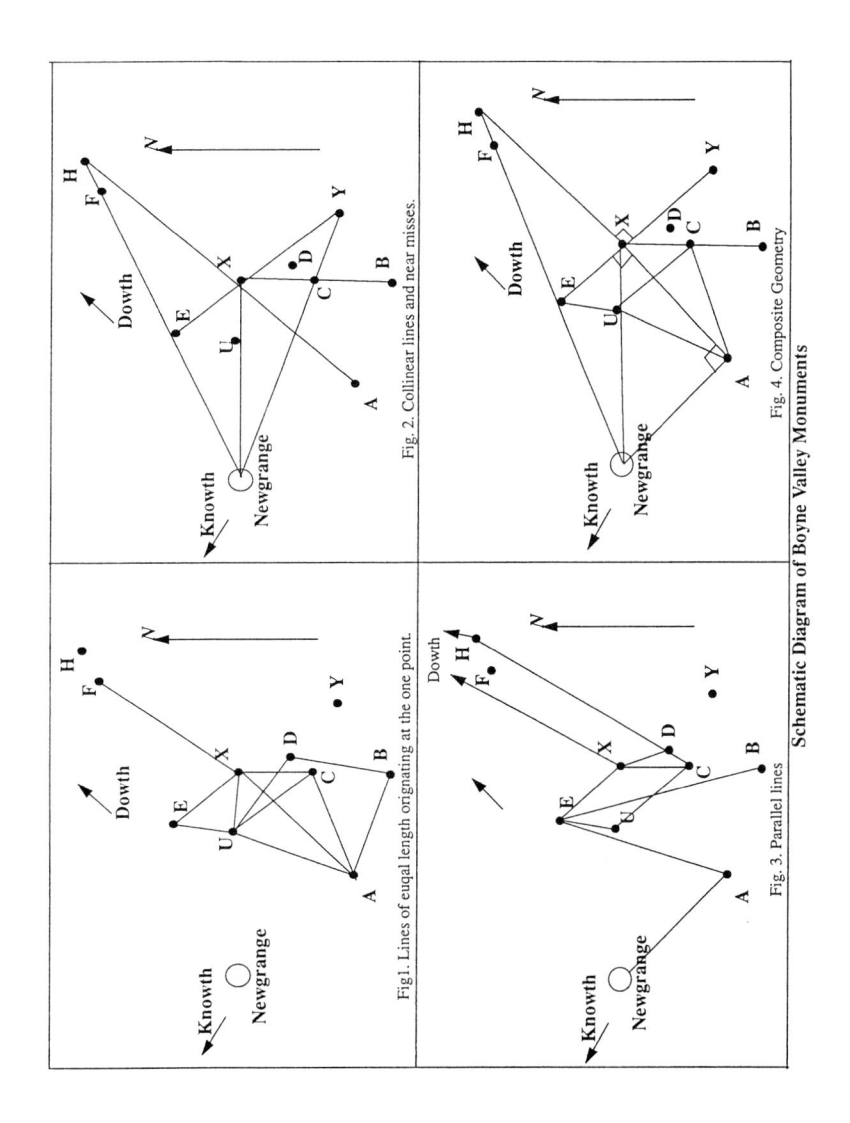

The astronomy and geometry of Irish passage grave cemeteries 205

Fig 1. Lines of euqal length orignating at the one point.

Fig. 2. Collinear lines and near misses.

Fig. 3. Parallel lines

Fig. 4. Composite Geometry

Schematic Diagram of Boyne Valley Monuments

1	CX	6°	19'	BX	6°	28'	BC	6°	34'	UE	7°	24'
2	HDt	22	58	AE	23	06						
3	FDt	30	08	AU	30	29	DDt	30	52	CDt	31	20
	DF	31	22	CF	32	05						
4	DH	35	16	CH	35	35	XDt	36	03			
5	ADt	40	10	XF	40	58						
6	UDt	43	45	AF	44	20	XH	44	26			
7	AH	45	59	AX	47	44						
8	EF	66	58	EH	67	17	NgF	67	17	NgH	67	50
	NgE	68	27									
9	NgU	88	07	NgX	88	08	UX	88	09			
10	KnH	92	31	KnF	93	26						
11	NgY	103	21	KnE	103	45	NgC	103	58			
12	KnU	111	32	AB	112	25	KnD	113	07	KnY	113	29
13	KnC	115	20	DY	116	02						
14	UD	122	21	KnNg	123	19						
15	EY	137	10	UC	137	50	NgA	138	32	EX	139	02
16	EB	171	15	XD	171	26						

TABLE 2. Boyne Valley: azimuths of all lines with nearly identical direction (within 1°).

In Table 2 there are sixteen groups of lines with similar azimuths. The third, fourth, fifth and sixth groups indicate a problem with the criteria, as over the long distance of some lines, the clustering of *F* with *H* and *C* with *D* appears appropriate yet over other short distances this action would be inappropriate. The eleventh group illustrates one of the problems that was sometimes difficult to resolve. Line *NgY* and *NgC* are 103° 21' and 103° 58' respectively while *CY* is 101° 45', thus not strictly fulfilling criterion 4; however, *CY* is 338m long while *NgC* is 890m long, so the points are accepted as collinear. Indeed, when viewed on the ground *YC* runs directly to the passage opening at Newgrange.

In the twelfth group, the extreme distance of Knowth from all these monuments means that the large variations in their positions has little effect on the bearings, creating a situation where essentially radial lines are classified as parallel. If either of the lines *YU* or *YD* is produced towards Knowth they miss the monument by at least two hundred metres.

Fig. 3 illustrates the sets of parallel lines (excluding straight lines) that are unequivocal.

Composite geometry. Fig. 4 is a composite drawing of the most significant geometric characteristics of the site. The most notable feature is the apparent focal position of *X* and the exclusion of Knowth and

Dowth. The two right angles at X and one at A add a surprisingly coincidental reliability to this geometry in the light of near miss alignment of YXE. Another coincidence came to light after this stage of the investigation was completed. The angles $EXNg$ and CXY are 50° 54′ and 50° 36′ respectively. Angle AXR has not been classed as a right angle because EXH is 85° 24′. The only gnomon angles generated by parallel lines are $CKnE$ and $KnCNg$ both of which are inadmissible.

Conclusions

1. There is evidence that some lines are deliberately oriented on solar phenomena. There is a good case for deliberate solar orientation of lines associated with monument X.

2. It is unlikely that any lines are deliberately oriented on lunar phenomena.

3. It is very unlikely that gnomon angles determined from the maximum altitude of the sun on the summer and winter solstices and the equinox were used deliberately to define all the positions of the monuments.

4. There is little evidence that the lower-order Pythagorean triples were deliberately used to define the plan of the monuments or that right angles were utilised as a predominant aspect of the plan.

5. The peculiar combinations of lines of equal lengths suggest that their occurrences in some cases are significant.

6. In five cases, three monuments or clusters are collinear, which seems on an intuitive basis to be much more than could be expected under random conditions. This argument is strengthened by two cases of near misses adding to a suggestion that X is a focal point of the plan.

7. Point 5 is further supported by the existence of two sets of parallel lines that form a parallelogram $EXCU$. The unusual feature of this figure is that one of the diagonals is nearly the same length as two of the sides.

8. Virtually every alignment centred on Newgrange would fit the investigated designs better if the entrance point to the monument was used instead of the centroid of the kerb.

The basic hypothesis that has been investigated is 'that the monuments in the Boyne Valley are laid out to a deliberate geometrical/astronomical design plan'. The null hypothesis is that they are positioned randomly or on a non-geometric design (e.g. based on topography). The evidence as presented strongly suggests that a geometrical and astronomical design was used to fix the positions of the monuments.

A possible procedure for laying out the sites would require the position of X with respect to Newgrange being selected first and then A positioned immediately so as to define the winter solstice and form a right angle between Ng and X. The existence of F for this proposed plan is unjustifiable as it is simply a redundant definition of the line $NgEH$. D does not seem to fit consistently with the proposed design.

The Carrowkeel study

The Carrowkeel cemetery is located in the south-west of County Sligo just west of Lough Arrow. The passage-graves are found on top of the flat-topped Bricklieve Mountains, a limestone upland area running 4 km north-west to south-east and about 2 km across. They are cut longitudinally by three vertical walled rifts 30m to 60m deep and 100m to 300m across. There are fourteen cairns labelled A to P and one standing stone S spread across the summits (see Fig. 5). The site is named Carrowkeel after one of the local townlands (Grid Ref. G 7612). Macalister (1912) excavated some of the cairns; his excavations yielded typical beads and bone pins and a form of pottery that has been identified at many other passage graves and which takes its name from this site, namely Carrowkeel ware. Cairns G and K are cruciform passage-graves and cairn F is a multiple-recess structure. Cairns B and H have passages but no recesses and are classified as undifferentiated. Cairns C, D, M and N are too badly destroyed for proper identification while A and P seem to have no internal structural features, O contains a cist and E is a long cairn with a ruined structure at its NW end.

The astronomy. There are 64 admissible alignments, of which three are indicated solar alignments and six are lunar alignments (see Table 3). Using the Binomial Theorem, these figures yield probability levels of 0·222 and 0·147 respectively, hence it is concluded no deliberate astronomical orientations are present. However, a χ^2 test of declinations yields ($\chi^2(k=10) = 23\cdot59; p < 0\cdot05$) giving strong evidence against a uniform distribution. An inspection of a histogram of azimuths shows a heavy concentration between 80° to 120°, confirming some skewness in the distribution.

Gnomon angles. The gnomon angles for Carrowkeel are slightly different from Newgrange, namely <30° 03', 59° 57'>, <53° 33', 36° 27'> and <78° 03', 11° 57'> for the summer solstice, equinox and winter solstice respectively. There are 155 admissible and 703 inadmissible angles and 12 and 64 gnomon angles have been found in each group. This

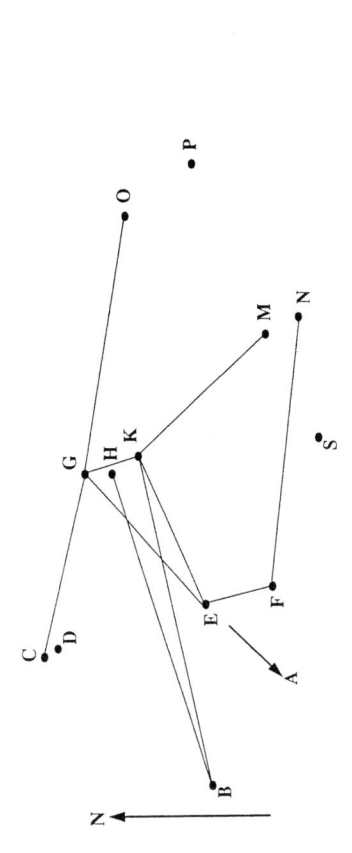

The astronomy and geometry of Irish passage grave cemeteries   209

Fig. 5. Schematic Diagram of Carrowkeel - Lines of Equal Length

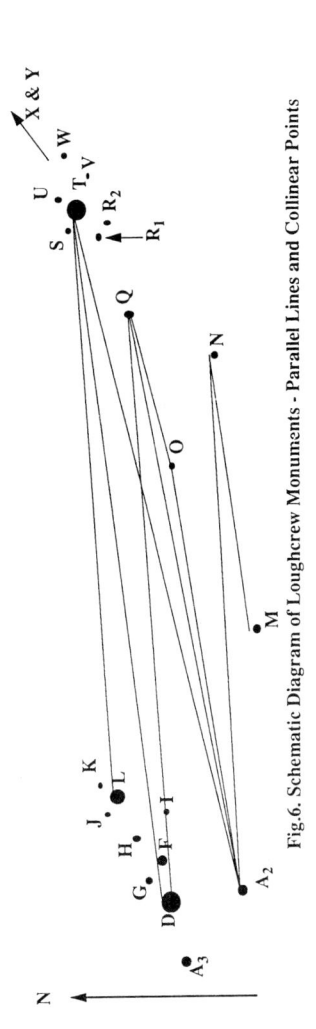

Fig. 6. Schematic Diagram of Loughcrew Monuments - Parallel Lines and Collinear Points

represents 7·7% and 9·9% respectively, suggesting that the gnomon angles occur no more often than would be expected by chance. Furthermore, no site has a predominance of gnomon angles over other sites.

Right angles. There is one admissible and seven inadmissible right angles, representing 0·6% and 1% of each group. These percentages are much lower than in the Boyne Valley. The only admissible triangle, *MSN*, closely matches the Pythagorean triple of 7-24-25 while the inadmissible *C-DMO* matches 8-15-17.

Line lengths. There are eleven groups of equal line lengths, of which five groups are admissible: *GK*(213), *EF*; *EK*(427), *KM*; *CG*, (510)*EG*; *FN*(651), *GO*; and *BH*(882), *BK*. Three of these form isosceles triangles.

Parallel lines. There are twenty different groups of parallel lines of which nine form admissible sets (Table 4). There are three parallelograms all of which are inadmissible: *BC-DON*, *BC-DHF* and *FHON*.

Conclusions

We have been unable to find any systematic astronomical or geometrical characteristics for the positioning of the monuments. However, the local topography may explain the few characteristics discovered. The site consists of three small elevated plateaux separated by rift valleys. Monuments *B*, *K* and *P* are sited in very conspicuous positions on the tops of ridges and near the edge of cliff walls that face to the NNW.

Monuments *E*, *F*, *G*, *H* and *O* are on less prominent positions but still on the tops of ridges. Monuments *C*, *D*, *M*, *N* and *S* are in flat areas of land where they do not occupy special topographic positions. The ridges run in a NNW direction and so the parallelism of *EF* and *GK* but no other groups of lines can be explained by the site topography. It seems unlikely that both *C-D* and *N* would occur in their respective positions if random choices were operating.

1	SH	3°	8′	SG	3°	29′			
2	MO	44	6	AK	44	53			
3	EG	48	24	FK	49	16			
4	NP	56	21	AO	56	32			
5	AN	64	31	MP	65	9			
6	SM	71	30	BH	72	5	EK	72	35
7	BF	113	36	CH	113	45			
8	HM	153	52	HN	154	47			
9	GM	157	33	GN	157	56			

TABLE 4. Carrowkeel: admissible parallel lines.

Monument	Accuracy	A2	D	I	F	H	L	M	N	O	Q	T	X
A2	2.5m		TC	TC	TC	NV	NV		NV	NV	NV		NV
D	2.0	TC		TC	TC	NV	TC				0.8R		NV
I	0.5	TC	TC		TC	TC	TC				0.3R		NV
F	0.5	TC	TC	TC		TC	TC				0.7R		NV
H	0.5	NV	NV	TC	TC		TC						NV
L	1.0	NV	TC	TC	TC	TC							NV
M	1.5									NV			NV
N	1.0	NV											NV
O	2.0	NV						NV					NV
Q	0.5	NV			−0.5S	0.4S	1.3S		−28.7S			NV	NV
T	1.0				NV	NV	NV	NV	NV	NV	NV		NV
X	3.0	NV	NV	NV	NV	NV	NV	NV	NV	NV	NV	NV	

TABLE 5. Loughcrew line accuracy, admissibility (blank squares) and solar and lunar declinations. NV–Not visible, TC–Too close.

The Loughcrew study

The passage grave cemetery on the Loughcrew Hills is situated in the north-west corner of Co. Meath (Grid Ref. N 5978). There are four main hills stretched roughly east-west over a distance of about 3 km. Most of the tombs are situated on the summits with three sites on the lower slopes (Fig. 6). The most westerly hill has eleven monuments on its summit, referred to as $A_2, B, C, D, E, G, H, I, J, K$, and L. Monuments B, C and G are so badly destroyed that it is impossible to be certain of their features, and so they are excluded from the analyses. Cairn M stands isolated on the summit of a lower hill, while O, N and Q are on lower slopes. On the central and highest summit, Sliabh na Calliaghe, are seven monuments R_1, R_2, S, T, U, V and W. Cairn T is very much larger and more impressive than any of the other sites. Farther to the east are two sites at X and Y. X is badly destroyed and only the monuments on the eastern summit (Sliabh na Calliaghe) are visible from it. Y is now in an area covered by a forestry plantation and it was not possible to find the site: hence it was excluded from the analyses. Most of these sites were excavated in the mid-nineteenth century by Eugene Conwell (1866a; 1866b). The finds include a distinctive range of grave goods such as beads, stone pendants and bone pins closely comparable to the small number of objects recovered from the Newgrange chamber. Cairns F, H, I, L, S and T show architectural features of cruciform or multiple recess chambers that are indistinguishable from Boyne Valley sites. Some sites have typical passage grave art.

Application of criteria. It was a difficult task to decide which monuments should be regarded as dominant and which should be regarded as a 'satellite'. After consideration of all the factors it was decided that under criterion 5, site G was subsumed by site F, sites J and K by site L and sites U and W by site T.

Orientation of alignments on astronomical phenomena. There are 64 admissible alignments that give four solar hits and one lunar hit. The declinations are shown in Table 5 along with monument accuracy and line classification. Line TN is somewhat unreliable, as N is about $4°$ below the horizon. Monument Q figures very prominently in these results but the azimuths given in Table 6 illustrate the difficulty of assigning monuments into clusters.

QD	267°	53'	QF	269°	04'	QH	270°	51'
QI	268°	40'	QL	272°	14'			

TABLE 6. Azimuths from Q to other monuments.

Lines QI and QF can be readily grouped under criterion 4 or 5 and the mean declination of the two lines would eliminate them as a significant alignment. Lines QH and QL are separated in azimuth yet they define exactly the same phenomenon. Monument L is much larger and more impressive than H and so it is considered fair to pair H and L and use them as one solar alignment. For the same reasons alignments QD, QF and QI should be grouped as a single line, giving a total of two solar hits in 58 alignments. It is very obvious that neither the solar nor lunar hits have any statistical significance.

A χ^2 test showed rejection of the rectangular distribution for the declinations with (χ^2 (n=64) = 772·16; $p < 0·005$). A histogram of the declinations reveals this fact quite immediately with the largest concentrations of values at about $-6°$ and $+6°$ and with all but four declinations between $-18°$ and $+18°$. These values represent an azimuth range of approximately 60° to 120° and 240° to 300°.

Gnomon angles. The gnomon angles at Loughcrew for the solstices and the equinox are <29° 45′, 60° 15′>, <77° 45′, 12° 15′> and <53° 15′, 36° 45′>. In the whole site there are 209 admissible and 451 inadmissible angles with hits on 13 (6·2%) and 37 (8·2%) gnomon angles respectively. Thus the number of admissible gnomon angles is less frequent than the inadmissible ones. No one site had subtended a significantly greater number of gnomon angles than any other site.

A feature that is not reflected in these results, as at Boyne Valley and Carrowkeel, is the dominance of peripheral monuments as the points with gnomon angles. The frequency histograms show distributions different from the other two sites, especially for the monuments M, N, O, Q, T and X. The reason for such discrepancies appears due to the topographical siting of most monuments in clusters on the tops of hills.

Right angles. There are seven right angles in the whole scheme, none of which is admissible. These are $DAQ, ADT, ADI, IHF, FIL, LIO, QLI$. There are 209 admissible and 451 inadmissible angles in the scheme, and so seven right angles represent 1·6% of the total inadmissible angles. A visual inspection of the angle distribution histogram shows that around 90° is the region of the smallest density of angles; thus it seems unlikely that these angles were set out deliberately. All the right angles occur at monuments on the west hill.

Line lengths. There are six groups of equal length lines, but some of these groups are meaningless as they represent pairs from very distant monuments to the monuments on the westerly hill that are close together. The two pairs of lines AF, DL and DF, LI are inadmissible. The other two pairs LM, MQ and OQ, QT would seem to have some importance, as L, M and T are on the tops of hills, whereas O and Q do not seem to be situated in any position of topographical importance.

Parallel lines and collinear points. The classification of lines into parallel and collinear categories proved to be the most difficult aspect of this work. Indeed, many of the criteria were formulated by attempting to make sense of the results from these data. There are six distinct groups of similar azimuths over the range from 0° and 100°. Some groups are distinguishable because their lines are radial from one station to other distant stations that are close together. The elimination of such lines as an artefact of any underlying deliberate choice is a purely subjective decision, yet without such a move there are so many alternative characteristics that a definitive geometric solution would be meaningless. The line groups amalgamated into single lines were *IT*, *DT* and *FT*; *HT* and *LT*; *DQ*, *IQ* and *FQ*; *IO* and *FO*; and *IF* and *FN*.

The line group of *NT* and *AI* is one of the few sets of lines not consistent with the topographical siting of the monuments. However, *A* and *I* are not intervisible by a strange quirk of the local topography and thus it is an inadmissible alignment. All the sets of parallel lines using only monuments on the west hill are unreliable, as the azimuths can not be determined to much better than 1°. This eliminates the groups of *QT*, *HL*; *DL*, *MO*, *NX*; *MT*, *DH*; *QX*, *MX*; and *MQ*, *DF*.

The remaining admissible groups are A_2T, *OQ*; *DT*, A_2Q; *LHT*, A_2O, *MN*; A_2N, *DQ*. No groups form any type of parallelogram as evidenced at the Boyne Valley and Carrowkeel, and they show no form of correlation with any other geometry. Indeed, most of these groups consist of lines radiating from the western group of monuments and this simply reinforces the strong grouping of the monuments.

Conclusions

All the astronomical and geometrical evidence illustrates strongly that the monuments are not positioned randomly. However, the site shows no simple geometric properties that encompass most of the monuments. The application of Occam's razor recommends adopting a simple alternative to a geometric design, that is, the siting of the monuments is based on topographic features. The monuments with a few exceptions were deliberately sited on the hilltops.

Overall conclusions

In the Boyne Valley there is a reasonable amount of evidence to suggest that monument *X* is the focal point of a design plan with astronomical and geometrical characteristics. This evidence could be strengthened if random models of the site were simulated and the frequencies of the previously specified geometry established. Excavation of monument *X* would lead to a better determination of its centre and yield valuable information about the character of the monument that was there.

At Carrowkeel, evidence drawn from this analysis is very inconclusive as far as defining a geometric design for the site. Four of the thirteen monuments are not incorporated in the predominant geometry. There seems to be an analogous lay-out of astronomical alignments between this site and the Boyne Valley, yet these alignments can be partially explained by the positioning of the monuments on prominent peaks. There is one unambiguous occurrence of three points being collinear and the deliberates of this arrangement is enhanced because it involves C-D, whose topographic position is inconspicuous. The number of gnomon angles in the plan is no greater than could be expected by a random distribution of the angles. Astronomical orientation of the site is not a dominant part of the plan even if the few alignments are deliberate.

Unlike the Boyne Valley there is no suggestion that one monument is a focal point for a design plan while the major similarity is the presence of parallelograms at both sites. The use or need for this type of geometry is not revealed in this analysis and it does not seem possible to offer a comprehensive solution with the existing information.

At Loughcrew the number of astronomical alignments that indicated theoretical declinations were so few as to be inconsequential. The gnomon angles were no more frequent than would be expected under random distributions, as was the case for right angles. There are very few lines of equal lengths and no geometry (although two isosceles triangles are apparent). The groups of parallel lines illustrate no extra geometric features and the azimuth density between 80° and 100° reflects the simple fact that half the monuments are grouped onto one hilltop. No comprehensive geometric design plan could be postulated and all the evidence supports the very apparent fact that the monuments were deliberately sighted on the hilltops.

It has been shown that the Boyne Valley monuments are probably deliberately laid out to a design plan. The evidence from Carrowkeel is very ambiguous with inconclusive results. At Loughcrew it is virtually certain that the monuments were deliberately sited on the tops of hills with little, if any, consideration for a geometrical or astronomical plan. Thus the research has not been able to fulfil the criterion that archaeologically similar sites should display similar astronomical and geometrical patterns. There are two courses of investigation that could clarify this situation: to seek sound evidence that the Boyne Valley monuments were built later than the Carrowkeel and Loughcrew sites; and to excavate monuments X and Y in the Boyne Valley to verify that they are man-made structures.

References

Barber, J. (1973). The orientation of the recumbent stone circles of the south-west of Ireland. *Journal of the Kerry Archaeological and Historical Society* **6**, 26–39.

Conwell, E.A., (1866a). Examination of the ancient sepulchral cairns on the Loughcrew Hills, County of Meath. *Proceedings of the Royal Irish Academy* **9**, 42–50.

Conwell, E.A., (1866b). On ancient sepulchral cairns on the Loughcrew Hills. *Proceedings of the Royal Irish Academy* **9**, 355–79.

De Valera, R. and Ó Nualláin, S. (1961). *Survey of the megalithic tombs of Ireland, vol. 1: Co. Clare.* Dublin.

Hawkins, G.S. (1968). Astro-archaeology. *Vistas in Astronomy* **10**, 45–88.

Heggie, D.C. (1981). *Megalithic science.* London: Thames and Hudson.

Macalister, R.A.S. (1912). Report on the exploration of Bronze Age cairns on Carrowkeel mountain, Co. Sligo. *Proceedings of the Royal Irish Academy* **29C**, 311–47.

O'Kelly, C. (1971). *Illustrated guide to Newgrange.* Wexford: J. English.

Patrick, J.D. (1971). Preliminary investigation of the astronomical and geometrical significance of Newgrange and its auxiliary mounds. Unpublished manuscript.

Patrick, J.D. (1974). *Investigation into the astronomical and geometrical characteristics of the passage-grave cemeteries at the Boyne Valley, Carrowkeel and Loughcrew.* MSc thesis, University of Dublin.

Patrick, J.D. (1979). A reassessment of the lunar observatory hypothesis for the Kilmartin stones. *Archaeoastronomy* no. 1 (supplement to *Journal for the History of Astronomy* **10**), S78–85.

Patrick, J.D. (1991). *Astronomical and geometrical designs of Irish passage grave cemeteries.* Technical Report 5/91, Department of Computing and Mathematics, Deakin University, Geelong, Australia.

Powell, T.G.E. (1938). The passage graves of Ireland. *Proceedings of the Prehistoric Society* **10**, 239–48.

Ruggles, C.L.N. (1984). *Megalithic astronomy: a new archaeological and statistical study of 300 western Scottish sites.* Oxford: British Archaeological Reports (BAR British Series, 123).

IV

CONTINUING RESEARCH: NEW RESULTS

19

Sun and sun serpents: continuing observations in south-eastern Utah

CAROL W. AMBRUSTER AND RAY A. WILLIAMSON

Introduction

The Holly group is one of several canyon-head ruin complexes at Hovenweep National Monument on the Colorado-Utah border. These masonry remains of the prehistoric Anasazi civilisation date largely to late Pueblo III times (AD 1200–1300). Location near the heads of canyons meant the availability of a dependable source of water in the form of run-off, seeps and springs. Various water control methods, such as check dams, had been developed, probably during Pueblo II times (AD 900–1100), to accommodate both an expanding population (Winter 1975: 105) and an increasingly dry climate (Winter 1977: 245; Weir 1977: 276–77).

The Holly ruins are located at the head of a canyon that opens approximately south; tree-ring dates range between AD 1200 and 1267 (Winter 1976: 288). There is also evidence in the immediate vicinity of occupation during Archaic and Pueblo I times that suggests a long history of (probably intermittent) human habitation (Winter 1975: 26, 85, 205).

Located some 150m south of the main Holly ruins, the petroglyph panel itself covers about half of the southern wall, some 7m long by 3m high, of a natural east-west rock corridor about 1m wide. Granaries, or walled-in natural rock cavities used for storage, are located at both ends of the great boulder forming the north side of the corridor; other granaries are found in the immediate vicinity. The west end of the corridor opens out onto a talus slope containing prehistoric farm fields and three check dams (Olsen 1985: 75). Prior to the recent arrival of large numbers of visitors, the ground a few metres downhill from the panel was littered with lithics and potsherds, proof that a variety of activities took place in the immediate vicinity.

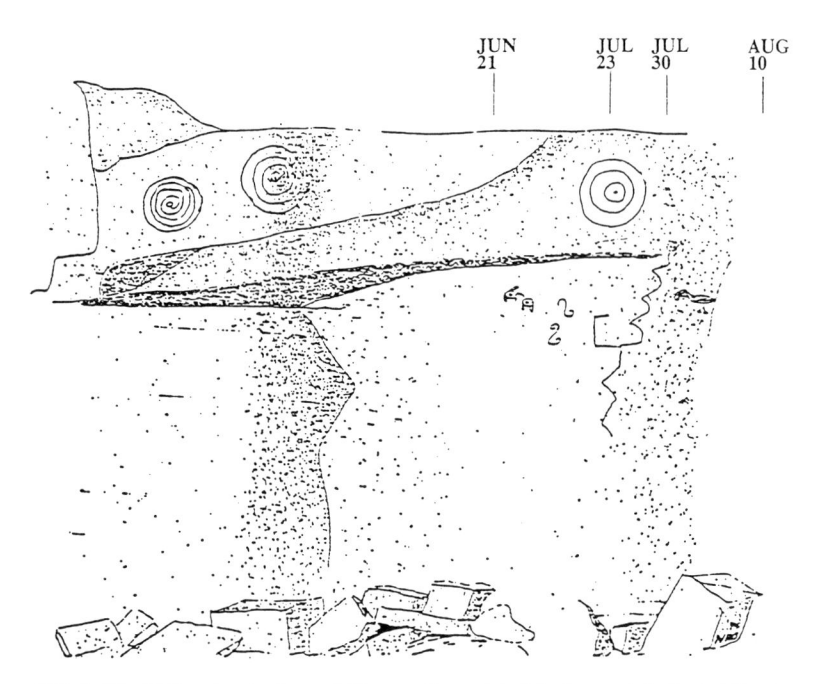

FIG. 1. The Holly petroglyph panel. The meeting point of the beams is indicated for various dates discussed in the text. Daily, or almost daily, changes are easily discernible, except at the solstice. (Adapted from Olsen 1985: 78.)

The panel, shown in Fig. 1, consists of the following pecked figures: two spirals, 32 cm and 36 cm in diameter, a concentric circle 33 cm in diameter, a vertical zig-zag glyph about 1m long, two tiny joined circles with central dots and vertical lines at the bottom, a bird, two 'S'-shaped glyphs, and a faint rectilinear abstract pattern. The two spirals and the concentric circle are about 2·5m above the ground, and were probably pecked from a ladder or some sort of scaffold; they appear to have been positioned intentionally to intersect the horizontal, serpent-like sun patterns. The other glyphs, with the exception of the top of the zig-zag glyph, could have been pecked by a person standing on the ground.

The summer solstice sunrise display is fairly well known and quite dramatic (Williamson 1984: 92–103; 1987). Narrow slits of light appear on the panel, coalesce into two 'sun-serpents', and then slowly edge toward each other, touch, and merge (Fig. 2). The meeting points range from about 92 cm east (left) of the concentric circle at summer solstice, to about 110 cm to the west (right) of the concentric circle 1·5 to 2 months on either side of the summer solstice. The entire sequence takes between seven and twenty minutes, depending on the date.

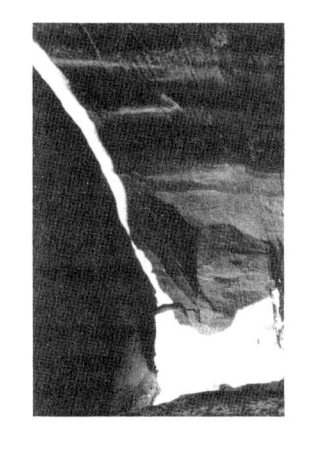

Fig. 2. Excerpts from the summer solstice sunrise sequence in 1984.
Upper left: Development of the left 'serpent' shortly after first light on the panel.
Upper right: Fragmentary beams 2–3 minutes later.
Lower left: Left and middle beams meet, right beam has just passed through the centre of the concentric circle.
Lower right: Meeting of the 'sun serpents', 92 cm left of the centre of the concentric circle. The complete sequence takes seven minutes.

The 'sun-serpents' are caused when the rising sun shines through a narrow gap between the overhang of the great boulder forming the south side of the corridor and the slightly lower top of the north boulder. For about two months on either side of the summer solstice, this sense of the meeting of two 'sun serpents' is maintained, though with intriguing and easily discernible variations. The sunrise display around the equinoxes is qualitatively very different, and will be discussed elsewhere.

A calendar?

For a site to function as a calendar, discernible changes over short time periods, at most a few days, must be apparent to the unaided eye; it has been suggested that, in practice, this means a linear motion (for one-day resolution) of 0·5 to 1 cm per day (Zeilik 1989: 152). We can now establish that this criterion is easily met by daily changes in the light pattern at Holly.

At the summer solstice, the light beams meet about 92 cm east (left) of the centre of the concentric circle. In contrast, on Jul 23, 32 days after summer solstice, the beams meet just below the central dot of the concentric circle (Fig. 3); the average daily motion between the two dates, then, is 2·8 cm per day. In this and other measurements below, we have taken into account the off-centre positioning of the central dot of the concentric circle.

Between Jul 23 and Jul 28 the meeting place of the beams moves from directly below the centre of the concentric circle to about 10 cm west (right) of its outer edge, resulting in a change of approximately 23 cm in five days, or about 4·5 cm per day, consistent with the more rapid horizon movement of the sun a month away from the solstice. We further confirmed this with our sequences of Jul 28 and Jul 30, two days apart, where the change in the meeting point can be seen clearly both in photographs and visually. We note that the sandstone of the panel surface is not perfectly smooth, so the meeting point of the beams does not progress smoothly from one day to the next.

The overall motion of the meeting point of the beams, then, is symmetric about summer solstice, moving from about 110 cm to the right of the centre of the concentric circle around May 1, to about 92 cm left of its centre at summer solstice. The meeting points then progress westwards (rightwards) again until about Aug 10, when the beams meet again about 110 cm to the west of the concentric circle.

By Aug 23, the meeting point has retreated east (left) a little (to about 90 cm west of the concentric circle) and the sunrise display ends with the little joined circles glyph below and east of the concentric circle cradled in an arc of light. By Aug 29, the display is in transition: the 'serpents' disappear; first light hits the panel in and to either side of the concentric circle, and spreads mostly east (left) in a shape most reminiscent of a large wing.

Sun and sun serpents 223

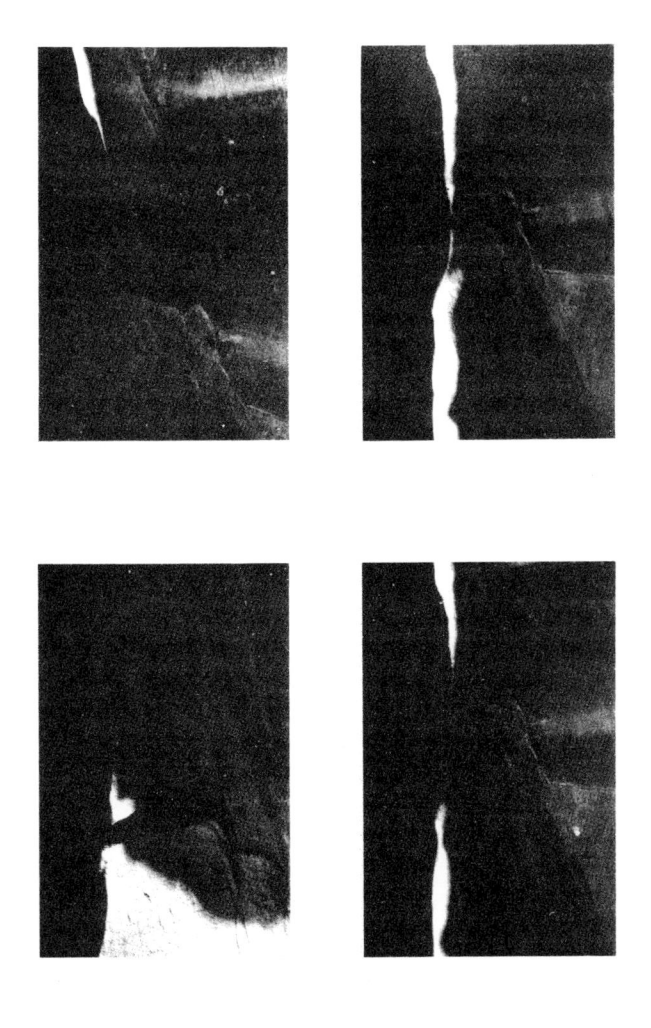

FIG. 3. Excerpts from the sunrise sequence for July 23, 1989.
Upper left: Shortly after first light on the east (left) part of the panel: the beam passes above the left spiral.
Upper right: Subsequent first appearance of the right beam to the west (right) of the concentric circle.
Lower left: The beams approach the concentric circle.
Lower right: The beams meet just below the centre of the concentric circle.

There can be little doubt that the patterns of sunlight on the panel today are, in their basic shapes and motions, the same as those the Anasazi saw in the thirteenth century. This is because the 'sun-serpents' are an artefact of the narrow gap between the overhanging southern boulder and the slightly lower top of the northern boulder. However, specific details of the shapes are probably not exactly the same because thin sheets of sandstone, roughly 10–20 cm on each side, have spalled off the underside of the overhanging southern boulder. Our observations suggest, for example, that the sharp angles suggestive of the wing tip of a large bird in the Aug 29 sunrise occur because of the interaction of the incoming sunlight with one of these separating sandstone sheets.

In subsequent seasons, we will investigate in depth the role the spalling sheets of sandstone play in modulating the basic shapes on the panel. In particular, it may be that the exact day at which the beams meet at a certain point is determined by the particular configurations of flaking sandstone. We also will test our suspicion that the main reason why the two beams meet 1–2 cm below, rather than exactly in, the centre of the concentric circle is a slight change in the height of the 'sun serpents' above ground level due to the cumulative effect of eight hundred years of spalling. If this is confirmed, it would also explain why, at other times, the 'sun serpents' always seem to pass the same 1–2 cm below the centre of the circle.

We emphasise, however, that the daily changes in the pattern of the sun on the panel, as well as the 2·8–4·5 cm per day daily progression of the meeting point of the 'sun serpents', would have been present in AD 1200 because they are determined by the basic shape and positioning of the two massive boulders which define the horizon as seen from the panel.

Another issue we have considered is the apparent absence of tick marks or markers of any sort on the panel to denote important dates. Extensive searching has produced no evidence for any marks other than those described above and indicated in Fig. 1. It is possible that there were painted marks which have since been destroyed by the water erosion visible in several places on the panel, most obviously in the eroded spiral. Or, and we lean towards this possibility, it may be that the interpretation of the panel depended not only, or even mainly, on the meeting point of the beams, but on daily changes in the overall patterning of the light display. That is, the determination of calendrical information was coded in the changing shapes of the display rather than in the specific meeting point of the beams. There would have been no practical way to mark the fluid, changing shapes on the panel, and the Anasazi, like their Pueblo descendants, may have passed down the interpretation of the panel to initiates by careful and extended apprenticeship and training.

The petroglyphs

The sunrise display at Holly is present throughout the agriculturally important spring, summer and autumn months. While the decision on when to plant a given crop was certainly made by taking into account other factors besides the position of the sun, such as temperature, weather, and the presence of certain wild flowers, the times of ceremonies to aid the success of the crops was likely to have been consistent from year to year.

The main glyphs on the Holly panel appear to be connected with the two essentials of agriculture, and of life itself: sun and water. The concentric circle in south-western rock art is widely accepted as a sun symbol (Ellis 1975), though it may not signify the sun in all contexts (Young 1986). And Olsen (1977) notes the association of spiral petroglyphs with water control features: spirals are present 75% of the time in eight Hovenweep sites with water control features, and absent 100% of the time from seven sites without water control features. The canyon-head location of the Holly ruins as well as the various water control features, a rim dam near the head of the canyon, and the three check dams on the slope to the west of the panel, testify to a concern about sufficient water. This association is strengthened by the fact that the climate at Hovenweep was drier between AD 1100 and 1300 than previously, and drier than today (Weir 1977: 277).

Despite uncertainty as to whether the light beams meet on the same dates as they did in AD 1200 (see above), it does seem reasonably clear that the concentric circle was carved in such a way that the beams would meet at its centre at some date. It thus seems a fairly safe assumption that this date had particular significance in the context of their calendar. Significant constraints on this date could serve to constrain the interpretation, at least within the context of a planting calendar. For example, if we are able to show that the meeting point in the concentric circle occurred within a few days of the present date, one possible interpretation might relate to the end of the planting season. For present-day conditions, McCluskey (1982: 37) has found for Hopi, which has approximately the same elevation and rainfall, that May 21—the symmetric date to Jul 23 about the solstice—is the latest time at which crops can be planted and harvested before there is a 50% probability of a hard frost in late summer.

Summary

Using only the most obvious feature of the sunrise sequence, the changing meeting point of the two 'sun serpents', we have documented an average daily motion of 2·8 cm per day over a one-month period on either side of the summer solstice. Except for the days immediately surrounding the solstice, the daily motion exceeds the 0·5–1·0 cm per day threshold of the eye to detect change. One month on either side of the solstice, this motion has increased to 4·5 cm per day. Daily changes in the entire sequence will inevitably occur even more rapidly towards equinox when the change in the sun's rising position on the horizon from day to day is greatest. Thus the panel could easily have been used as a calendar. A better understanding of the context of the panel must wait until the site is excavated.

References

Ellis, F.H. (1975). A thousand years of the Pueblo sun-moon-star calendar. In *Archaeoastronomy in pre-Columbian America*, ed. A.F. Aveni, pp. 59–87. Austin TX: University of Texas Press.

McCluskey, S.C. (1982). Historical archaeoastronomy: the Hopi example. In *Archaeoastronomy in the New World*, ed. A.F. Aveni, pp. 31–57. Cambridge: Cambridge University Press.

Olsen, N.H. (1977). Hovenweep rock art and agriculture. In *Hovenweep 1976*, ed. J.C. Winter, pp. 285–87. Archaeological report no. 3, Anthropology Department, San José State University, San José CA.

Olsen, N.H. (1985). Hovenweep rock art—an Anasazi visual communication system. Occasional Paper 14, Institute of Archaeology, University of California at Los Angeles.

Weir, G.H. (1977). Paleoenvironment. In *Hovenweep 1976*, ed. J.C. Winter, pp. 246–78. Archaeological report no. 3, Anthropology Department, San José State University, San José CA.

Williamson, R.A. (1984). Living the sky: the native American cosmos. Boston MA: Houghton Mifflin.

Williamson, R.A. (1987). Light and shadow, ritual, and astronomy in Anasazi structures. In *Astronomy and ceremony in the prehistoric Southwest*, eds. J.B. Carlson and W.J. Judge, pp. 109–14. Albuquerque NM: Maxwell Museum Press (Papers of the Maxwell Museum of Anthropology, no. 2).

Winter, J.C., ed. (1975). *Hovenweep 1974*. Archaeological report no. 1, Anthropology Department, San José State University, San José CA.

Winter, J.C., ed. (1976). *Hovenweep 1975*. Archaeological report no. 2, Anthropology Department, San José State University, San José CA.

Winter, J.C., ed. (1977). *Hovenweep 1976*. Archaeological report no. 3, Anthropology Department, San José State University, San José CA.

Young, M.J. (1986). The interrelationship of rock art and astronomical practice in the American Southwest. *Archaeoastronomy* no. 10, (Supplement to *Journal for the History of Astronomy* **17**), S43–58.

Zeilik, M. (1989). Keeping the sacred and planting calendar. In *World archaeoastronomy*, ed. A.F. Aveni, pp. 143–66. Cambridge: Cambridge University Press.

20

The origin and meaning of Navajo star ceilings

VON DEL CHAMBERLAIN AND POLLY SCHAAFSMA

Introduction

The rock art motifs of interest in this paper are pictographic depictions of stars, in the form of crosses, on the ceilings of rock overhangs and rock shelters. Some are very simple, containing only one or a few faded star-crosses. Others are impressive and complex with large numbers of stars.

Star ceilings have been described in numerous publications (Britt 1975; Chamberlain 1974; 1978; 1983; 1989a; 1989b; De Harport 1951; 1953; 1959; Grant 1978; Hester 1962; Jett 1984; Schaafsma 1963; 1966; 1975; 1980). To date, we have studied sixty star ceilings, and undoubtedly many more exist. They are found in scattered locations throughout the area of the Colorado Plateau occupied by Navajos in the eighteenth century. The greatest concentration is in the Canyon de Chelly area in the upper Chinle drainage in north-eastern Arizona. They also occur on an individual basis in various other localities in north-western New Mexico, southern Colorado and extreme south-eastern Utah.

The Navajo origin of these star ceilings is implied by oral traditions and ceremonial practices of Navajos (Britt 1975; Jett 1984), and by the contexts of star ceilings with other Navajo ritual rock art.

In this paper, we feature three star ceiling locales of particular significance in our attempt to understand the origins of star ceilings and the traditional significance they might have.

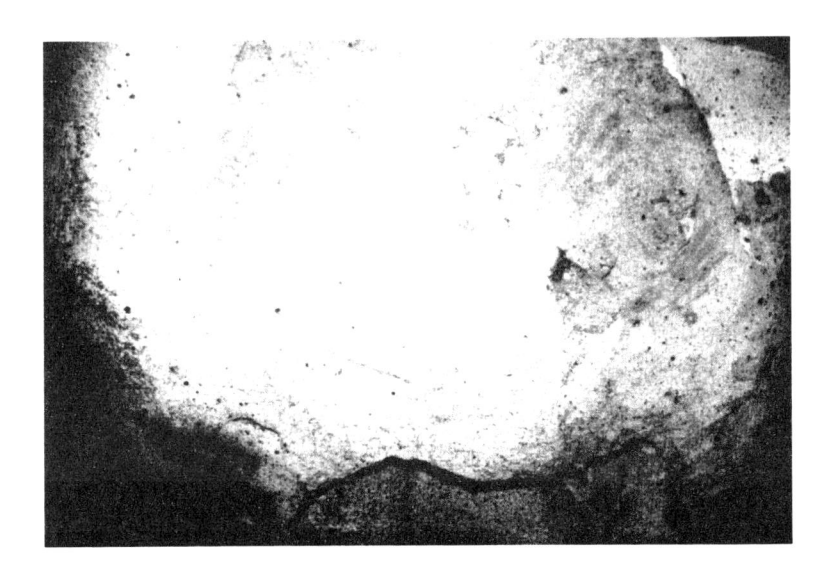

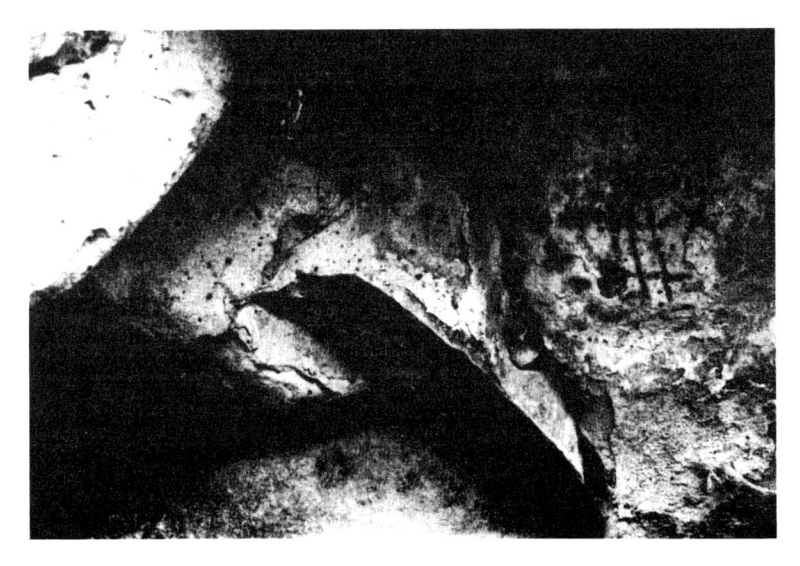

FIG. 1. The Dulce, New Mexico star ceiling. (a) Red hand-prints and a single star-cross, about 7cm across, decorate the ceiling standing over a large sun symbol (not shown). (b) Several positive and negative crosses are found on a section of the wall at the back of the rock shelter. (Photograph by the authors.)

The star ceilings

The Gobernador Phase star ceilings

The Navajo Reservoir District in the upper San Juan drainage in northwestern New Mexico was occupied by Gobernador Phase Navajos during the Pueblo Refugee Period between AD 1696 and 1720 (Carlson 1965: 100–1). The Navajo rock art made during that period was ceremonial in content. The hallmark of the Gobernador Representational style is the Navajo supernatural, or *ye'i*. One of two star ceilings, now under the waters of the Navajo Reservoir, was in Todosio Canyon. Immediately beneath the overhang painted with star-crosses and flower forms were some of the finest *ye'i* in the District (Schaafsma 1963: 42–44; pls. 1–3 and fig. 42), an association that leaves no doubt about the ethnic origins of the star ceiling.

Another star ceiling in the upper San Juan area near Dulce, New Mexico, has both positive and negative stencilled red crosses along with a Born-For-Water hour-glass shaped sign, or scalp-knot, the symbol of one of the Navajo twin war heroes (Fig. 1). The ceiling, standing over a large sun or sun shield, contains several hand prints and one star-cross.

The Penistaja star ceiling

The Penistaja star ceiling is in a deep south-west–facing rock shelter located about 30 km outside of the town of Cuba, New Mexico. The site has been used for a long time as a shelter and watering place for animals, and the growing amount of multi-coloured graffiti on the lower rocks indicates that it remains a popular place among current area residents.

The star ceiling consists of a variety of elements painted on various levels of the roof of the cave, a more varied set of pictographs than in any other star ceiling we know of (Fig. 2). There are the usual star-crosses in red and black. There are single bars in black and red; black crosses with ends tapered towards one edge, resulting in decidedly birdlike forms; black circular 'loops'; double-barred crosses resembling dragonflies; and red triangular birds with voided centres. Finally, there are tiny red figures of people on horseback carrying long lances and a rather generalised animal that could represent a bear. The men on horseback confirm the historic origins of these paintings.

Unlike the other elements, the complex horsemen, located on a relatively low ceiling in the rear of the shelter, are hand painted. Most of the other figures, including the birds with voided centres, are on high

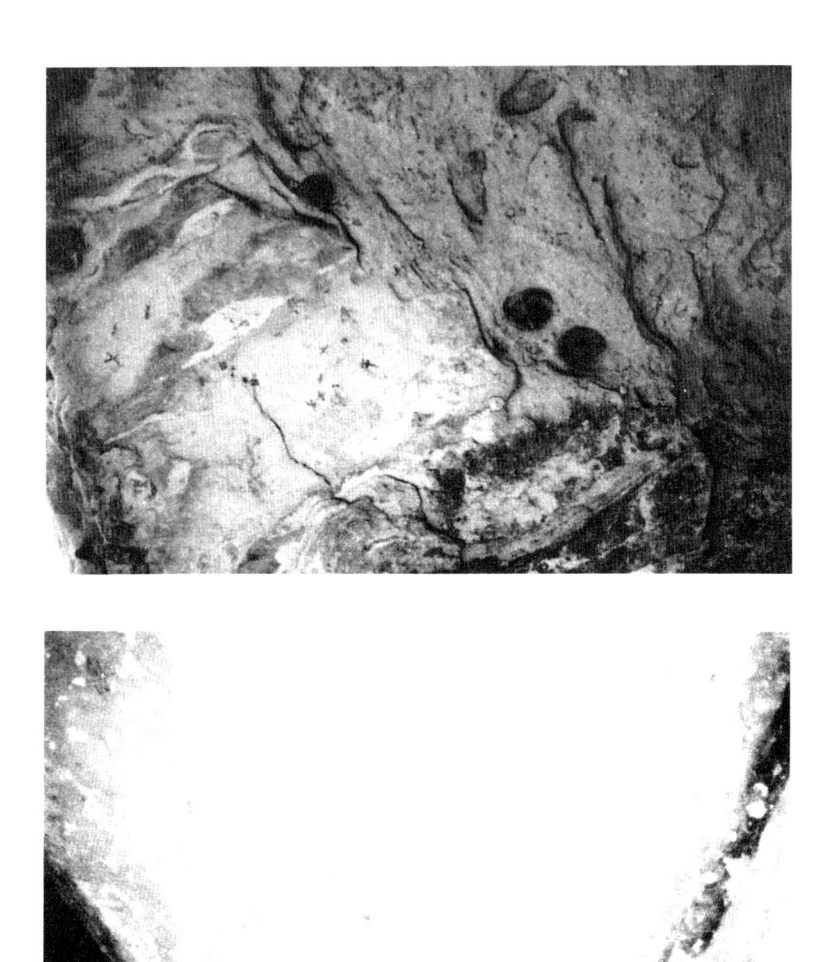

FIG. 2. The Penistaja star ceiling. (a) A portion of the high ceiling in this large rock shelter cave has black and red figures including crosses, dragonfly forms and birds. The figures vary in size from about 6cm to 12cm. (b) A lower section of the ceiling has crosses and other forms, including tiny red figures of people with lances on horseback (upper right). (Photograph by the authors.)

inaccessible surfaces and appear to have been printed on these sandstone overhangs by means of stamps shot as arrows (Jett 1984). The total inaccessibility of most of these figures, twenty metres high and far inside the lip of the overhang without any possible means of support for scaffolding, suggests the projectile method of placing the figures on the rock.

The San Cristobal star ceiling

A single known significant exception to the star ceilings of apparent Navajo origin is located in a rock outcrop just north of the ruin of San Cristobal Pueblo in the Galisteo Basin near Santa Fe, New Mexico. This star ceiling occurs in the context of a large Rio Grande–style rock art site in a locality in which star iconography is emphasised. Many of the petroglyphs are four-pointed stars. The rock art in question was made between AD 1325 and 1680 by the Southern Tewa people, also known as Tanos, who resided at San Cristobal.

Under a shallow overhang with a south-eastern exposure on the pale sandstone ceiling near the lip, at least seven simple red crosses occur in a scattered arrangement (Fig. 3). Some of these are weathered to the point of near invisibility, a factor that suggests that originally there might have been more. The clearest figures with their crisp edges suggest that they were printed with a stamp in the fashion of the Navajo star ceilings. At least one of the crosses has a small voided centre like many of the Navajo crosses. The star-crosses on the San Cristobal star ceiling are virtually identical to those of Navajo origin. Tucked under the same overhang, on vertical surfaces, are seven or more painted masks, or faces, some of which have tall head-dresses of red feathers, giving the distinct impression of Kachina masks. Facial features are pecked, and one has stars for eyes. There is one large feathered white star face with a dark centre outlined in red.

Of particular importance is a nearby large boulder, the sloping face of which is pecked with small sky symbols (Fig. 4). Stars as simple crosses, a star with expanded points and circular centre, spread-winged birds, and a star with a 'tail' that resembles a meteor or a comet are spaced in the manner of a star ceiling. Indeed, the small birds and star crosses here are strongly reminiscent of the Penistaja star ceiling just described.

The star ceiling and boulder covered with sky symbols present convincing evidence for a historical and cultural continuity between Pueblo and Navajo star ceilings. Furthermore, the San Cristobal star-crosses occur in the context of Pueblo IV Rio Grande style rock art, a style ancestral to the Gobernador Representational style of the eighteenth-century Navajo (Schaafsma 1963: 57–60; 1980: 306–7). This is especially

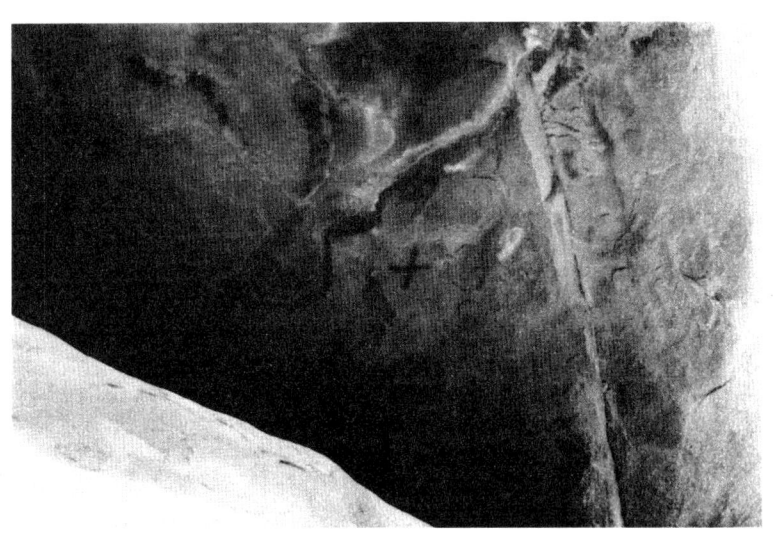

FIG. 3. The San Cristobal star ceiling, located in a rock shelter with Pueblo IV historic rock art. (a) Several faces, or masks, include a star face (lower left-centre) and a face with star-cross eyes (left centre). (b) Ceiling surfaces contain several star-crosses, including the two shown here that are much more distinct than the others. The crosses are about 7cm across and the masks are about life-sized. (Photograph by the authors.)

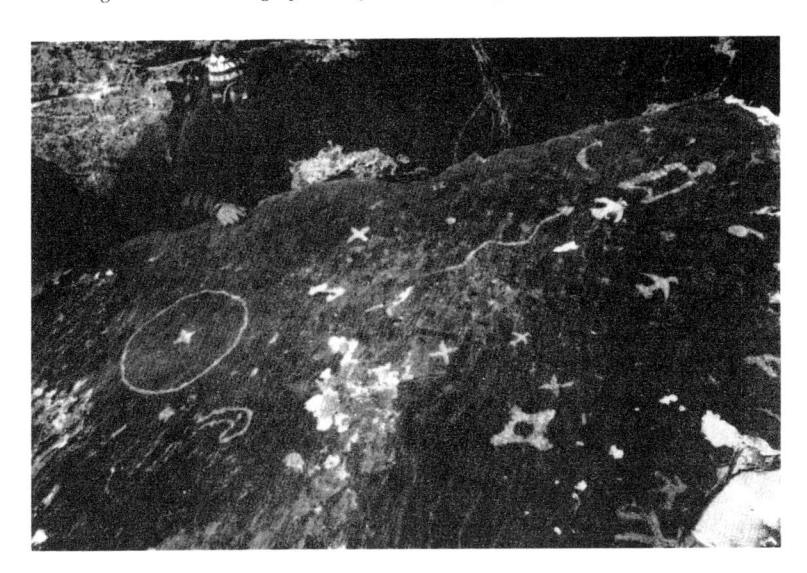

The origin and meaning of Navajo star ceilings 233

FIG. 4. The San Cristobal sky symbol boulder. The top of a large rock, located near the San Cristobal star ceiling, displays sky symbol petroglyphs. Features range from a few up to about 20cm across. (Photograph by the authors.)

interesting in view of stated opinions (Forbes 1960: 263–73; Hester 1962: 22–28) that the people from San Cristobal Pueblo fled to Navajo country in 1696–97.

Significantly, the people who are believed to have painted the Penistaja star ceiling located west of the San Cristobal site are thought to have been a mixed population of Chacra Mesa Navajos and Pueblo Refugees (Carlson 1965: 100; Forbes 1960: 270–2). It is reasonable to propose that southern Tewa contributed to this Pueblo Refugee group. The southern Tewa, or Tanos, had abandoned their Galisteo Basin pueblos of San Cristobal and San Lazaro at the time of the Pueblo Revolt in 1680 and moved to the northern Tewa area north of Santa Fe. During the second Pueblo Revolt of 1696 they were described as having crossed the Rio Grande with loaded horses in the vicinity of Santa Clara as they fled westwards (Espinosa 1988: 275–88). It is known that many of them went to Hopi where they founded the Tewa village of Hano on First Mesa. It is not unlikely, however, that some lingered with the intervening Navajos and possibly even ultimately became part of that population.

To summarise the significance of the Penistaja and San Cristobal star ceilings, we note that clusters of star depictions of Pueblo origin, either pecked as petroglyphs or painted on the ceilings of rock shelters in the manner of Navajo star ceilings, have not been described before. The presence of both in one section of the extensive rock art site near the southern Tewa Pueblo IV village of San Cristobal in the Galisteo Basin south of Santa Fe, is of considerable interest and was the initial motivation leading to this paper. The presence of a star ceiling at San Cristobal and the similarity between the San Cristobal boulder with the stars and birds and the Penistaja ceiling, which was near the eastern border of Navajo occupation around AD 1700, suggest that San Cristobal refugees specifically were amongst the Refugee population in north-western New Mexico. These factors also suggest, of course, that the practice of making star ceilings started with Pueblos and was adopted by Navajos.

Summary of research on the meaning of star ceilings

The San Cristobal panel indicates that Navajo star ceilings fit into the ceremonial context of other Navajo rock art that was strongly influenced by Pueblo contact following the Pueblo Revolts at the end of the seventeenth century. At this time, large numbers of Pueblo Indians fled the Rio Grande Valley to escape the Spanish, and joined their Navajo neighbours living in the wilds of north-western New Mexico. Although Pueblo star ceilings are extremely rare, the practice of painting—or, more correctly, stamping—stars on sandstone overhangs and ceilings of rock shelters persisted amongst the Navajos. While the easternmost examples date from the early 1700s, or the Refugee Period, those in Canyon de Chelly probably date from the late eighteenth to early nineteenth centuries for the most part. The concentration of star ceilings in the Canyon de Chelly area suggests that ritual practices involving the making of star ceilings assumed greater importance amongst Canyon de Chelly Navajos than among Navajos in other regions. In order to understand the meaning of these rock paintings, we have explored Pueblo star-cross symbolism, along with Pueblo IV and protohistoric rock art and other star-cross symbolism for features that are similar to star ceiling motifs.

The star-cross motif is very common in Pueblo as well as Navajo iconography. It is found on Hopi pottery (Fewkes 1898: 724, pls. CXXVI, CXXX, CXXXVI, CXXXIX, CL, CLVI, CLXII, CLXVII; 1919: 242, 257–58), on many Kachina figures (Fewkes 1903: pls. IV, V, VI, XXVIII, XXXI, XXXVII, XL, XLVIII, XLIX; Colton 1959: 18, 20, 21, 44, 54, 55, 66, 67, 69, 116, 118, 119, 126, 131, 139; Wright 1977: 32–33, 40–41, 44, 94–95, 112–14; Dockstader 1985: 39, 47, pls. V, VI, VII, VIII), on woven items (Dockstader 1987: 38, 43, 67, 68, 70; James 1974: vi, 2, 97, 98, figs. 22, 23, 32, 36, 201, 233; Rodee 1981: pls. 1, 10, 11, 12, 13, figs. 11, 14, 49; Kaufman and Selser 1985: 33, 42, 57, 58, 62, 68, 81; Blomberg 1988: 42, 43, 59, 63, 67, 80, 104, 106, 107, 111, 113, 117, 130, 138, 161, 172, 189, 191, 201, 217, 219, 220, 239, 240), on sand paintings (Newcomb and Reichard 1975: figs. 5, 9, pl. XVII; Reichard 1977: fig. 2), and on ceremonial kiva art (Hibben 1975: fig. 45); and it is also found in Anasazi rock art (Chamberlain: slide files).

There are many Pueblo associations between birds, stars and war (Schaafsma 1992), and certain continuities of thought between Navajo and Pueblo concepts of stars in general. There are numerous indications from mythology and ceremonial practices that, for the Navajo, stars have a great deal of protective symbolism. Thus it is suggested that stars painted or stamped on the ceilings of sandstone rock shelters were perceived to have the power of protective magic (Chamberlain 1989a: 337–39). One of the characteristics that the vast majority of them seem to have in common is that they occur in places which were used by Navajos for such things as storage and for animals (Hester 1961: 101).

Stars could have been placed on ceilings for any or all of the following protective reasons.

- Some star ceilings might have been made for protection of shrines.

- Some of the most extensive star ceilings overlook Anasazi ruins, and it is well known that Navajos tend to be wary of places where ghosts or *chindi*, malevolent spirits of the dead, might lurk. The effects of these spirits would be perceived as dangerous to either animals or people, so the star ceilings might have been made to protect from such influences.

- Navajos do not typically reside in the rocks as the Anasazi did, yet they might have found this necessary at times, and the practice of making star ceilings might have 'made it safe' for them to be there.

- Contemporary Navajos have said that star-crosses keep the rock from falling (Chamberlain 1989a: 337).

- The ceilings might have been involved in ceremonies relating to the threat of enemy invaders, both ancient and contemporary, into the canyons where the people lived.

- The placement of stars on ceilings might have been ceremonially related to the mythological placement of stars in the sky by Holy People, an act that had a great deal of protective symbolism. Indeed, in referring to Black God, creator of the stars, and the other Holy People involved, Ethelou Yazzie (1971: 21) wrote, '...they instructed the stars to guard the sky and man'.

There are possible relationships between the star ceilings and the Great Star Chant (McAllester 1956), a Navajo Evil Way ceremony (Sandner

1979: 46, 147; McAllester 1956: 121) seldom practised today. The Chant had its origin in dangerous conflicts and in the need for returning balance and harmony to those who had engaged in unsafe activities. It was and is practised for the divination of evil influences, and the casting out of evil powers resulting from such things as handling the dead, fear of the night and darkness, the powerful influences of the stars, and the presence of aliens (Sandner 1979: 46, 50, 146; McAllester 1956: 38, 121). One of the recent uses of the ceremony has been for veterans returning from service in World War II. There are also instances in which the Great Star Chant was held to banish any sort of bad luck. McAllester (*ibid.*: 118) described this ceremony being held after a rock shelf had fallen and killed some sheep. The protection against outsiders (aliens) provided by the Great Star Chant and the use of this ceremony to combat the bad luck signified by a rock fall are functions compatible with those suggested earlier for the star ceilings. In other words, stars on the sandstone overheads are said to *prevent* the rock from falling in the first place. They may also have served a more generalised protective function against outsiders: Utes or Caucasian invaders into the canyon system, or the previous residents, the Anasazi, whose remains are everywhere in evidence, often within the same rock shelters and alcoves.

The prayers in the Great Star Chant request protective help from the birds and from the stars and for safety and escape (McAllester 1956: 47, 79–80, 90). The combination of symbolic elements—stars, birds, rock, danger, threat and war—and protection suggests that star ceilings might well have been involved in early renderings of ceremonies similar to, or even related to, the Great Star Chant. Perhaps the stars were put up on ceilings, sometimes shot there on star-tipped arrows, following the example of Star People in the Great Star Myth, shooting their arrows of light, for the purpose of making people safe from potential illnesses that might be acquired in the rocky canyons where they resided, canyons previously inhabited by the Anasazi.

The key to understanding star ceilings is in the symbols involved and the places where they are found. Reichard (1983: 81; 152–53: 307) discussed the importance of places, noting that '...No one can be sure a place is safe unless he knows its history'. Indeed, she said that place itself is an outstanding symbol. She emphasised (*ibid.*: 112) that the chanter's 'ultimate goal is to identify the patient with the supernaturals being invoked' and that the purpose of symbols is to allow the patient to absorb the powers depicted. Recognising that Navajos seem to have been obsessed with rocks, rock formations and rocky places, we wonder if star ceilings might represent cases where the places are as much the 'patient' as any human participants might have been. Knowing that, in Navajo ritual, identity between patient and the supernatural through symbols, based on myth, is at the heart of the system (Sandner 1979: 17), perhaps it was most important for the places where star ceilings are found to absorb the powers of the depicted stars, to remove unwanted influences, and protect them, so that they would be safe for use.

Star ceilings are most abundant in and around Canyon de Chelly, which functioned as home as well as a place of refuge for Navajos during the late eighteenth and through much of the nineteenth centuries. This was a period of stress, during which time the Navajos were under pressure from the outside by various hostile groups. At the same time, they were undoubtedly exploring ways of dealing with the Anasazi presence and other problems within the canyons themselves. It is hypothesised that all of these factors might have led to the development of rituals aimed at protection, including protection of the type they believed would come from the stars. Such an hypothesis would account for the large number of star ceilings there.

Stars are important symbols in Pueblo and Navajo cosmologies. Newcomb *et al.* (1956: 34) stressed the importance of symbols in the context of sand paintings and ritual:

> When a sand painter wishes to show who or what lived on a certain mountain, or walked in a certain direction, then tracks of that person, animal, or bird will be drawn in the proper position. These track symbols are considered as powerful for exorcism as the sketch of the person, animal or bird would be. Those most often used are pollen tracks of the immortals, star crosses of the sky forces, ...

Finally, as Reichard (1983: 149) put it, for the Navajo '...a symbol that stands for a power is the power'. In other words, a graphic image assumes the power of the object it represents. The star 'tracks' on star ceilings were meant to have the power of the stars themselves, and the meaning of these star symbols is related to the places where they are found.

Conclusions

In conclusion, the equal-armed cross, as a star symbol, is prominent in both Pueblo and Navajo symbolism, and there appears to be a continuity of star symbolism in ritual contexts between these peoples. The San Cristobal star ceiling and the nearby petroglyphs are strong evidence for Pueblo origins of the Navajo practice of making star ceilings. More specifically, the concept of star ceilings could have originated with the Tewa who communicated the practice to the Navajo after 1696. The Penistaja star ceiling, with motifs astonishingly similar to those at the

San Cristobal site, seems to be a transition between the Pueblo IV ceiling at San Cristobal and those in the upper San Juan and Canyon de Chelly.

We have suggested that star ceilings were made in connection with ceremonies or rituals which were performed to make these places safe for various kinds of use by the Navajo. The desired protection might have been from things as specific as the collapse of fragile overhangs, or as broad as protection from alien influences such as the Anasazi, or Mexican, Ute, and Anglo intruders into the canyon systems. The lingering concept of protective symbolism seems to be present in the contemporary Navajo point of view that just as the stars hold the sky together, the power of the stars painted on the ceilings, hold the rocks in place.

Acknowledgements

We wish to thank our many colleagues and friends who have shared their ideas and opinions about rock art related to star ceilings. They are too many for us to name them all. We especially want to mention here the kind and thoughtful discussions we have had with Harry and Anna Walters, Mike Mitchell, Dave Wilson and his family, Johnson John, Chauncey Neboyia, Robert McFearson, Max King and Keith Franklin.

References

Blomberg, N.J. (1988). Navajo textiles: the William Randolph Hearst collection. Tucson AZ: University of Arizona Press.

Britt, C. (1975). Early Navaho astronomical pictographs in Canyon de Chelly, northeastern Arizona, U.S.A. In *Archaeoastronomy in pre-Columbian America*, ed. A.F. Aveni, pp 89–107. Austin TX: University of Texas Press.

Carlson, R.L. (1965). Eighteenth century Navajo fortresses of the Gobernador District. Boulder CO: University Press of Colorado (Earl Morris Papers, no. 2; University of Colorado Studies Series in Anthropology, no. 10).

Chamberlain, Von Del (1974). American Indian interest in the sky as indicated in legend, rock art, ceremonial and modern art. *The Planetarian* **3**, 89–106.

Chamberlain, Von Del (1978). Sky symbol rock art. In *American Indian rock art. Papers presented at the fourth annual American Rock Art Research Association symposium, vol 4*, pp. 79–89. El Toro CA: American Rock Art Research Association.

Chamberlain, Von Del (1983). Navajo constellations in literature, art, artifact and a New Mexico rock art site. *Archaeoastronomy* (Center for Archaeoastronomy) **6**, 48–58.

Chamberlain, Von Del (1989a). Navaho Indian star ceilings. In *World archaeoastronomy*, ed. A.F. Aveni, pp. 331–40. Cambridge: Cambridge University Press.

Chamberlain, Von Del (1989b). Rock art and astronomy: Navajo star ceilings. In *Rock art of the Western canyons*, ed. J.S. Day, P.D. Friedman and M.J. Tate, pp. 30-45. Denver CO: Denver Museum of Natural History and Colorado Archaeological Society (Colorado Archaeological Society Memoir, no. 3).

Colton, H.S. (1959). *Hopi Kachina dolls*. Albuquerque NM: University of New Mexico Press.

De Harport, D.L. (1951). An archaeological survey of Canyon de Chelly: preliminary report of the field season of 1948, 1949, and 1950. *El Palacio* **58**(1), 35–48.

De Harport, D.L. (1953). An archaeological survey of Canyon de Chelly: preliminary report for the 1951 season. *El Palacio* **60**(1), 20–25.

De Harport, D.L. (1959). *An archaeological survey of Canon de Chelly, northeastern Arizona: a Puebloan community through time*. Unpublished PhD dissertation, Harvard University, Cambridge MA.

Dockstader, F.J. (1985). The Kachina and the White Man: the influence of white culture on the Hopi Kachina cult. Albuquerque NM: University of New Mexico Press.

Dockstader, F.J. (1987). Song of the loom: new traditions in Navajo weaving. New York NY: Hudson Hills Press, in association with Montclair Art Museum.

Espinosa, J.M. (1988). The Pueblo Indian Revolt of 1696 and the Franciscan missions in New Mexico. Norman OK: University of Oklahoma Press.

Fewkes, J.W. (1898). Sikyatki and its pottery. In *Seventeenth annual report of the Bureau of American Ethnology to the Secretary of the Smithsonian Institution 1895–96*, pp. 631–728 (excerpt from an accompanying paper by J.W. Powell entitled 'Archaeological Expedition to Arizona in 1895'). Washington DC: US Government Printing Office.[1]

Fewkes, J.W. (1903). *Hopi Katcinas. Twenty-first annual report of the Bureau of American Ethnology to the Secretary of the Smithsonian Institution, 1899–1900*. Washington DC: US Government Printing Office. Reprinted in 1985 by Dover Publications, Inc., New York NY.

---

[1] Fewkes (1898) and Fewkes (1919) are combined in Fewkes (1973).

Fewkes, J.W. (1919). Designs on prehistoric Hopi pottery. In *Thirty-third annual report of the Bureau of American Ethnology to the Secretary of the Smithsonian Institution 1911–12*, pp. 207–84 (an accompanying paper). Washington DC: US Government Printing Office.[2]

Fewkes, J.W. (1973), *Designs on prehistoric Hopi pottery*. New York NY: Dover.

Forbes, J.D. (1960). *Apache, Navajo and Spaniard*. Norman OK: University of Oklahoma Press.

Grant, C. (1978). *Canyon de Chelly: its people and rock art*. Tucson AZ: University of Arizona Press.

Hester, J.J. (1961). *Early Navajo migrations and acculturation in the Southwest*. PhD dissertation 61-3846, University of Arizona.

Hester, J.J. (1962). *Early Navajo migrations and acculturation in the Southwest*. Santa Fe NM: Museum of New Mexico Press (Papers in Anthropology no. 6).

Hibben, F.C. (1975). *Kiva art of the Anasazi at Pottery Mound*. Las Vegas NV: KC Publications.

James, G.W. (1974). *Indian blankets and their makers*. New York NY: Dover.

Jett, S.C. (1984). Making the 'stars' of Navajo 'planetaria'. *The Kiva* **50**, 25–40.

Kaufman, A. and Selser, C. (1985). *The Navajo weaving tradition: 1600 to the present*. New York NY: E.P. Dutton.

McAllester, D.P., ed. (1956). *The myth and prayers of the Great Star Chant and the myth of the Coyote Chant*, recorded by M.C. Wheelwright. Santa Fe NM: Museum of Navajo Ceremonial Art (Navajo Religion Series, vol. IV).

Newcomb, F.J., Fishler, S. and Wheelwright, M.C. (1956). *A study of Navajo symbolism*. Cambridge MA: Peabody Museum.

Newcomb, F.J. and Reichard, G.A. (1975). *Sandpaintings of the Navajo Shooting Chant*. New York NY: Dover.

Reichard, G.A. (1983). *Navajo religion: a study of symbolism*. Tucson AZ: University of Arizona Press. Reprinted from the 1974 edition by Princeton University Press.

Reichard, G.A. (1977). *Navajo Medicine Man sandpaintings*. New York NY: Dover.

Rodee, M.E. (1981). *Old Navajo rugs: their development from 1900 to 1940*. Albuquerque NM: University of New Mexico Press.

---

[2] Fewkes (1898) and Fewkes (1919) are combined in Fewkes (1973).

Sandner, D. (1979). *Navaho symbols of healing*. New York NY: Harcourt Brace Jovanovich.
Schaafsma, P. (1963). *Rock art in the Navajo reservoir district*. Santa Fe NM: Museum of New Mexico Press (Papers in Anthropology, no. 7).
Schaafsma, P. (1966). *Early Navaho rock paintings and carvings*. Santa Fe NM: Museum of Navaho Ceremonial Art.
Schaafsma, P. (1975). *Rock art in New Mexico*. Albuquerque NM: University of New Mexico Press (published for the Cultural Properties Review Committee, State Planning Office).
Schaafsma, P. (1980). *Indian rock art of the Southwest*. Santa Fe and Albuquerque NM: School of American Research and University of New Mexico Press.
Schaafsma, P. (1992). War imagery and magic: petroglyphs at Comanche Gap, Galisteo Basin, New Mexico. In *Archaeology, art and anthropology: papers in honor of J.J. Brody*, ed. M.S. Duran and D.T. Kirkpatrick, pp. 157–74. Albuquerque NM: The Archaeological Society of New Mexico (no. 18).
Wright, B. (1977). Hopi Kachinas: the complete guide to collecting Kachina dolls. Flagstaff AZ: Northland Press.
Yazzie, E., ed. (1971). *Navajo History, vol. 1*. Many Farms AZ: Navajo Community College Press.

21

Organisation of large settlements of the northern Anasazi

JOHN MCKIM MALVILLE AND JAMES WALTON

Introduction

During the twelfth and thirteenth centuries, the major concentration of people on the northern frontier of the Anasazi culture area was in the fertile lands of the Montezuma Basin north-west of Mesa Verde. By AD 1100, the population in the Montezuma Basin may have exceeded 30,000 distributed amongst at least eight major settlements and many smaller ones (Rohn 1983; 1985). By contrast, the largest population estimated for Chaco Canyon is 5700 (Hayes 1981), while the population of Mesa Verde at its peak may not have exceeded 2500 (Rohn 1985).

Table 1 provides estimates of the numbers of kivas, towers, and rooms in nine of the largest sites in the Montezuma Basin. In none of these sites is there evidence that they were constructed by a ranked society with a high degree of centralisation of socio-economic institutions. Models that have been applied to the social organisation of the Anasazi are those of aggregated and nucleated societies (Irwin-Williams 1980). The nucleated pattern involves centralisation of all elements of the community around a dominant socio-economic core of institutions and unique personnel. The Chaco phenomenon, which reached its florescence between AD 1050 and 1100, is the primary Anasazi example of a nucleated society. Aggregated communities such as the secondary occupation Salmon Ruin lacked the tight social control and specialisation of Chacoan communities. The nucleated structure of the Chacoan communities was short-lived and fragile, while the aggregated pattern appears to have been more flexible and resilient.

Aggregated communities are capable of elaborate community projects such as that of Mummy Lake on Mesa Verde, which indicates both

	Kivas	Great kivas	Estimated rooms	Towers
Yellow Jacket	165	1	250–700	22
Sand Canyon	90	1	300–400	17
Goodman Point	86	2	240	6
Yucca House	84	1		
Mud Springs	80	–		
Lowry	80	1		
Wilson	74	–		
Lancaster	45	–		
Reservoir Community	39	1		

TABLE 1. Estimates of the numbers of kivas, towers, and rooms in nine of the largest sites in the Montezuma Basin.

technological skill and organisation of labour in the Pueblo II time period (AD 950–1100). In the Pueblo III Mesa Verde populations that shifted from the mesa top to the canyons there is little evidence of social hierarchy. There are, however, two special purpose sites, Fire Temple and Sun Temple, which indicate village integration and ceremonial organisation, but still not hierarchies of society.

In aggregated communities, festivals and group rituals involving more than one household are necessary to achieve social integration. The integrative facilities employed by the northern Anasazi appear to have been medium- and large-sized kivas, plazas, and towers. Astronomy appears to have played a role in public ceremony, as several of the large sites indicate large-scale architectural planning associated with observational astronomy.

Two examples which provide evidence for astronomically initiated planning are Yellow Jacket and Goodman Point, neither of which have been scientifically excavated. The large site of the Montezuma Basin that has been most carefully excavated is Sand Canyon Pueblo, which experienced most of its growth during the half century just before the general abandonment of the region near AD 1300. Sand Canyon, however, has not yet revealed any evidence of astronomically initiated planning.

Yellow Jacket

The largest settlement in the Montezuma Basin was that of 5MT-5, near Yellow Jacket Spring (Lange *et al.* 1986; Wheat 1983; Ferguson and Rohn 1987; Malville and Putnam 1989; Malville 1990) (Fig. 1a). The ruins occupy a large flat mesa and contain evidence of approximately

165 kivas, 22 towers, a great kiva, and a great tower. Only the outlying communities to the south-west of Yellow Jacket, 5MT-1, 5MT-2 and 5MT-3, have been excavated. The main site, 5MT-5, remains today essentially unexcavated, except for the mischievous activities of pot hunters.

Extensive excavation by J.B. Wheat and his colleagues during the past three decades in the Yellow Jacket area provides the most complete record of settlement patterns. The area appears to have been occupied from about AD 500 to the time of abandonment near AD 1300, except for a hiatus in Pueblo I times (AD 700–900) when there was apparently a movement of people to the Dolores area to the east. Much of the construction at 5MT-5 in Pueblo II times (AD 900–1100) appears to have been jacal (mud and wattle) rooms and sub-surface kivas placed in east-west rows. Associated with the Pueblo II occupation there are middens indicating extensive residential activity.

In Pueblo III times (AD 1100–1300), pecked masonry room-blocks, kivas, and towers replaced jacal structures. Scattered through the middens are sandstone slabs, most of which may have covered burials and which were apparently discarded during the past century by pot-hunters. Our survey of the site revealed 140 of these burial slabs, primarily in identifiable middens and largely contained in east-west scatter fields. The use of middens for burial may simply have been expedient owing to the ease of digging, or the east-west organisation of those spaces may have been symbolically significant.

Rohn (1985) has estimated that the population of 5MT-5 may have been as high as 2500, based upon a person-to-kiva ratio of 15:1. Wheat proposes a smaller population based upon his estimate of a room-to-kiva ratio as small as 1:1. The extraordinary number of kivas remains a puzzle. The small kivas may been locations for residential as well as ceremonial activities (Cater and Chenault 1988), while the larger ones were probably exclusively ceremonial (Ferguson 1989).

Caves are sacred places in Pueblo cosmology (Ellis and Hammack 1968), and two caves in the southern and eastern cliffs may have been the primary features at Yellow Jacket involved in the initial architectural planning of the site. The southern mound, the remains of the tallest structure in that portion of the ruin, may have been intentionally located north of the southern cave and west of the much larger eastern cave. The floor of the southern cave contains fine, white ash and may have been the sacred repository of ash from fires of the many kivas on the mesa top.

The eastern cave is associated with a tower and several kivas. Above the cave, a small semi-circular rock enclosure opens to the east and may have served as a solar shrine. From the cave, a line of monoliths and walls, oriented upon summer solstice sunrise, culminates in the standing solar monolith. The top of the standing monolith has been worked

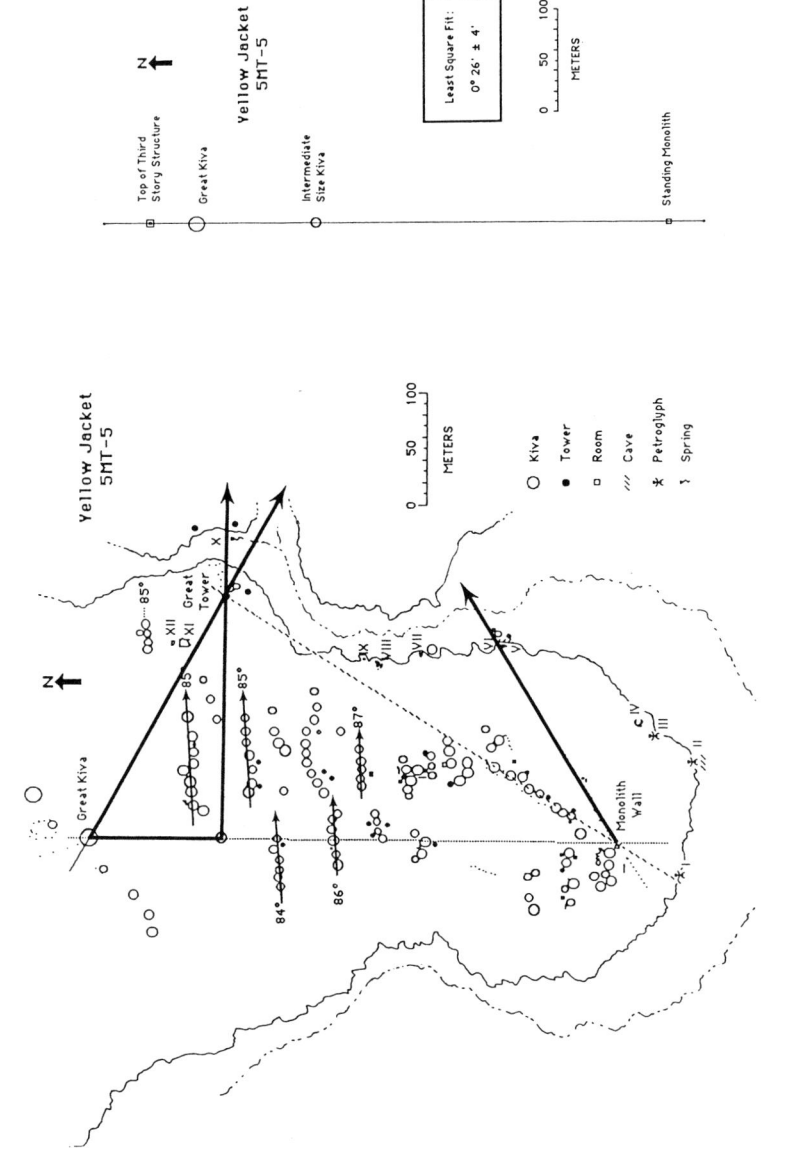

Organisation of large settlements of the northern Anasazi 245

FIG. 1. (a) Map of Yellow Jacket, 5MT-5. (b) North-south line.

(marks are visible) into a wedge such that it provides a sighting device to the location of summer solstice sunrise. A line of regression through the features of the line departs from the position of the first gleam of summer solstice sunrise of AD 1200 by approximately 10'.

From the monolith two additional lines diverge. One line, approximately perpendicular to the direction towards winter solstice sunrise, terminates in the north-east with the great tower and in the south-west with a major ceremonial feature on the adjacent mesa, the Dance Circle of 5MT-3. The second line (Fig. 1b) heads north to the Intermediate Kiva, the great kiva, and the northern mound with an azimuth departing from north by 26'. The northern mound may be the Yellow Jacket counterpart of a Chacoan great house. We propose that the sequence of planning events for the ceremonial structures was as follows: (1) the eastern cave established the location of the standing monolith in alignment with summer solstice sunrise; (2) major ceremonial structures, i.e. the Intermediate Kiva, the great kiva, and the northern house, were placed north of the monolith; (3) the line perpendicular to the winter solstice sunrise was established; and (4) the great tower was located relative to the great kiva.

The Intermediate Kiva has a diameter of approximately 9·7m, intermediate in size between the majority of the kivas and the great kiva. From the centre of the Intermediate Kiva, the centre of the great kiva is approximately 1' from true north and the centre of the great tower is 58' from east. The great tower, great kiva, and Intermediate Kiva lack middens, indicating that they were primarily ceremonial, non-residential structures.

The rim of the great tower has a diameter of 8·4m and lies 1·9m above the surrounding landscape. Based upon our measurements of rubble volume, a filling factor of 0·7, and a wall thickness of 0·6m, we estimate that the additional height of the tower was 11·4m, yielding an original total height of 13·3m. As viewed from the great kiva, the present rim of the Great Tower is 10·7m below the line of sight to the horizon at the point of winter solstice sunrise, and our estimate of the height of the tower thus places it slightly above the south-eastern horizon viewed from the great kiva.

As it would have been seen from the centre of a platform built on top of the great kiva, the centre of the great tower lies at an azimuth of 120° 20', approximately 10' to the south of the position of the centre of the sun on the horizon at winter solstice near AD 1200 (Fig. 2). The positioning of the two greatest ceremonial structures at Yellow Jacket towards winter solstices supports our hypothesis, generated by ethnographic analogy, that winter solstice at Yellow Jacket should have been the most important of solar festivals.

WINTER SOLSTICE
YELLOW JACKET

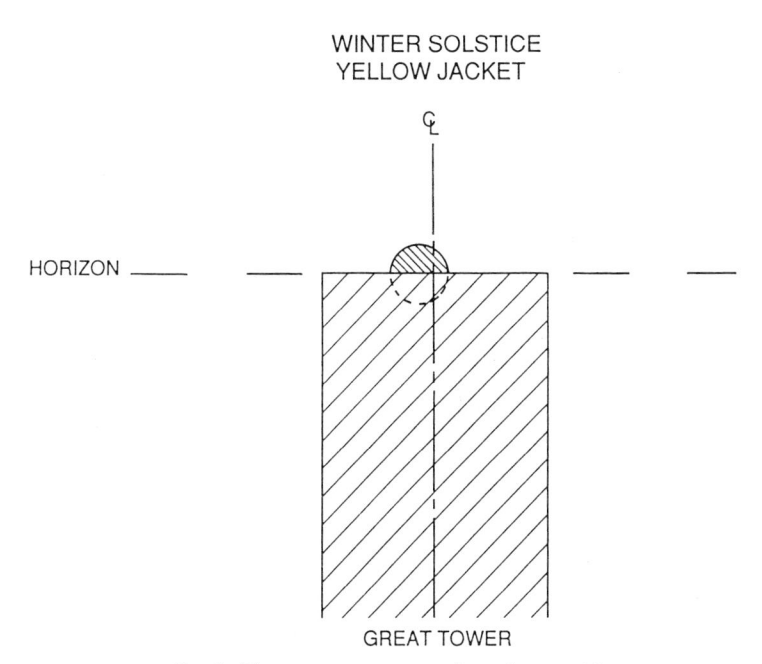

FIG. 2. The great tower as seen from the great kiva.

Goodman Point, Sand Canyon and Yucca House

The ruins of Goodman Point, 5MT-3287, 12 km south of Yellow Jacket, contain evidence of some 86 kivas, 240 rooms, 6 towers, and two great kivas (Fig. 3). Walls enclose all but a south-western sector and include an approximately square area with sides 200m long. Containing two great kivas, the site may have been a major ceremonial complex. The second great kiva is separated from the main ruin and lies 900m to the south-west.

Our data include our theodolite survey of major features of the site in 1987–88 and an aerial map made for the National Park Service. The northern kiva house—140m long with a long northern wall, and containing 30 kivas and 90 rooms—is an impressive feature, which has an orientation of 89° 15′. The eastern wall has an azimuth of approximately 2° 15′. The cardinality of these features appears to be intentional as they depart from the local topography. The great kiva shows four foundations for posts, which are approximately cardinal.

Our discovery of the winter solstice association of the great kiva and the great tower at Yellow Jacket led us to propose that a similar line of sight may have involved the great kiva at Goodman Point. Indeed, from

FIG. 3. (a) Map of Goodman Point, 5MT-3287. The dashed line is the north wall of the kiva house. GK = great kiva; T = tower. (Courtesy of Mesa Verde National Park).
(b) Artist's reconstruction of Goodman Point (Jean Kindig). Reproduced with permission.

the centre of the great kiva a tower situated just at the edge of the canyon has an orientation of approximately 120°. Furthermore, there are an adjacent wall and a second non-cardinal wall associated with a tower and a kiva in the north-east quadrant with similar orientations.

Three miles to the east of Goodman Point is Sand Canyon Pueblo. Based upon tree ring dates, its major period of occupation was between AD 1240 and 1280 (Lipe 1989). The site is enclosed by a wall and contains some 90 kivas, 300–400 rooms, 17 towers, and a bi-wall or tri-wall structure. There is no evidence for the orientation of structures along cardinal or solstitial directions.

Yucca House is strategically located between the eastern slopes of Sleeping Ute Mountain and the Mesa Verde at the southern entry into Montezuma Basin. With a view across this two-mile-wide passage, Yucca House could have dominated the flow of people and goods from the Chaco basin. The site was probably chosen partly because of the presence of a spring, but its location may also have been influenced by the sunset on winter solstice that occurs at Towaoc, the spectacular spire which is the toe of Sleeping Ute.

Discussion

The non-ranked societies of Yellow Jacket and Goodman Point appear to have engaged in activities normally attributed to urban communities, such as the planning and construction of major ceremonial structures associated with large scale cardinality. Each presents evidence of astronomical ceremonialism. In both of these sites different types of kivas may have served different purposes. The great kivas and the Intermediate Kiva of Yellow Jacket may have been primarily integrative facilities while the smaller kivas may have been associated with the ceremonies of residential kin groups.

Social integration is probably a prerequisite for the use of astronomy in large-scale community planning. Other communities in which astronomy has played a significant role are Chaco Canyon and Chimney Rock, both of which have great kivas. A common feature of the northern Anasazi sites is attention to winter solstice, which appears to be evidenced at the Sun Temple of Mesa Verde, Chimney Rock, Yucca House, Yellow Jacket and Goodman Point.

Acknowledgements

We thank Jack Smith for the map of Goodman Point and John Jacobs for use of his measurements of the great tower of Yellow Jacket.

References

Cater, J.D. and Chenault, M.L. (1988). Kiva use reinterpreted. *Southwestern Lore* **54**(3), 19–32.

Ellis, F.H. and Hammack, L. (1968). The inner sanctum of Feather Cave, a Mogollon sun and earth shrine linking Mexico and the Southwest. *American Antiquity* **30**, 25–44.

Ferguson, T.J. (1989). Comment on social integration and Anasazi architecture. In *The architecture of social integration in prehistoric Pueblos*, ed. W.D. Lipe and M. Hegmon, pp. 169–73. Cortez CO: Canyon Archaeological Center (Occasional Papers no. 1).

Ferguson, W.M. and Rohn, A.H. (1987). *Anasazi ruins of the Southwest in Color*. Albuquerque NM: University of New Mexico Press.

Hayes, A.C. (1981). A survey of Chaco Canyon archaeology. In *Archaeological surveys of Chaco Canyon, New Mexico*, ed. A.C. Hayes, D.M. Brugge and W.J. Judge, pp. 1–68. Washington DC: National Park Service (Archaeological Series 18A: Chaco Canyon Series).

Irwin-Williams, C. (1980). Investigations at Salmon Ruin: methodology and overview. In *Investigations at the Salmon site: the structure of Chacoan society in the northern Southwest*, ed. C. Irwin-Williams and P.H. Shelley, pp. 107–70. Portales NM: Eastern New Mexico University.

Lange, F., Mahaney, N., Wheat, J.B., Cater, J. and Chenault, M.J. (1986). *Yellow Jacket: a Four Corners Anasazi ceremonial center*. Boulder CO: Johnson Books.

Lipe, W. D. (1989). Social scale of Mesa Verde Anasazi kivas. In *The architecture of social integration in prehistoric Pueblos*, ed. W.D. Lipe and M. Hegmon, pp. 53–71. Cortez CO: Crow Canyon Archaeological Center (Occasional Papers no. 1).

Malville, J.M. (1990). Prehistoric astronomy in the American Southwest. *Astronomical Quarterly* **7**, 192–232.

Malville, J.M. and Putnam, C. (1989). *Prehistoric astronomy in the Southwest*. Boulder CO: Johnson Books.

Rohn, A.H. (1983). Budding urban settlements in the northern San Juan. In *Proceedings of the Anasazi symposium*, ed. J. Smith, pp. 175–80. Mesa Verde CO: National Park Service.

Rohn, A.H. (1985). Prehistoric developments in the Mesa Verde region. *Exploration: Annual Bulletin of the School of American Research*, 1985, 3–10.

Wheat, J.B. (1983). Anasazi who? In *Proceedings of the Anasazi symposium*, ed. J. Smith, pp. 11–15. Mesa Verde CO: National Park Service.

22

Summer solstice: a Chumash basket case

EDWIN C. KRUPP

Indians and ancient astronomy are probably not what first come to mind when California—the home of Hollywood and Disneyland and the Gold Rush—commands our attention. The State's reputation for colourful behaviour, glamorous lifestyles, and exotic landscapes suggests that we would expect to encounter there reports of UFO landings, celebrity liaisons, and Elvis sightings—not accounts of skywatching indigenous Californians. There is, however, ample archaeological and ethnographic evidence to document California Indian interest in the sky (Benson and Hoskinson 1984; Hudson 1984; Hudson et al. 1983; Hudson et al. 1979; Hudson and Underhay 1978; Krupp 1983a; 1983b; 1983c; 1987; 1991; Schiffman 1988).

California Indian astronomical traditions are also important in the consideration of the character of native astronomy throughout North America. At the time of European contact, California was the most populated region in the US and Canada (Kroeber 1939: 153). It was also one of the most linguistically diverse areas in the entire world. Evaluated in terms of population, then, California Indian astronomy was a mainstream tradition.

Astronomical components of California Indian life also shed light on the development and cultural function of astronomy among hunters and gatherers. California Indians did not need to practice intensive cultivation. The State's celebrated climate provided amply for their needs, and their systematic 'harvest' of the wild produce of nature was naturally tempered by familiarity with the celestial signals of seasonal change. Practical observations of the sun, moon, and stars, celestial myths, sacred astronomical symbols, shamanic quests for celestial power, and seasonal ritual were all features of traditional life in indigenous California. We know, for example, that the solstices were observed throughout the State (Hudson et al. 1979), and several reviews have summarised field studies that have linked solstice observation with shamanism and indigenous rock art (Benson and Hoskinson 1984; Hudson 1984; Hudson et al. 1979; Krupp 1983a; 1983b: 129–37; 1987; 1991; Schiffman 1988).

Modern studies of California Indian astronomy were greatly stimulated in 1978 by the publication of *Crystals in the sky*, Travis Hudson's and Ernest Underhay's compilation and analysis of the astronomical and cosmological lore of southern California's Chumash Indians. Chumash culture had a maritime focus and was centred in the Channel Islands, but Chumash groups also occupied two hundred miles of California coastline, from Malibu to the vicinity of San Luis Obispo. Their territory reached inland to the west edge of the San Joaquin Valley.

The Chumash were relatively wealthy. Craft specialisation, a moneyed economy, extensive trade with their neighbours, and a class structure added complexity and richness to the Chumash way of life. Although traditional Chumash beliefs, like most of the native culture of southern California, were nearly extinguished by decimation of the Indian population through contact with outsiders—first Europeans, then Mexicans, and finally Anglos, there were still myths, songs, dances, and ceremonies to be recorded in the final decades of the last century and the first decades of this century. John Peabody Harrington prepared voluminous notes from accounts provided by surviving Chumash, and much of Hudson's and Underhay's work was based upon Harrington's unpublished notes. Most of these notes are now held at the University of California, Berkeley, and in the National Anthropological Archives of the Smithsonian Institution in Washington DC.

Elaborate polychrome rock paintings are found throughout the mainland territory of the Chumash. These pictographs have been called 'the finest rock paintings in North America' (Grant 1967: 107) and so stand out even in a state that possesses one of the largest and most varied concentrations of indigenous rock art in the entire world. Since 1975, some California Indian rock art has been linked with native astronomical traditions (Brandt et al. 1975; Hedges 1986), and it was against this backdrop of ethnographic evidence of Chumash astronomy and field studies of connections between astronomy and Chumash rock art, that the present investigation was conducted.

Although the Chumash calendar was based upon the phases of the moon, the solstices were determined through horizon observations of the sun (Hudson and Underhay 1978: 53). A major ceremony dedicated to the New Year renewal of the sun, and to a felicitous outcome of the great, year-long, high-stakes celestial gambling game played by supernaturals in the sky, was conducted at the time of the winter solstice (Blackburn

Summer solstice: a Chumash basket case 253

FIG. 1. Chumash shamans attempted to procure power from the supernatural beings they believed populated their universe. The sun, recognised as a source of supernatural power, is probably represented by the rayed disc shown here, and the figure interacting with the sun may represent a skywatching, power-seeking shaman. This pictograph is located on the Carrizo Plain, near the northern limit of Chumash territory. The figure is about 10cm high. (Photograph: E.C. Krupp)

1975a: 91). The officials who presided over this event were members of the *'antap*, an elite religious cult. The winter solstice was considered to be a time of acute cosmological crisis when guided human intervention was required to maintain the delicate balance of nature. The 'antap organisation was responsible for achieving the desired outcome. Its membership included shamans, and in the nineteenth century, two Chumash shamans, identified by name, are known to have gone into the mountains at the time of winter solstice and painted pictographs in a cave (Blackburn 1975b: 127). The Chumash called the sun *Kakunupmawa*, and his name meant 'the radiance of the child born on the winter solstice' (Hudson and Underhay 1978: 51).

Fernando Librado Kitsepawit, the Ventureño Chumash who provided Harrington information on the winter solstice activities, also described what happened at summer solstice. A sun-stick was inserted into the earth by the sun priest and his twelve 'antap assistants. At winter solstice, the same thing occurred. The sun-stick was intended to tether the sun and induce it to move back north.

A 'big basket' was filled with valuable offerings for the sun at summer solstice with the hope that such respect would inspire a rich harvest in the autumn (*ibid.*: 66). This reference to a ritual container may endorse the validity and clarify the meaning of summer solstice light-and-shadow effects observed at a few Chumash painted shrines.

Hoskinson and Cooper (1988) have described the formation of an 'arrow' of sunlight that appears at the time of summer solstice inside a pictograph shelter (SBa-526) on the Sierra Madre Ridge. During the afternoon, the 'arrow' moves across the shelter floor and crosses and partly illuminates a five-sided hole that was carved intentionally into the natural rocky pavement. Although this site is known popularly as Painted Rock (they are all known as Painted Rock), it has been identified (Lee and Horne 1978) as a place described to Harrington by María Solares, an Inezeño Chumash. She mentioned a cave known as *Sapaksi* , the 'House of the Sun,' and described its paintings of animals and its 'big painting of the sun.'

FIG. 2. The Chumash and other southern California tribes fabricated 'sun-sticks' by attaching stone discoids, perforated with a central hole, to wooden shafts. This sun-stick, about 40cm long, is one of a pair found in 1884 in Bowers Cave, Los Angeles County. A pattern of eight radial lines is painted on the top of the stone, and shell beads are attached to the stone with asphalt. Chumash sun-sticks were erected during winter and summer solstice ceremonies. (Collection: Peabody Museum, Harvard University. Photograph: E.C. Krupp)

There is a rayed disc, about 49 cm in diameter and painted mostly in red, on the ceiling of the shelter, along with many other pictographs, including several animals, on the walls and ceiling. *Kakunupmawa* was said to live in a crystal house in the sky (Blackburn 1975a: 36), and he also

Summer solstice: a Chumash basket case 255

FIG. 3. A large rayed disc, about 49cm in diameter, painted inside a Chumash rock shelter on the Sierra Madre Ridge, above the Cuyama Valley, helped identify the site as a place known to the Chumash as *Sapaksi*, the 'House of the Sun'. (Photograph: E.C. Krupp)

visited another special house every night to play the cosmic gambling game. *Sapaksi* may have been regarded by the Chumash shaman as a terrestrial and symbolic counterpart for either or both of these solar-powered homes. Its natural architecture—flat floor and domed ceiling—also makes it a small-scale version of the earth and sky. The seasonally significant interaction of sunlight with the receptacle inside the shelter (first observed in 1980) suggests that a summer solstice gift was directed to the sun in a private shamanic ritual that paralleled the public basket offering.

Another summer solstice light-and-shadow involvement with a ritual container was first observed at Burro Flats (Ven 151-161) in 1979 (Krupp 1983c: 19; Romani 1981: 173–76; Romani *et al.* 1988: 116; Romani *et al.* 1985: 95–97). Here, in the north-west San Fernando Valley, there is an elaborate and well-known panel of pictographs that hosts a winter solstice lighting effect. South-east of the main shelter, a curious complex of natural rock formations, bedrock mortars, and cupule patterns provide a stage for the summer solstice event. The largest mass of rock looms above a 'bear paw' pattern of five deep bedrock mortars. Their size, location, and arrangement suggest ceremonial—rather than domestic—use, and an obviously domestic mortar site is relatively nearby. Natural features of the large boulder suggest a face. It may have been noticed by the Chumash, too, for apparent references to simulacra—faces and figures in the natural landscape—are known from Harrington's notes (Lee and Horne 1978: 220) and observed near other Chumash sites.

From an isolated mortar, the rising sun first appears in a high natural hook in the nearby sandstone cliff, in line with a cup-marked boulder. The 'bear paw' pattern of mortars in the bedrock is just a few feet north of this isolated mortar, and an upright boulder slab in front of the 'bear paw' carries a line of nineteen cupules on its top surface. The cupule row is aligned roughly with the direction of the summer solstice sunrise, and when the sun climbs over the top of the cliff, the slab casts a shadow upon the 'bear paw' arrangement of mortars. The tip of the shadow falls into the central mortar, and the shadow bisects one of the mortars on the rim so that it is half in shadow and half in light. As the sun continues to rise, the shadow recedes from the bedrock. This shadow has been monitored thirty days after the solstice, and its behaviour at that time is dramatically different from its behaviour in June. In fact, observable differences have been reported as soon as five days after the solstice (Romani *et al.* 1985: 98). At summer solstice, then, another 'stone basket'—this time at Burro Flats—is tagged by sunlight, and in this case, the event takes place below the gaze of whatever supernatural being that might have been imagined in the face on the big rock.

FIG. 4. An afternoon 'arrowhead' of sunlight moves across the floor of the Sapaksi rock shelter around the time of summer solstice. Its trajectory carries it across a five-sided cavity carved into the floor. The cavity is 3 to 5cm across and about 7·5 cm deep. (Photograph: E.C. Krupp, Jun 21 1982)

Summer solstice: a Chumash basket case

Field observations undertaken from 1987 to 1989 at Mutau Flat (Ven-51), a Chumash site in the Los Padres National Forest with several delicately painted shelters, provided another example of a ritual container associated with light-and-shadow effects at summer solstice (Krupp and Wubben 1990). Prior to this work, the significance of the containers and their possible relation to the summer solstice offering basket had not been recognised.

Although the pictographs at Mutau Flat have been known since 1889 (Eberhart and Babcock 1963: 11), an archaeological survey of the area was not conducted until 1963, when Hal Eberhart and Agnes Bierman recovered material that prompted them to guess the site was a seasonal camp (*ibid.*: 13). At an elevation of nearly five thousand feet, it provided a cool and pleasant highland haven in summer. Heavy snow and prohibitive cold probably emptied the site in winter.

In 1984, Bob Wubben discovered a new feature at Mutau Flat—a dish carved from an upturned projection of a boulder on the floor in the middle of the main shelter. Once noticed, the shallow bowl is hard to miss, but none of the previous site reports (Steward 1929: 99–100; Fenenga 1949; Eberhart and Babcock 1963; Grant 1965) mentions it. When Wubben ret-

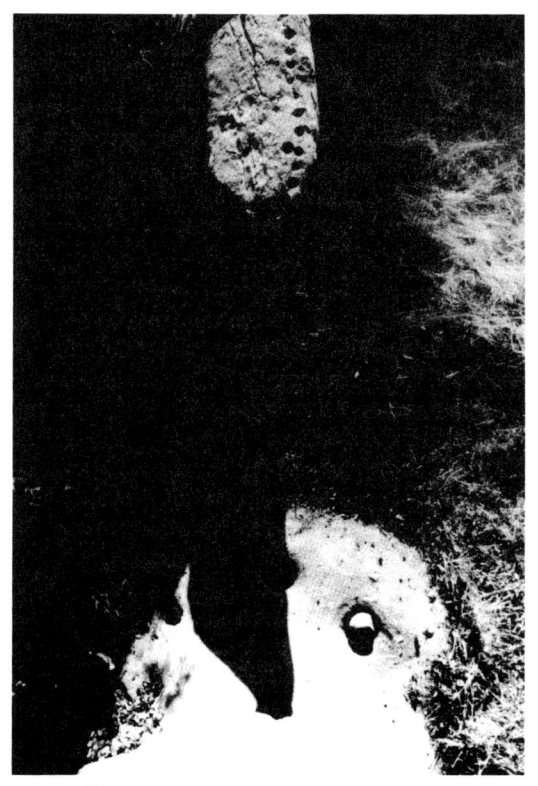

FIG. 5. The rising summer solstice sun at Burro Flats casts the shadow of a cupule-marked slab of rock upon an arrangement of bedrock mortars. The tip of the shadow coincides with the central mortar while the shadow's straight edge bisects one of the four mortars on the rim of the 'bear paw'. The bisected mortar is about 12cm across. (Photograph: E.C. Krupp, Jun 21 1980)

FIG. 6. This mass of sandstone dominates the open grassland at Mutau Flat in the Los Padres National Forest south of Mount Pinos. The main shelter is located in the central split in the rock. (Photograph: E.C. Krupp)

FIG. 7. A shallow bowl was carefully cut and polished on the top of a boulder firmly lodged in the floor of the main shelter of Mutau Flat. This feature of the site was first discovered by Bob Wubben in 1984. Although a portion of the bowl is broken away, enough remains to confirm its original shape. It was about 23cm in diameter. The considerable mass of the boulder suggests that the dish is still in, or close to, its original position. (Photograph: E.C. Krupp)

Summer solstice: a Chumash basket case 259

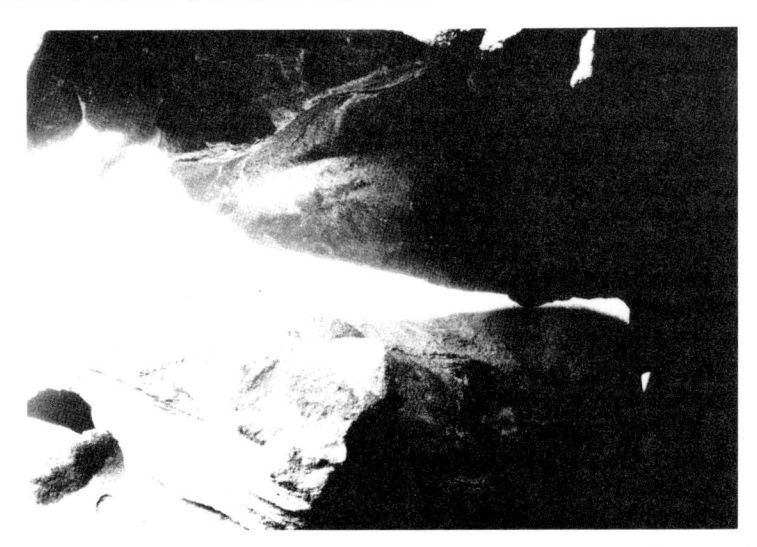

FIG. 8. At summer solstice, two triangles of sunlight knife across the shelf in the main shelter and meet at a precisely painted red line that is split when the blades touch. The shelf is 1·37m wide and 0·41m deep. (Photograph: E.C. Krupp, Jun 20 1988)

urned to the site in 1985, on the summer solstice, he noticed that the dish was partially illuminated by the sun in the morning and completely awash in light at about noon. Subsequently, he brought photographs of the dish to my attention, and we began a programme of systematic observation in 1987.

Although there are several painted shelters in the outcrop at Mutau Flat, only the main shelter hosts an eye-catching light-and-shadow performance at summer solstice. It involves a small natural ledge a few feet south of the dish and a few feet above it. The wall behind the shelf is painted with a number of typical Chumash elements along with a figure that resembles a shaman in ceremonial dance skirt. Just after 9:00 Pacific Daylight Time, two triangles of sunlight form on opposite sides of the shelf and head towards each other. After a few minutes, their points touch, and where the two blades of light first meet, there is a finely painted red line. As the two sunlight knives merge, they open into a wide band of light that gradually moves north through the shelter.

At 9:45, sunlight reaches the dish. In a few minutes, it becomes fully illuminated, and it remains that way until 10:55, when a shadow edges over the south-west rim. The shadow progresses across the dish until half of the dish is covered. Then the shadow slowly rotates clockwise, and this keeps the dish half in shadow, half in light, until 11:53, when the shadow retreats completely from the container. Later, at 12:20, a second shadow overtakes the dish, which remains in darkness until the next day.

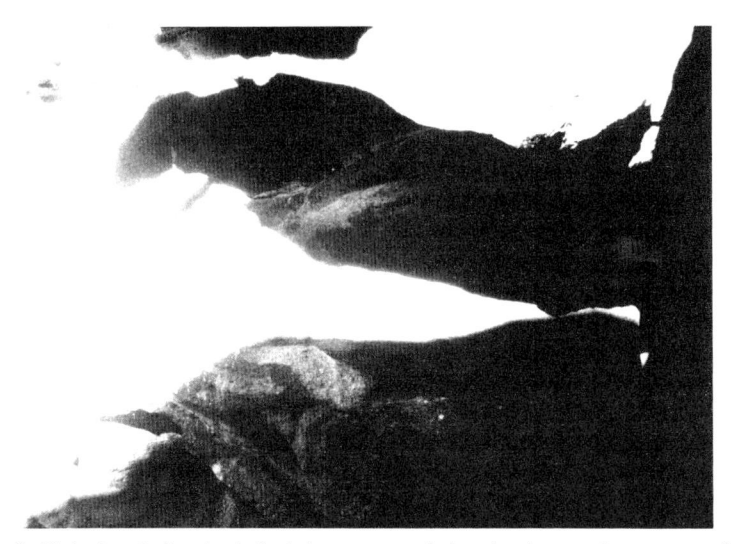

FIG. 9. Eight days before (and after) the summer solstice, the shape and movement of the two blades over Mutau Flat's rocky shelf are discernibly different from what is observed on the solstice itself. Specifically, a 'shoelace' of light forms between the blades. On the solstice itself, the triangles themselves touch, and no 'shoelace' is seen. Four days before (and after) the solstice, however, the event is indistinguishable from what takes place on the solstice. (Photograph: E.C. Krupp, Jun 13 1989)

A carefully formulated sequence of subsequent visits to the site revealed that it is possible to see a distinct difference in the behaviour of the two knives of light eight days before or after the summer solstice. Four days before or after the solstice, however, the lighting effect on the shelf is indistinguishable from what happens on the solstice. The summer solstice effect at Mutau Flat is, then, consistent with other solstitial light-and-shadow effects at rock art sites in California. In general, they 'work' for about a week on either side of the solstice, and this argues that they functioned as shrines and not as observatories.

The play of light and shadow on the dish at Mutau Flat is not as well resolved in time as the splitting of the red line on the shelf. Twelve days before and after the summer solstice, the dish effect still looks the same. It does gradually change, and certainly a month and a half from the solstice the effect is entirely different.

At Mutau Flat, then, we have a third example of a ritual container that is split by light and shadow at summer solstice. Fernando Librado Kitsepawit provided some additional commentary about this time of the year that has a bearing on what has been observed at Mutau Flat, Burro Flats, and Sapaksi. The month in which the summer solstice falls was known as

Summer solstice: a Chumash basket case 261

the 'month when things are divided in half' (Hudson and Underhay 1978: 128). At this time of year, two daggers of light split the rocky shelf at Mutau Flat in half, and where they meet, they split a painted line in half.

The summer solstice was regarded by the Chumash as the mid-point of the year, and sacred containers at three Chumash shrines are divided in half by sun and shadow when the year is divided in half. This encourages the prediction that similar containers will be found at other sites and will be discovered to interact with summer solstice sunlight that divides things in half.

FIG. 10. Rock formations above the main shelter at Mutau Flat cast a summer solstice shadow that divides the boulder bowl in half at the same time the Chumash said the year is divided in half. (Photograph: E.C. Krupp, Jun 20 1988)

Fernando Librado Kitsepawit also said that the month of summer solstice was the time 'to go out in different directions.' He added that things would in time be cyclically reunited. The sun, of course, heads in a different direction at summer solstice, and its migration south ends six months later, at winter solstice, when the Great Gambling Game in the Sky is wrapped up with the rebirth of the sun and the promise of a new year.

Symbolic displays that appear to turn Chumash shrines into halfway houses for the year provide graphic confirmation that summer solstice really was the time when things were divided in half and when offerings to the sun were made in special containers. Furthermore, these features of the summer solstice rock art sites seem to argue that they functioned not as observatories but as shrines where astronomical knowledge served Chumash shamanic ritual and where the dish ran away with the sun.

References

Benson, A. and Hoskinson, T., eds. (1984). *Earth and sky—proceedings of the first western regional conference on archaeoastronomy.* Thousand Oaks CA: Slo'w Press.

Blackburn, T.C. (1975a). *December's child—a book of Chumash oral narratives.* Berkeley and Los Angeles CA: University of California Press.

Blackburn, T.C. (1975b). Further information on the de Cessac photograph. *Journal of California Anthropology* 2(1), 127–28.

Brandt, J., Maran, S., Williamson, R.A., Harrington, R., Cochran, C., Kennedy, M., Kennedy, W. and Chamberlain, Von Del (1975). Possible rock art records of the Crab Nebula supernova in the western United States. In *Archaeoastronomy in pre-Columbian America*, ed. A.F. Aveni, pp. 45–58. Austin TX: University of Texas Press.

Eberhart, H. and Babcock, A.B. (1963). *An archaeological survey of Mutau Flat, Ventura county, California.* Berkeley CA: University of California, Archaeological Research Associates (Contributions to California Archaeology, no. 5).

Fenenga, F. (1949). Methods of recording and present status of knowledge concerning petroglyphs in California. *University of California Archaeological Survey* 3, 1–19.

Grant, C. (1965). *The rock paintings of the Chumash.* Berkeley and Los Angeles CA: University of California Press.

Grant, C. (1967). *Rock art of the American Indian.* New York NY: Promontory Press.

Hedges, K. (1986). The sunwatcher of La Rumorosa. In *Rock art papers, vol. 4*, ed. K. Hedges, pp. 17–32. San Diego CA: San Diego Museum of Man.

Hoskinson, T. and Cooper, R.M. (1988). Sapaksi: archaeoastronomical investigation of an inland Chumash site. In *Visions of the sky—archaeological and ethnological studies of California Indian astronomy*, ed. R. Schiffman, pp. 31–40. Salinas CA: Coyote Press.

Hudson, T. (1984). California's first astronomers. In *Archaeoastronomy and the roots of science*, ed. E.C. Krupp, pp. 11–81. Washington DC: Westview Press.

Hudson, T., Labbe, A., and Moser, C. (1983). *Skywatchers of ancient California.* Santa Ana CA: Bowers Museum (Booklet and catalogue to accompany the 'Skywatchers of Ancient California' exhibit).

Hudson, T., Lee, G. and Hedges, K. (1979). Solstice observers and observatories in native California. *Journal of California and Great Basin Anthropology* 1, 39–63.

Hudson, T. and Underhay, E. (1978). *Crystals in the sky: an intellectual odyssey involving Chumash astronomy, cosmology, and rock art* Socorro NM: Ballena Press.

Kroeber, A.A. (1939). Cultural and natural areas of native North America. *University of California Publications in American Archaeology and Ethnology* **38**, 1–242.

Krupp, E.C. (1983a). Emblems of the sky. In *Ancient images on stone: rock art of the Californias*, ed. J.A. van Tilburg, pp. 38–43. Los Angeles CA: University of California (Rock Art Archive, Institute of Archaeology).

Krupp, E.C. (1983b). *Echoes of the ancient skies*. New York NY: Harper and Row.

Krupp, E.C. (1983c). Light and shadow. *Griffith Observer* **47**(6), 12–20.

Krupp, E.C. (1987). Saluting the solstice. *News from Native California* **1**(5), 10–13.

Krupp, E.C. (1991). Hiawatha in California. *The Astronomy Quarterly* **8**(1), 47–64.

Krupp, E.C. and Wubben, R. (1990). When things are divided in half. In *Rock art papers, vol. 7*, ed. K. Hedges, pp. 41–48. San Diego CA: San Diego Museum of Man.

Lee, G. and Horne, S. (1978). The Painted Rock site (SBa-502 and SBa-526): Sapaksi, the House of the Sun. *Journal of California Anthropology* **5**(2), 216–24.

Romani, J. (1981). *Astronomy and social integration: an examination of astronomy in a hunter and gatherer society*. MA. thesis, California State University, Northridge CA.

Romani, J., Larson, D., Romani, G., and Benson, A. (1988). Astronomy, myth, and ritual in the west San Fernando valley. In *Visions of the sky—archaeological and ethnological studies of California Indian astronomy*, ed. R. Schiffman, pp. 109–34. Salinas CA: Coyote Press.

Romani, J., Romani, G., and Larson, D. (1985). Archaeoastronomical investigations at Burro Flats: aspects of ceremonialism at a Chumash rock art and habitation site. In *Earth and sky—proceedings of the first western regional conference on archaeoastronomy*, eds. A. Benson and T. Hoskinson, pp. 93–108. Thousand Oaks CA: Slo'w Press.

Schiffman, R.A., ed. (1988). *Visions of the sky—archaeological and ethnological studies of California Indian astronomy*. Salinas CA: Coyote Press.

Steward, J.H. (1929). Petroglyphs of California and adjoining states. *University of California Publications in American Archaeology and Ethnology* **24**(2), 47–239.

23

Counting and sky-watching at Boca de Potrerillos, Nuevo León, Mexico: clues to an ancient tradition

WM. BREEN MURRAY

> 'All these people [of the Texas Gulf Coast] had no notion of time by the Sun or the Moon, nor do they count the month and the year, and they know more about the different times when the fruits ripen, and the time when fish die, and the appearance of the stars, in which they are very skilled and practised.' (Cabeza de Vaca 1942[1537]: ch. 22.)

Amongst the most intriguing motifs found at prehistoric rock art sites in north-eastern Mexico (Nuevo León and Coahuila states) are numerous pecked dot configurations and lines of tally marks. A decade ago at the Oxford 1 conference, I proposed (Murray 1982b) that many of these petroglyphs could be explained as simple cumulative (monosymbolic, or tally) counts. This notion derived primarily from an analysis of two extraordinarily complex examples, a 207-tally petroglyph at Presa de La Mula, N.L. and a unique dot configuration at Boca de Potrerillos, N.L., thirty kilometres to the east, which registers the same total sum. Both of them can be explained, I contend, as quasi-observational records of seven lunar synodic months (206·7 days). In 1981, the chronological span, geographical extent, and cultural function of this petroglyphic counting tradition could only be guessed, but the initial discoveries provided an especially telling case for numeracy, since they refer to unchanging natural cycles whose countable sequences are independent of place, time, or cultural context. They also defined a working paradigm by which further examples of counting could be identified iconographically, whether they refer to lunar time-reckoning or not.

Since then, fieldwork at more than thirty rock-art sites in our study area has uncovered many more dot and tally petroglyphs, and clarified considerably the motif complexes most often associated with them (Murray 1985; 1986; 1992). The artefactual basis and geographical range of the tradition has also been re-defined by the discovery of similar marking systems on engraved bison scapulae from the Texas Gulf Coast (Hester 1980; Murray 1984a) and wooden calendar sticks from the Winnebago and other North American Indian tribes (Marshack 1985; 1989; Murray 1989), while recently discovered petroglyphs at Xihuingo in the Valley of Mexico (Wallrath and Rangel 1984; Aveni 1989) raise new questions about the tradition's possible Mesoamerican connections. Most importantly, fieldwork still in progress at Boca de Potrerillos, which is documenting the largest sample of petroglyphic counts at any single site in our study area, has uncovered a horizon orientation system which appears to be associated with the counts. In this study I will try to summarise briefly how these new perspectives modify my earlier hypotheses about petroglyphic counting, particularly in relation to lunar time-reckoning and sky-watching in general.

The most fundamental interpretative change involves the separation of dot and tally counting into two archaeologically distinct (though related) traditions. The exact nature of the separation is still not clear; since the two symbols occur at the same sites, and have the same cognitive meaning, they can not be totally divorced. Nevertheless, once a larger sample of counts was detected, a look at glyphic frequencies, spatial distribution, relative rock wear, and examples of super-positioning established very different profiles for each motif, effectively separating them in archaeological terms in spite of their apparent similarities.

Dot configurations, for example, outnumber tally counts by a ratio of at least 10:1, and occur at virtually every site in the region. At the larger sites, they are often densely concentrated at specific points or on particular panels, as if they reflect some special activity which had to be carried out at a certain place. They show a wide range of patinas from quite fresh (possibly even historic) examples to heavily repatinated ones, often extensively fractured, which appear to be among the oldest petroglyphs at their respective sites. Several of the most spectacular examples have been so over-carved or obliterated that numerical reconstruction is no longer possible. The typical configurational pattern is columns or rows of dots, but no two petroglyphs are exactly alike, and a wide variety of patterns is in fact represented, including arcs, serpentine lines, fan shapes, cruciforms, and rectangles amongst others. This configurational variability may have some cognitive significance, but since our sample usually includes only one or a few examples of each type, this is difficult to demonstrate. A unique triple spiral configuration at Cerro

La Bola, Coah (Murray 1984b), however, replicates a petroglyph at Xihuingo in central Mexico, and suggests that a broader multi-site comparison might yield significant results. In general, dot counting seems to be an old and widely distributed tradition which undergoes considerable diversification over time. The north-eastern Mexican examples are certainly a spectacular manifestation of this tradition, but they are by no means unique, and must ultimately be explained, in my opinion, as a local variant of a widely diffused system.

Tally counts, on the other hand, have a more restricted distribution in our study area, and are usually limited to a few examples at each site. They are systematically associated with a hunting motif complex which includes representations of atlatls, projectile points and scraper blades of various types, animal paw and hoof prints, and deer antlers. Whenever superpositions occur, the tallies and hunting motifs are carved over the other glyphs (including dot configurations). Tally counting, then, appears to be an intrusive tradition of somewhat later date, and to have co-existed alongside the older dot counting tradition in a more specific cultural niche.

A comparison of deer antler representations at various sites (Murray 1992) allowed a deeper look into this cultural context, and suggested that tally counting might even be an independently invented local tradition. This study revealed an iconographic continuum ranging from highly realistic depictions of antler architecture to progressively more abstract ones which appear to encode antler size or number in some way. This process of abstraction culminates in an antler glyph less than a hundred metres from the La Mula tally count in which the beam has been reduced to a nearly straight baseline, and the tine count (29 or 30) is compatible with a lunar synodic month. From this angle the tally itself took on a new meaning when it was noted that the 207 tallies would also be a very good estimate of the gestation period of the female whitetail deer, which is 205–212 days (Leopold 1977: 574). Whether the tally is a record of real animal observation, a warning against out-of-season poaching, or a ritual symbol clothed in numerical guise is still impossible to say, but precise knowledge of animal gestation cycles has obvious adaptive advantages for any hunting people, and as a motive for counting, it tightens the fit with the associated glyphs.

The gestation cycle explanation neither invalidates nor contradicts the lunar features noted earlier in the La Mula count, but it does convert them into a means to an end, rather than the end itself, and raises the following questions. What kind of astronomical knowledge does the lunar time-keeping system used imply in and of itself? Was sky-watching ever done for its own sake?

A re-analysis of the La Mula tally's counting pattern revealed a hitherto unsuspected facet of the tradition's astronomical basis. The pattern

Counting segments	Segment total	Accumulative total	Observation postulated
1/ 11 + 11 + 4 + 1	27	27	Sidereal month
+ 2	2	29	Synodic adjustment
2/ 14 + 13	27	56	Sidereal month
3/ 12 + 3 + 3 + 11	29	85	Synodic month
4/ 7 + 8 + 7 + 5 + 1	28	113	Synodic month? Sidereal month?
+ 5	5	118	Synodic adjustment
5/ 10 + 6 + 6 (+ 2 + 3)	27	145	Sidereal month
+ 2	2	147	Synodic adjustment
6/ 4 + 12 + 6 + 4 + 1	27	174	Sidereal Month
+ 3	3	177	Synodic adjustment
7/ 9 + 2 + 5 + 5 + 9	30	207	Synodic month

TABLE 1. Counting pattern of the tally stone at Presa de La Mula, N.L. (Mexico) in relation to lunar periodicities.

and its total sum show clearly that the lunar/synodic relationship is being noted, but when its monthly segments are isolated (Table 1), one can see that the sidereal month (27·34 days) is also being noted in some cases. In fact, the La Mula tally seems to be a complex combination of the two. Sidereal lunar notation requires a set of star markers, and an awareness of sidereal motion independently of lunar and solar motion. Such knowledge of the annual sidereal cycle seems to have been shared by both the tally and the dot counters. With this clue, a number of dot configurations can be explained as sidereal notations, among them the 'drip lines' which cross the left side of the La Mula tally, whose total number (54±1) is an acceptable approximation of two sidereal months (54·68 days), and the unique 'jar' count at Boca de Potrerillos, whose three loops sum to 191+1, or seven sidereal months (191·38 days), with a fifteen-day 'post-fix' added in the middle to reflect the accumulated synodic/sidereal discrepancy.

The site of Boca de Potrerillos also provides the key to the sources of this knowledge, because here petroglyphic counting forms part of a unique horizon orientation system which brings together sun, moon, and stars. Boca is a large site with more than three thousand petroglyphs extending for almost a kilometre on both sides of a canyon mouth which cuts through a ridge crest running due north-south. Its most peculiar feature is that even though perfectly suitable rock is available for carving on the west-facing slope, all of the petroglyphs face east. The eastern horizon is spiked by jagged mountain crests 7-20 km distant whose long sightlines permit almost any heliacal rising to be pinpointed with relative ease and accuracy. In March 1979, field observations from a rock promontory directly behind the canyon mouth identified several circle petroglyphs which aligned extremely well with the equinox sunrise position (Murray 1982a). Subsequently, other heavily carved monoliths on the North Crest were detected with the same due east orientation. On these stones a motif complex could be identified consisting of dot configurations, circles, and zigzags which appears to relate systematically to horizon observation/representation.

Once the basic principle of horizon orientation was mastered, times and directions other than the equinox could be easily established by reference to heliacal star risings. Further away from the canyon mouth, other monolithic rocks with the same horizon motifs mark both cardinal east and cardinal north, which can be established observationally by simply noting circumpolar star motion, perhaps with the aid of the north-south oriented crest itself. With north and east established, south and west can be inferred, and intercardinal directions could be derived by using star risings in the appropriate directions at any given time of year. My present hypothesis is that Boca de Potrerillos displays the elements of a terrestrial navigation system much like that described by Cabeza de Vaca in the sixteenth century, which would have allowed a hunter-gatherer group to orient themselves in space and time, and enhance their chances of survival by opportune scheduling of movements to highly localised food resources.

Much remains to be learned about how petroglyphic counting relates to terrestrial navigation by the stars. Nevertheless, the pieces to the puzzle continue to multiply, and a more coherent picture of this ancient astro-numerical tradition gradually emerges. Its further elucidation may eventually contribute to a new and clearer picture of the cultural ideas which guided Amerindian peoples in their adaptation to the New World.

Acknowledgements

The author's participation in Oxford 3 was made possible by a special grant from the Office of Cultural Affairs, Municipio of Garza Garcia, N.L., and he wishes to thank Sra. Carolina Sada de Viesca, its director, for this support. Thanks are also due to Sr. Conrado Hidalgo (Monterrey, N.L.) for his invaluable photographic assistance; Phyllis Pitaluga (Adler Planetarium, Chicago) and Edwin C. Krupp (Griffith Observatory, Los Angeles) for access to their respective planetaria and their patient explanations of sky phenomena; Ing. Armando Buentello and the members of the Exploradores de México (San Nicolás de los Garza, N.L.) for their aid in field documentation; and Prof. Anthony Aveni (Colgate University) for his comments on some of the hypotheses presented here.

References

Aveni, A.F. (1989). Pecked cross petroglyphs at Xihuingo. *Archaeoastronomy* no. 14 (supplement to *Journal for the History of Astronomy* **20**), S73–115.

Cabeza de Vaca, Alvar Nuñez (1942[1537]). *Naufragios y comentarios*. Madrid: Espasa Calpe.

Hester, T.R. (1980). Digging into South Texas prehistory. San Antonio TX: Corona Publishing Co.

Leopold, A.S. (1977). *Fauna silvestre de México*. Mexico City: Editorial Pax-México.

Marshack, A. (1985). A lunar-solar year calendar stick from North America. *American Antiquity* **50**(1), 27–51.

Marshack, A. (1989). North American calendar sticks: the evidence for a widely-distributed tradition. In *World archaeoastronomy*, ed. A.F. Aveni, pp. 308–24. Cambridge: Cambridge University Press.

Murray, W.B. (1982a). Rock art and site environment at Boca de Potrerillos, N.L. In *American Indian rock art, vol. 7–8*, ed. F. Bock, pp. 57–68. El Toro CA: American Rock Art Research Association.

Murray, W.B. (1982b). Calendrical petroglyphs of northern Mexico. In *Archaeoastronomy in the New World*, ed. A.F. Aveni, pp. 195–204. Cambridge: Cambridge University Press.

Murray, W.B. (1984a). Numerical characteristics of three engraved bison scapulae from the Texas Gulf Coast. *Archaeoastronomy* (Center for Archaeoastronomy) **7**, 82–88.

Murray, W.B. (1984b). A spiral pecked dot petroglyph from eastern Coahuila, México. *La Pintura* **11**(2), 11 and 14.

Murray, W.B. (1985). Petroglyphic counts at Icamole, Nuevo León, México. *Current Anthropology* **26**(2), 276–79.

Murray, W.B. (1986). Numerical representations in North American Rock Art. In *Native American mathematics*, ed. M. Closs, pp. 45–70. Austin TX: University of Texas Press.

Murray, W.B. (1989). A re-examination of the Winnebago calendar stick. In *World archaeoastronomy*, ed. A.F. Aveni, pp. 325–30. Cambridge: Cambridge University Press.

Murray, W.B. (1992). Antlers and counting in northeast Mexican Rock Art. In *American Indian Rock Art*, ed. K. Sanger, vol. 15, pp. 71–79. San Miguel CA: American Rock Art Research Association.

Wallrath, M. and Rangel, A. (1984). *Xihuingo (Tepeapulco): centro de cómputo de la astronomía Teotihuacana*. Paper presented at the symposium on archaeoastronomy and ethnoastronomy in Mesoamerica, Universidad Nacional Autónoma de México, México City.

24

Venus orientations in ancient Mesoamerican architecture

IVAN ŠPRAJC

Recent archaeoastronomical research in Mesoamerica has shown that some orientations in Prehispanic architecture most probably refer to the planet Venus. The purpose of this paper is to

- call attention to certain astronomical facts concerning Venus extremes;
- summarise the evidence about those Venus orientations that have already been described in the literature;
- report on new orientations discovered by the author;
- show that all Venus orientations in Mesoamerica known so far most probably relate to the Evening Star; and
- indicate the possible symbolic significance of these orientations.

With the exception of the alignments incorporated in Temple 22 of Copan (Closs *et al.* 1984; Šprajc n.d.), all Venus orientations known so far relate to the extremes. It is therefore necessary to point out some characteristics of the apparent motion of the planet, which must be taken into account in any consideration of the alignments to Venus extremes.

Venus extremes

Since Venus is a planet, its declination changes and, consequently, its rising and setting points move along the eastern and western horizon, respectively, reaching northerly and southerly extremes. Dates and magnitudes of the extremes vary considerably, but exhibit eight-year patterns (like other Venus phenomena). Because the plane of Venus's orbit is slightly inclined to that of the ecliptic, some eight-year cycle extremes are greater than the solstitial extremes of the sun.

The importance of Venus extremes in Prehispanic Mesoamerica has been discussed by Aveni (1975), Aveni *et al.* (1975), and Closs *et al.* (1984). The last authors (*ibid.*: 234 ff.) noticed that all great northerly extremes (when the planet attained a declination in excess of 25°·5) in the eighth and ninth centuries were visible in late April or early May, i.e. they were seasonally fixed, coinciding approximately with the onset of the rainy season in Mesoamerica. Further research (Šprajc 1990a) has shown that

- *all Venus extremes are seasonal phenomena*; and

- *the maximum extremes of the morning star and evening star are asymmetric*, i.e. maximum extremes visible on the eastern horizon differ in magnitude from those visible on the western horizon.

During the Classic and Postclassic eras in Mesoamerica, Venus attained absolute declinations in excess of about 24° 10′ (up to about 27°·5) exclusively when it was visible as *evening star*, and always some time *before* the solstices: between April and June (northerly extremes) and between October and December (southerly extremes). The *maximum extremes*, occurring at eight-year intervals, always fell between May 2 and May 6 (north) and between Nov 2 and Nov 7 (south), Gregorian. On the other hand, the extreme declinations of Venus as *morning star* were always attained *after* the solstices, between late December and February (south) and between late June and August (north), but they never exceeded ±24° 10′ (Šprajc 1990a: table 1; for the astronomical explanation of these phenomena see *ibid.*: Appendix).

In view of the rainfall data for various parts of Mexico (E. García 1987[1964]: 22–30, 62–70), the evening star extremes seem particularly interesting, since they coincide approximately with the start (northerly extremes) and with the end (southerly extremes) of the rainy season in Mesoamerica. As we shall see, Venus orientations in Mesoamerican architecture refer to the *maximum extremes*, which, as was noted above, are particularly accurate time-markers.

Orientations to Venus extremes

Chichen Itza

While exploring the astronomical properties of the Caracol of Chichen Itza, Aveni *et al.* (1975) discovered that some of the lines with a possible astronomical significance point to the northerly and southerly Venus extremes on the western horizon. The azimuths of the perpendiculars to the base of the lower platform and to the base of the stylobate platform agree with the azimuth of the maximum northerly extreme. Furthermore, maximum northerly and southerly extremes could have been observed along the diagonal lines of windows 1 and 2 respectively (*ibid.*: table 1, figs. 2, 4–6). Some evidence is quoted to demonstrate that these alignments were most probably intentional (*ibid.*: 980).

Uxmal

Aveni (1975: 184ff., fig. 6, table 5) observed that a straight line taken from the western doorway of the House of the Magician and passing through the centre of the ball court, the centre of the northern plaza of the south group and the principal doorway of the west group lies within 1° of Venus maximum southerly setting point around AD 750. Beyond this correspondence there is, however, no evidence indicating that this alignment was intentional (Aveni 1990).

Venus orientation was also suggested for the House of the Governor, skewed about 15° relative to the common orientations at Uxmal. The line from the principal doorway towards the south-east, perpendicular to the front face of the Palace, passes almost exactly over a distant mound, believed to be the Great Pyramid of Nohpat (Aveni 1975: 183). Field research in 1989 revealed that this identification was erroneous and that the bump visible on the south-eastern horizon is the main pyramid of Cehtzuc, a relatively small site situated about 4·5 km from Uxmal (Šprajc 1990b). According to Aveni (1975: 184), 'the alignment from the Governor's Palace to Nohpat points almost exactly to the azimuth of Venus rise when the planet attained its maximum southerly declination around AD 750'. The azimuth of the line from the principal doorway to Nohpat (=Cehtzuc) is given as 118° 13' (*ibid.*: Table 5). However, considering what has been stated above about the asymmetry of the morning star and evening star extremes, Venus rising as morning star on the eastern horizon of Uxmal could never reach an azimuth greater than 115° 40'. It seems improbable that the Maya would have made such a great error, if indeed they had wanted to incorporate this Venus alignment in their architecture.

The alignment was more likely intended to work in the opposite direction: an observer standing at Cehtzuc would have been able to see Venus *set* behind the House of the Governor, when the planet attained its greatest *northerly* extremes.[1] Such an hypothesis has iconographic support in the decoration of the façade of the Palace: almost four hundred Venus (or star) glyphs adorn the cheeks of the Chac (rain god) masks. It should be recalled that maximum northerly extremes, occurring around May 3, coincided with the beginning of the rainy season. A relationship between

---

[1] Since the angular width of the Palace, as viewed from Cehtzuc, is 1° 14', Venus would actually have set behind the Governor's Palace at two or three of the five northerly extremes visible in an eight-year cycle.

the structure and the planet is also implied by the presence of the numeral eight on the masks at both northern corners (Aveni 1982: 15, fig. 3).

Santa Rosa Xtampak

The Great Palace of this Chenes site in Mexican federal state of Campeche may also be oriented to the Venus extreme, since Aveni (1982: 14) mentions that it possesses the same orientation as the Governor's Palace at Uxmal.

Nocuchich

Structure 2 of this Chenes site in the state of Campeche (Pollock 1970: 45 ff., fig. 54) faces south-west. Transit measurements in 1989 showed that it may have been oriented to the maximum southerly extreme of Venus as evening star, since it is skewed approximately 28° south of west.[2] Unfortunately, there is no other evidence indicating that this was an intentional Venus orientation. It can only be pointed out that Venus may be hinted at in the name of the site: *Nocuchich*, meaning 'Great Eye', reminds us of *Nohoch Ich*, which is one of the names for Venus among the Maya of Belize (Thompson 1930: 63).[3]

Huexotla

One of the structures at Huexotla, a Postclassic site south of Texcoco in the state of México, is the so-called *Circular*. The building faces east and has a substructure of which the northern balustrade crops out (García García 1987: 79, pl. 4). The two phases of construction do not have the same orientation (*ibid.*: 19). Results of transit measurements carried out in November 1988 by A. Ponce de León and myself suggest that the substructure was oriented to the maximum northerly extreme of Venus as evening star. The hypothesis is supported by the following evidence.

The only visible balustrade of the substructure yields an azimuth of 118° 16'. Viewing east, the azimuth does not seem to be astronomically significant. However, the declination of 26° 52', corresponding to the azimuth in the opposite direction, agrees with the value of the maximum northerly declination of Venus in the mid-twelfth century. Archaeological dating places the construction of El Circular-sub between AD 1150 and 1350 (García García 1987: 82, 100).

---

[2] I was not able to determine the accurate orientation, since the measurable walls of this tower-like structure are relatively short. If Venus orientation was involved, it could only have functioned in relation to some other building, placed in the required direction, to form an alignment appropriate for observations.

[3] A local informant said, however, that the site owes its name to the bulbous eyes of the face that was modelled in stucco on Structure 1 (Pollock 1970: 44, fig. 53), now ruined.

In Pollock's (1936: 147, table 5) classification of round structures, El Circular of Huexotla belongs to Type 5, which appears amongst the types associated with the cult of Quetzalcóatl (*ibid.*: 159 ff.). The wind jewels, prevalent motives on the pottery found in the vicinity by Batres (1904: 6), confirm this association which, in view of the relationship between Venus and Quetzalcóatl, has obvious implications for the orientational hypothesis.

The proposition that the orientation referred to the phenomena on the western horizon is reinforced by the way in which the structure relates to the surrounding topography. The axis of the balustrade prolonged eastwards does not point to any significant natural feature; it passes over the southern slope of the Tlapanco hill, relatively close to the site. In the opposite direction, however, the same axis leads to Pico Tres Padres, the highest mountain visible on the NW horizon, lying north of Mexico City and about 30 km from Huexotla.

A few recent studies have demonstrated that a number of Prehispanic temples are oriented upon prominent mountains in their neighbourhood; in many cases these orientations are also astronomically significant (cf. Ponce de León 1983; Tichy 1983; Aveni *et al.* 1988). It means that the localities for the erection of ceremonial buildings must have been premeditated, in accordance with certain principles of 'sacred geography' or geomancy, in which astronomical and calendrical considerations, beliefs concerning local topographic features, and probably many other factors were involved (Carlson 1981; Broda 1993). Though we are far from understanding how the whole combination of these rules really functioned, it is a fact that mountains had an important role in the Prehispanic Mesoamerican world view (Broda 1982; 1987; 1991b).

Specifically, we know that Pico Tres Padres was of particular importance for the Mexica. In the month *Atlcahualo* children were sacrificed on various hills in the Valley of Mexico; one of these was Quauhtepetl (Sahagun 1985: 98–b.2, ch. 20; Broda 1971: 273), which can be identified with Pico Tres Padres (Broda 1991a). Fragments of Prehispanic pottery found on its summit, as well as piles of stones visible on a plain near the summit, corroborate the historical reports.[4]

Considering the evidence presented, it seems highly probable that the substructure of El Circular was erected on a carefully selected place, from which the line to the prominent mountain on the NW horizon marked the maximum northerly setting point of Venus.

Final remarks

Many years ago Seler (1963[1904], vol. II: 117) observed that the temples consecrated to the planet Venus face west. This statement should now be corrected: the façades of the round temples of Quetzalcóatl are oriented

---

[4] The place was already recognised as an archaeological site by Sanders *et al.* (1979: map 18).

towards the west on the Yucatan peninsula, whereas in central Mexico they look east (Pollock 1936: 160). The Governor's Palace at Uxmal, evidently associated with Venus, also has its access from the east. In spite of these dissimilarities, it can be argued that all Venus alignments examined so far refer to the extremes of the evening star, visible on the western horizon. Thus, the side of the building (east or west) on which the entrance or stairway was placed is not apparently an indication of the direction of astronomical importance.

What, then, can be said about the beliefs underlying these orientations? Closs *et al.* (1984) disclosed a conceptual relationship between the planet Venus and rain and maize in the ancient Mesoamerican world view. My own research (Šprajc 1989) has revealed that their conclusions can be substantiated by a vast amount of archaeological, ethnographic, iconographic and historical data, which suggest, moreover, that the so-called Venus-rain-maize complex was associated primarily with the Evening Star. I have also argued that the most important observational basis for these concepts must have been the concomitance of Evening Star extremes with two crucial annual climatic changes in Mesoamerica.

The fact that the maximum extremes always occur on almost the same dates of the tropical year, delimiting the rainy season and agricultural cycle, probably accounts for their importance attested in orientations. The significance of these dates is implied in a number of other orientations, which most probably refer to the sunrise and sunset points: days around May 3 and Nov 3 are recorded by the so-called 17°-family of orientations, very common in ancient Mesoamerican architecture (Aveni 1980: 237, 311 ff.). The great popularity of the Holy Cross festival and of All Saints' Day among the present-day Indians may also reflect the importance of these dates in Prehispanic times. If the Evening Star was regarded as one of the agents responsible for important cyclic changes in natural environment, the orientations to its extremes can be viewed as one more manifestation of the attempts of ancient Mesoamericans to reproduce and perpetuate the cosmic principles and heavenly order in their earthly environment.

References

Aveni, A.F. (1975). Possible astronomical orientations in ancient Mesoamerica. In *Archaeoastronomy in pre-Columbian America*, ed. A.F. Aveni, pp. 163–90. Austin TX: University of Texas Press.

Aveni, A.F. (1980). *Skywatchers of ancient Mexico*. Austin TX: University of Texas Press.

Aveni, A.F. (1982). Archaeoastronomy in the Maya region: 1970–1980. In *Archaeoastronomy in the New World*, ed. A.F. Aveni, pp. 1–30. Cambridge: Cambridge University Press.

Aveni, A.F. (1990). The real Venus-Kukulcan in the Maya inscriptions and alignments. In *Sixth Palenque Round Table, 1986*, ed. V.M. Fields, pp. 309–21. Norman OK: University of Oklahoma Press.

Aveni, A.F., Calnek, E.E. and Hartung, H. (1988). Myth, environment, and the orientation of the Templo Mayor of Tenochtitlan. *American Antiquity* **53**(2): 287–309.

Aveni, A.F., Gibbs, S.L. and Hartung, H. (1975). The Caracol tower at Chichen Itza: an ancient astronomical observatory? *Science* **188**, 977–85.

Batres, L. (1904). *Mis exploraciones en Huexotla, Texcoco y montículo de 'El Gavilán'*. Mexico City.

Broda, J. (1971). Las fiestas aztecas de los dioses de la lluvia: una reconstrucción según las fuentes del siglo XVI. *Revista Española de Antropología Americana* **6**, 245–327.

Broda, J. (1982). El culto mexica de los cerros y del agua. *Multidisciplina* **3**(7), 45–56.

Broda, J. (1987). Templo Mayor as ritual space. In *The Great Temple of Tenochtitlan: center and periphery in the Aztec world*, eds. J. Broda, D. Carrasco and E. Matos Moctezuma, pp. 61–123. Berkeley CA: University of California Press.

Broda, J. (1991a). The sacred landscape of Aztec calendar festivals: myth, nature and society. In *To change place: Aztec ceremonial landscapes*, ed. D. Carrasco, pp. 74-120. Niwot CO: University Press of Colorado.

Broda, J. (1991b). Cosmovisión y observación de la naturaleza: el ejemplo del culto de los cerros. In *Arqueoastronomía y etnoastronomía en Mesoamérica*, eds. J. Broda, S. Iwaniszewski and L. Maupomé, pp. 461–500. Mexico City: Instituto de Investigaciones Históricas, Universidad Nacional Autónoma de México.

Broda, J. (1993). Astronomical knowledge, calendrics and sacred geography in ancient Mesoamerica. In *Astronomies and cultures*. eds. C.L.N. Ruggles and N.J. Saunders. Niwot CO: University Press of Colorado. In press.

Carlson, J.B. (1981). A geomantic model for the interpretation of Mesoamerican sites: an essay in cross-cultural comparison. In *Mesoamerican sites and world-views*, ed. E.P. Benson, pp. 143–215. Washington DC: Dumbarton Oaks (Trustees for Harvard University).

Closs, M.P., Aveni, A.F. and Crowley, B. (1984). The planet Venus and Temple 22 at Copán. *Indiana* **9**, 221–47.

García, E. (1987[1964]). *Modificaciones al sistema de clasificación climática de Köppen*, 4th edn. Mexico City: published by the author.

García García, M.T. (1987). *Huexotla: un sitio del Acolhuacan*. Mexico City: Instituto Nacional de Antropología y Historia (Colección Científica, 165).

Pollock, H.E.D. (1936). *Round structures of aboriginal Middle America*. Washington DC: Carnegie Institution of Washington (Publication no. 471).

Pollock, H.E.D. (1970). Architectural notes on some Chenes ruins. *Papers of the Peabody Museum of Archaeology and Ethnology* **61**, 1–87.

Ponce de León H., A. (1983). Fechamiento arqueoastronómico en el altiplano de México. In *Calendars in Mesoamerica and Peru: native American computations of time*, eds. A.F. Aveni and G. Brotherston, pp. 73–99. Oxford: British Archaeological Reports (BAR International Series 174).

Sahagun, B. de (1985). *Historia general de las cosas de Nueva España*, 6th edn., ed. A.M. Garibay. Mexico City: Editorial Porrúa.

Sanders, W.T., Parsons, J.R. and Santley, R.S. (1979). *The basin of Mexico: ecological processes in the evolution of a civilization*. New York NY: Academic Press.

Seler, E. (1963[1904]). *Comentarios al Códice Borgia*, 2 vols. Mexico City: Fondo de Cultura Económica. (First edn. in German.)

Šprajc, I. (1989). *Venus, lluvia y maíz: Simbolismo y astronomía en la cosmovisión mesoamericana*. MA thesis, Escuela Nacional de Antropología e Historia, Mexico City.

Šprajc, I. (1990a). Venus, lluvia y maíz: el simbolismo como posible reflejo de fenómenos astronómicos. In *Memorias del segundo coloquio internacional de Mayistas: Campeche, 17-22 de agosto de 1987*, pp. 221–48. Mexico City: Universidad Nacional Autónoma de México (Instituto de Investigaciones Filológicas, Centro de Estudios Mayas).

Šprajc, I. (1990b). Cehtzuc: a new Maya site in the Puuc region. *Mexicon* **12**(4), 62–63.

Šprajc, I. (n.d.). Venus and Temple 22 at Copán: revisited. *Archaeoastronomy* (Center for Archaeoastronomy) **10**, in press.

Thompson, J.E.S. (1930). *Ethnology of the Mayas of southern and central British Honduras*. Chicago IL: Field Museum of Natural History (Publication 274: Anthropological Series, vol. 17, no. 2).

Tichy, F. (1983). El patrón de asentamientos con sistema radial en la meseta central de México: ¿'sistemas ceque' en Mesoamérica? *Jahrbuch für Geschichte von Staat, Wirtschaft und Gesellschaft Lateinamerikas* **20**, 61–84.

25

Mesoamerican geometry combined with astronomy and calendar: the way to realise orientation

FRANZ TICHY

1 Introduction

Observations of the orientation of the groundplans of settlements and field patterns as well as of archaeological structures in the high basins of Mexico, Puebla-Tlaxcala and Oaxaca since 1973 (Tichy 1981) were originally intended to show their distribution and how they developed. However, it soon became necessary to learn more about the results and methods of archaeoastronomy, a discipline which was being established at the time. Co-operation with Anthony Aveni made it possible for me to do so without great difficulty. This was accompanied by an exchange of research results and views between the present author and Johanna Broda in the field of ethnohistory, especially the Mesoamerican calendar. Soon the geographer's first questions were answered. Clear links were established with the position of the sun on the horizon and, in addition, the agrarian calendar and a hypothetical solar calendar. There were also observations on the alignments between ceremonial centres or settlements and the most important mountains and between the settlements themselves, independently of the orientation of the groundplans.

A revision of the research results already reported led to new considerations and approaches. How was the astronomically measured direction translated—that is to say, how was the direction of the point where the sun rose or set on a relevant day in the calendar used in the planning and construction of buildings at another site? To what extent might a knowledge of geometry and its methods have been involved, in addition to experience gained during observation of the sun and by making calendars? To my knowledge this question has not yet been asked; though it is an obvious one, since astronomical observations cannot be recorded without measuring angles or determining certain radii of circles. Yet, we still know very little about the geometrical knowledge of the peoples of Mesoamerica. There are no written records, hence we are dependent on the interpretation of evidence from archaeology, sculpture and graphic representations in codices. A few rather meagre results have been gained by showing that right angles were formed and how space was divided up in primitive wall paintings. Vinette (1986) summarised the current state of research on Mesoamerican geometry. She discusses Aveni's archaeoastronomical publications as well as Hartung's on the planning of ceremonial centres, ending with the words: 'There is little doubt that further investigations directed towards an evaluation of Mesoamerican knowledge, will include geometry as a part of the amalgam of Mesoamerican science and religion'.

I shall call upon the body of data described and interpreted elsewhere (e.g. Tichy 1983; 1988; 1991), but using it now for a different purpose. The archaeoastronomical basis is still valid but, in my opinion, a simpler order can now be perceived, one that can be described as geometry.

2 The geometry of the Mesoamerican world order with the four solstitial points in the Olmec sign *olin*

It is well known that the peoples of Mesoamerica based their cosmological views on a horizon-zenith system, in which the fixed points of the horizon are not the main directions familiar to us, but the four points of sunrise and sunset on the solstitial days, the extreme positions of the sun on the horizon. These points of standstill and return are especially noteworthy and easy to observe in the field. One has no difficulty in marking this solar alignment.

FIG. 1. Olmec engraving on a stone celt from La Venta, Tab. Mexico, identified as a complex representation of the cosmic ideogram (after Köhler 1982: fig. 4).

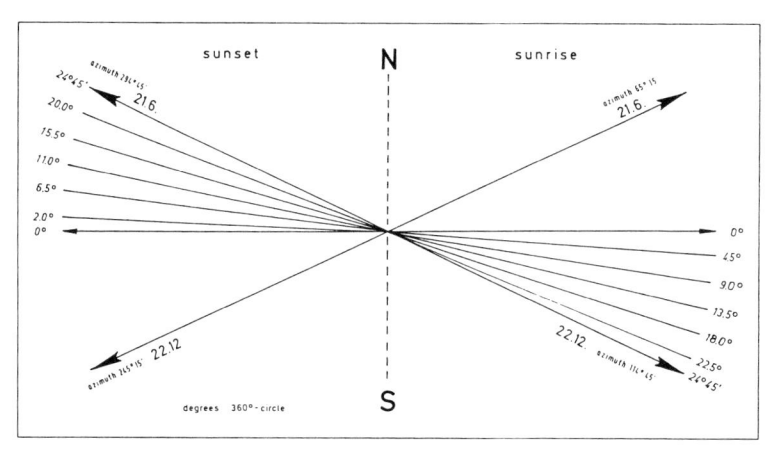

FIG. 2. Diagram of the Mesoamerican orientation system. The alignments are in accordance with the hypothetical solar orientation calendar with thirteen-day weeks. The clockwise deviations from east represent the 'winter sunrise Maya type' and the deviations from west the 'summer sunset central Mexico type'.

The Olmec sign *olin* (Fig. 1) represents the basic plan of their relative positions at the horizon in a very natural way. The ideogram with the observer's own position as the central point and with the sightlines leading to the solstitial points is rectangular in shape with sides in the ratio 1:2. The diagonal lines of a rectangle of this type subtend an angle of 54° at the centre; in the figure represented, the angle measured is 52°. This is a very good approximation to the actual distance between the solstitial points, which, determined astronomically at a latitude of 19°N, is approximately 49°·5. As a cosmological sign *olin* is, of course, not an astronomical geometrical sketch, but the expression of elementary links between astronomy, cosmology, religion, the calendar and, as we have seen, geometry.

It appears that the positions of sunrise between due east and the winter solstice, and of sunset between due west and the summer solstice, provided the astronomical basis for the orientation not only of ceremonial buildings but also of settlements including the field patterns of small and large territorial units (Fig. 2). Such orientations most frequently follow directions 4°·5 apart, as demonstrated by histograms of the orientations of churches in the Mexico basin. The sequence 7° – 11°·5 – 16° – 20°·5 – 25° south of east/north of west was found there: the final value in this sequence clearly corresponds to the solstitial point. The well-known Teotihuacan bearing of 16° is the most frequent. Aveni and Hartung (1986) report on data for the Maya area which have a similar sequence but which begin in the east or the west, i.e. in an equinoctial point (or rather in the sun's position on mid-quarter days);

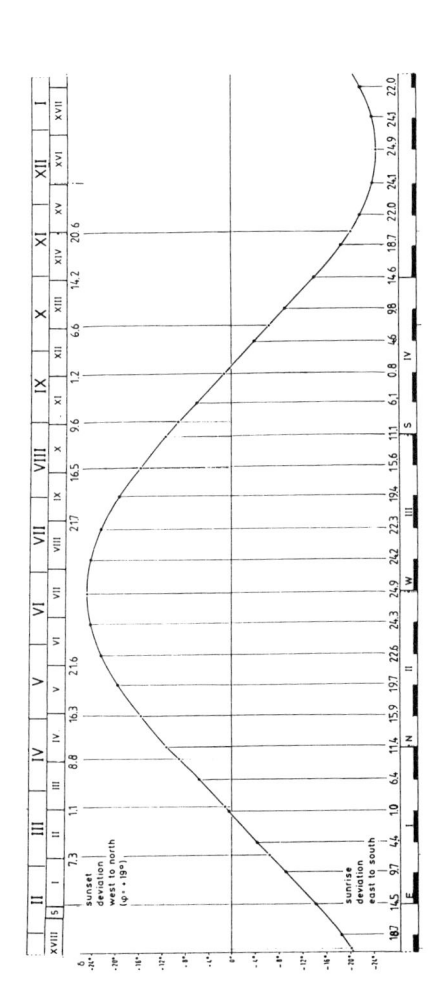

FIG. 3. Hypothetical solar orientation calendars with twenty-day periods (above) and thirteen-day periods (below). The vertical axis shows the declination of the sun. The horizontal axis shows the horizon azimuth for latitude 19° N.

the series is 4°·5 – 9° – 13°·5 – 18° – 22°·5. There is clearly a geometrical principle at work here, namely the division of the right angle into twenty units of 4°·5. The whole sweep between the two solstice points in the east or west contains eleven such units within 49°·5; that in the *olin* sign, i.e. in the rectangle with its sides in the ratio 1:2, contains twelve of these units within 54°. If the long axes of buildings are oriented towards the position of the sun, it must also be possible to find their geometry in the movement of the sun throughout the year, or rather in the calendar.

3 Geometry, the movement of the sun and the calendar

The passage of the sun as the year progresses is described using the change in the declination of the sun, which can be represented as a sine curve (Fig. 3). If the orientation sequences found for central Mexico and the Maya area are assigned to the curve, and one adds the calendar with thirteen-day periods, the connection between geometry, the passage of the sun and the calendar becomes evident. This representation is the final result that I have reached in my work. After realising that an ordering principle for a sequence of angles for the orientation axes existed in central Mexico, I first proposed a fixed solar calendar with twenty-day periods, which was regulated in some way, following the festival calendar of the Mexica. However, only a few of the monthly festivals could be connected with the position of the axes. This attempt was finally successful when I postulated a thirteen-day calendar, one which was essentially well known, although it was not fixed either. What is significant is that it contains the solstices and the mid-quarter days and, for some latitudes, the days of the passages of the sun through the zenith. In the periods between the solstices, in which the branches of the sine curve are more or less rectilinear, the declination values fall or rise by almost the same values every day. At thirteen-day intervals they change on average by 4°·5, and in the Tropics the change in the azimuths of the positions of the sun on the horizon in this period is also approximately 4°·5.

We can assume that this relationship had already been recognised in early pre-Hispanic times. The angular intervals to the nearest solstitial point for the position of the sun at the limits of the thirteen-day periods must have been known. Geometry could be used to fix a suitable axial position for a building where the sunrise or sunset could not be observed near the horizon because of high mountains. Of course, some sort of measurement was required for this.

How can a unit for the measurement of angles be arrived at? Perhaps people proceeded as they did for units of length, i.e. they found appropriate units on their own bodies. Outspread fingers at arm's length seen

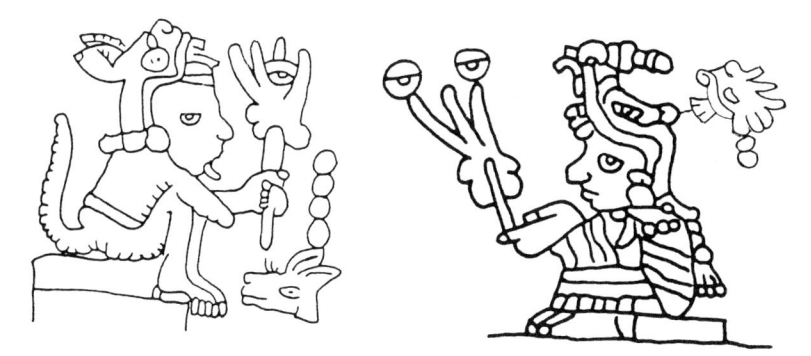

FIG. 4. Person with an implement which can be interpreted as a device for astronomical observation.
   Left: Man 'Four Deer', Codex Selden p. 14–IV, holding implement with four prongs ('Jaguar-torch with eye'). After Caso (1979: 164).
   Right: Woman 'Two Water', Codex Bodley p. 17–IV, holding implement with three prongs ('Fire snake torch stars'). After Caso (1960: 17) and Tichy (1988: fig. 5).

with one eye are a sort of instrument for observing angles. Using two fingers, 4°·5 can just be covered; the angular separation between the index finger and the little finger is three times this, or 13°·5. As this method is not accurate enough for astronomical measurements, a device will have been made resembling that represented in the Selden and Bodley codices and which I interpret as the man 'Four deer' on the left and the woman 'Two water' on the right, each holding out in front of them a tendril with three or four prongs or a corresponding bundle of twigs. The tops of the twigs must be equidistant from each other. The stars—drawn in the figure like eyes—suggest that the device was used to observe stars. In this way it was possible to determine the altitude of stars above the horizon and the angular distance between individual stars. One observation may have been of special significance: the fact that below latitude 19°N at noon on Jun 21, the summer solstice, the sun reaches its highest point—around 4°·5 from the zenith towards the north. This is an example of the application of geometry with a circle divided into eighty parts in Mesoamerican astronomy.

4 Determining the axial positions for orientation using geometry

A question that is often asked is whether observations of the sun or the stars were always used for determining the orientation of building axes when ceremonial centres were planned. It is often difficult to prove this;

incorrect dates have frequently been given and repeated, as the example of Teotihuacan shows. However, in the case of solstitial orientations, such as one of the axes of the Cholula pyramid upon winter solstice sunrise, the astronomical relationship can easily be determined.

Let us assume that the people of Mesoamerica knew a geometry of the circle divided into eighty units, that is to say that they conceived and made practical use of a standard angle of $4°·5$. If this was so, then with simple aids—just by using their open hands—they could indicate directions quite accurately and reproduce them. If the hypothesis were proved that a regulated calendar was used, i.e. one that was always adjusted to the passage of the sun, either with twenty-, or even better, thirteen-day periods, then the close connection between the passage of the sun, the calendar, calendar festivals, agrarian rites, the agricultural cycles and, very importantly, the axial positions of ceremonial buildings would always have been guaranteed. Under these conditions it would be understandable that a principle of orientation such as that of Teotihuacan could be transferred with the spread of Teotihuacan culture, together with all the constituent elements mentioned above. It was not necessary to wait until the beginning of May, and the start of the rainy season with its corresponding ceremonies, to fix the orientation of a building. It was sufficient to measure the angular distance of $9°$ from a well-known solstitial direction using the distance between three fingers. I believe that in the Maya area the direction of the sun on the mid-quarter days formed a firm basis for this. In this way people became independent of the influence of mountains and mountain ranges. This explains why the orientations of ceremonial centres deviate very little from the sequences of angles that have been discovered.

5 Orientation on the whole horizon

In recent years following the observations begun by the architect González Aparicio (1973), it has become much more evident that the planning of architecture and settlements in Mesoamerica was not only carried out according to the principles of the astronomical calendar discussed so far. Individual places were imagined as being linked to mountains or other places by straight lines. It is clear now that the connecting lines found by González Aparicio, and many other lines, were also oriented upon azimuths belonging to the eighty-part circle. In addition, planned arrangements of small dependent places and ceremonial sites evidently surrounded some large central places, with the satellites being situated on a small number of radii. The geometry of the whole circle very probably provided the basis for such arrangements. Comparisons with the Ceque system of Cuzco have met with criticism, however. In this case astronomy, including observations of the sun, plays hardly any role. Yet there are several striking connecting lines which lead to solstitial points across and beyond individual mountains, running, for example, from the Mexico basin across the Sierra Nevada to the Puebla basin and on to Cholula.

6 The geometry of the rectangle and its diagonals

The discovery of the Mesoamerican angular unit comprising one twentieth of the right angle resulted from frequency counts of orientations taken from aerial photos and maps as well as from measured data. These angles have so far been found as azimuths of the horizontal circle. It can be shown, however, that this angular unit plays a part in the geometry of rectangles with sides in different ratios, i.e. with the properties of its tangent. The Olmec sign *olin* gave a first indication that this might be the case. With sides in a ratio of 1:2 its diagonals intersect at an angle of 54°, or 12 units, to the west and east. I have examined rectangles with their sides in different ratios which are found in the groundplan of courtyards in Teotihuacan, of the pyramid of the plumed serpent in Xochicalco and in the frontispiece of the Mendoza codex (all 6:7) and also in representations of stelae from Tikal, described by Vinette (1986: 391) quoting Clancy (5:12 and 5:7 or 3:4). Table 1 gives one or more appropriate ratios of sides for every number of Mesoamerican units from 1 to 10.

Ratio of sides (tan α)	Angle α in degrees	Angle α in units of $4°\cdot 5$
1 : 13	4°·40	1
3 : 19	8°·97	2
1 : 4	14°·04	3
1 : 3	18°·44	4
7 : 17	22°·38	5
5 : 12	22°·62	5
1 : 2	26°·57	6
8 : 13	31°·61	7
5 : 7	35°·54	8
13 : 18 *	35°·84	8
3 : 4	36°·87	8
6 : 7	40°·60	9
1 : 1	45°·00	10

TABLE 1. The geometry of the rectangle and its diagonals in Mesoamerica. The asterisk denotes the golden section.

7 Conclusion

Proceeding from the Olmec day sign *olin*, which has a clear geometrical structure, I have provided further data suggesting that geometrical procedures were known and used in Mesoamerica. The idea of an angular unit of measurement has received further support from the relationship between the solar declination and horizon azimuth in the course of the thirteen-day periods of a hypothetical calendar, from the possible use of an observation device for the measurement of angles, and from a geometry of the rectangle and its diagonals. Though no new data on archaeoastronomy and orientation in Mesoamerica have been presented, I suggest that it would be worthwhile to reconsider current opinions in the light of whether an assumed geometry of the eighty-part circle and of the rectangle with various proportions could not contribute towards a better understanding of these interrelated matters.

Acknowledgements

The author is grateful to David Heath MA for the translation of the German manuscript. The paper contains data and results of studies carried out in the Puebla-Tlaxcala/Mexico region between 1962 and 1980 with the support of the Deutsche Forschungsgemeinschaft.

Bibliography

Aveni, A.F. and Hartung, H. (1986). Maya city planning and the calendar. *Transactions of the American Philosophical Society* **76**(7), 1–87.

Caso, A. (1960). *Interpretación del Códice Bodley 2858*. Mexico City: Sociedad Mexicana de Antropología.

Caso, A. (1979). *Reyes y reinos de la Mixteca*, 2nd. edn. Mexico City: Fondo de Cultura Economica.

González Aparicio, L. (1973). *Plano reconstructivo de la región de Tenochtitlan*. Mexico City: Instituto Nacional de Antropología e Historia.

Köhler, U. (1982). On the significance of the Aztec day sign 'Olin'. In *Space and time in the cosmovisión of Mesoamerica*, ed. F. Tichy, pp. 111–28. Munich: Wilhelm Fink Verlag (Lateinamerika-Studien 10).

Tichy, F. (1981). Order and relationship of space and time in Mesoamerica: myth or reality? In *Mesoamerican sites and world-views*, ed. E.P. Benson, pp. 217–45. Washington DC: Dumbarton Oaks.

Tichy, F. (1983). Observaciones del sol y calendario agrícola en Mesoamérica. In *Calendars in Mesoamerica and Peru: native*

American computations of time, eds. A.F. Aveni and G. Brotherston, pp. 135–43. Oxford: British Archaeological Reports (BAR International Series 174).

Tichy, F. (1988). Measurement of angles in Mesoamerica: necessity and possibility. In *New directions in American archaeoastronomy*, ed. A.F. Aveni, pp. 105–20. Oxford: British Archaeological Reports (BAR International Series 454).

Tichy, F. (1991). Direction lines in central Mexico and in the Maya area as elements of hypothetical orientation calendars. In *Colloquio Internazionale Archeologia e Astronomia*, ed. G. Romano and G. Traversari, pp. 123–29. Rome: Giorgio Bretschneider Editore (Supplementi alla RdA, 9).

Vinette, F. (1986). In search of Mesoamerican geometry. In *Native American mathematics*, ed. M.P. Closs, pp. 387–407. Austin TX: University of Texas Press.

26

Mesoamerican cross-circle designs revisited

STANISŁAW IWANISZEWSKI

In this paper I offer some observations on factors that might have contributed to the emergence and development of the cross-circle design in Mesoamerica.

Aveni (1988; 1989) has recently summarised the present state of our understanding of the meaning and function of the figure. I see no need to discuss the literature. For the purposes of the present paper I have utilised the material from the ceremonial centre at Teotihuacan and various places in the Basin of Mexico.

The cross-circle design as a counting device

It is known that cross-circle figures were pecked in stucco floors in ceremonial centres or engraved on boulders or rock outcrops. Sharer (1985) suggested that several markings, such as dots, cupules and other depressions found on monuments, altars and boulders and dated to Late Preclassic times (500–100 BC) could serve notational purposes. He observed that dots were often arranged in columns or two-, three-, four- and five-dot clusters, and even that they were associated with grooved cross-like figures. In Sharer's opinion they might serve as counting devices of some sort. The oldest cross-circle figures (e.g. those at Tlalancaleca) date from this period. On the other hand, Aveni (1989) and Aveni *et al.* (1978) had already noticed that counts of cross-circle dots closely oscillate around such numbers as 20 or 260 which were intimately related to the Mesoamerican calendar. Thus, cross-circle dots may have originated from such earlier notational traditions.

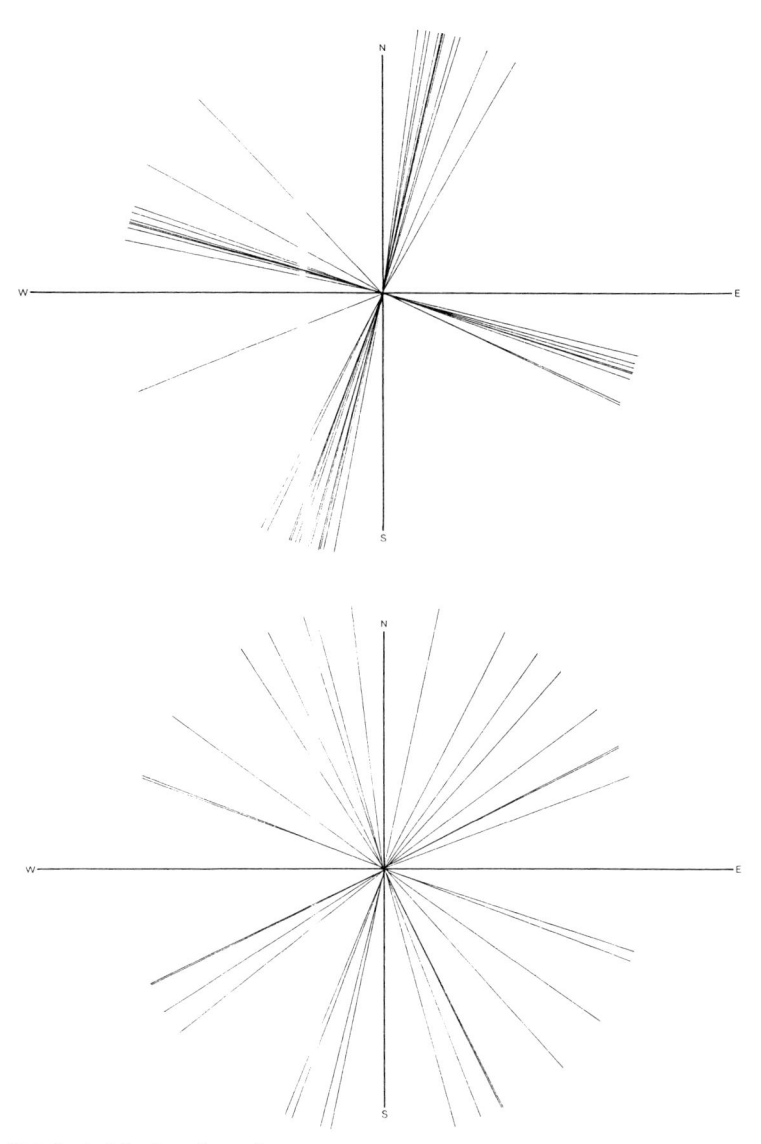

FIG. 1. Axial orientations of
(a) Teotihuacan figures (TEO 1, TEO 2, TEO 3, TEO 4, TEO 9, TEO 10, TEO 12, TEO 17, TEO 18, TEO 19, TEO 20, TEO 21 and TEO 22);
(b) Basin of Mexico figures (TEO 5, TEO 6, TEO 7, TEO 11, TEO 13, TEO 14, TEX, ACA and TOM).

Quadrant	Mean azimuth relative to cardinal direction		Standard deviation		Orientation of Street of the Dead (E of N)	Orientation of E–W Avenue (S of E)
Teotihuacan figures						
NE	15° 36'		6°·85		15° 25'	
SE	16° 36'		5°·98			16° 30'
SW	21° 18'	*17° 24'	14°·96	*5°·38	15° 25'	
NW	20° 05'	*16° 52'	10°·69	*4°·89		16° 30'
Basin of Mexico figures						
NE	45° 52'		19°·96			
SE	50° 18'		21°·35			
SW	38° 25'		23°·06			
NW	53° 10'		24°·56			

TABLE 1. The azimuths of cross-circle figures. The figures included are those listed in Fig. 1. An asterisk indicates the value when TEO 20 is excluded.

The cross-circle figure as an architectural marker

While the axial orientations of the Teotihuacan designs are concentrated about a few directions, the azimuths of the rest are widely dispersed (see Fig. 1 and Table 1).

The mean azimuthal values of the cross-circle figures at Teotihuacan follow very closely the city's general layout defined by the orientations of the Street of the Dead and the East-West Avenue. Exceptions are TEO 4 and TEO 17 which refer to the winter solstice sunrises.

Several authors have suggested that the artefacts engraved on rocks could have served as bench-marks for the builders of Teotihuacan. The evidence (see Iwaniszewski 1991; Ruggles and Saunders 1984) suggests rather the reverse: it is the city's layout that influenced the orientation of the pecked figures. Most of the artefacts comes from the Miccaotli-Tlamimilolpa periods (AD 150–450), that is, they were pecked some two or three centuries after the building of the Sun Pyramid had been initiated.

The cross-circle figure as an astronomical/calendrical marker

Since both easterly and westerly orientations fall within the range of solar risings and settings (in Teotihuacan these range from 65° to 115° and from 245° to 295° respectively) it is natural to suggest that the cross-circle figures were intended as markers indicating specific solar rising and setting points.

Aveni and Hartung (1986; 1989) and Tichy (1980; 1983) have suggested that the Mesoamericans had a horizon calendar involving the marking of solar divisions at regular intervals. It seems reasonable, then, to search for such a calendar at Teotihuacan (cf. Drucker 1977). The solar dates indicated by cross-circle figures cluster around the first half of February, the end of April/beginning of May, the first part of August and the end of October/beginning of November (Iwaniszewski 1989; 1991). I suggested that they could be related to the important climatological cycles (such as precipitation, temperature, frost) and to the main agricultural activities (with maize as a principal staple and rainfall-based agricultural techniques). If so, then it is possible to associate the figures with specific agricultural cycles.

The cross-circle figure and Tlaloc

The average dates when sunrise or sunset is indicated by the Teotihuacan orientations are Feb 9, Apr 29, Aug 12 and Nov 2. These are all close to dates on which some of the Tlaloc feasts were celebrated by the end of the Late Postclassic period (see Broda 1971; 1982; see also 1993). The first of the dates corresponds to the beginning of the month of *I Atlcahualo* (Feb 12, Gregorian) during which rain-related ceremonies were held in the mountains. The second date coincides with the great ceremony at the end of the month of *IV Huey Tozoztli* when the ritual was performed in the sanctuary at the top of Mt. Tlaloc (May 9, Gregorian). The third of the dates (mid August) can be related to maize deities ceremonies and the last with the feast held in honour of the mountain gods at the end of the month of *XIII Tepeilhuitl* (end of October).

The cult of a Tlaloc-like god in Teotihuacan is widely attested. The first god images belong to the Late Tzacualli phase (AD 100–150) and are related to the ceremonial cave under the Sun Pyramid (Millon *et al.* 1965). From the very beginning, the Tlaloc A figure was associated with the solar year symbol (Pasztory 1974; Von Winning 1988, I: 66). Some of the cross-circle figures are linked to Tlaloc-like images, especially in Tlalancaleca (TLA 3—Aveni *et al.* 1978: fig 1i), Xihuingo (TEP 27—Aveni 1989: fig. 2.23 and Table 3); and Cerro Maravillas (TEO 11—Aveni 1989: fig. 5).

Finally, the 'sacred' Teotihuacan orientation is repeated by that of Mt. Tlaloc–La Malinche (Iwaniszewski n.d). We know that at both places ceremonies took place in honour of rain-sending mountain gods.

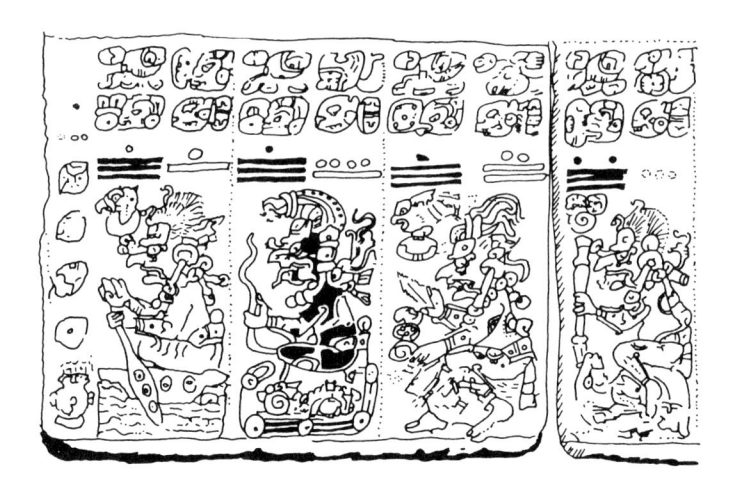

FIG. 2. The agricultural Almanac on pages 29c–30c of the Dresden Codex (after Thompson 1972).

The cross-circle figure and agricultural ceremonies

Although a great many of the figures at Teotihuacan have been partially damaged, it is still possible in some cases to establish counts of dots in adjacent quadrants. For example, in the inner circle of TEO 1, the classic two-circle figure, we find a 16–16–16–17 dot sequence starting with the SE quadrant. On the other hand, the outer circle is partially destroyed and the number of dots is 30, 20, 24, and a number between 27 and 29. This gives a total of 65 dots for the inner circle and 101–103 for the outer one. It is easy to observe that $4 \times 65 = 260$ days, or one *tonalpohualli/tzolkin* cycle. If we add 101–103 days we shall arrive at 361–363 days, a total that could refer to the 360, 364 or 365-day periods known in Mesoamerica.

We deal with similar sequences of days in Mesoamerican codices. For example, a series of 16–16–16–17 is found in Almanac 64 on pages 29c–30c of the Dresden Codex (see Fig. 2). According to Thompson (1972) it is a chant of praise which refers to 'the food offerings to the four Chacs associated with world directions and colors'. The so called Farmers' Almanacs cover pages 29a–44a, 43b and 45c and deal with the agricultural ceremonies scheduled in different series of days which tend to reach a total of 52 or 65 days or their multiples.

The series of 4×65 days refers to the Chacs and/or Xibs, four colours and directions, and employs the following verbs: T588 'to sidle'

(Bricker 1986), T765b 'to enter' (Grube and Stuart 1987: 4–6) and T667 'tongue sacrifice' (Kelley 1976: 139, 144).

The series of 5 × 52 days also refers to the Chacs and the world direction, but employs a richer vocabulary: T588, T667, T44:110 'to burn' (Grube and Stuart 1987: 8), T563a 'to fire' (Bricker 1986: 149) and T190 'to extrude, pop out' (Bricker 1986: 158–60), together with three references to eclipses.

All this indicates that special ceremonies in honour of Chacs were held in previously established intervals and consisted of food offerings and auto-sacrificial rites. They were established to assure good weather and crops. At the same time they could serve to make prognoses of the coming weather.

Which offertory and divinatory aspects could, however, be associated with cross-circles? Two examples will answer this question. These are the figure from Sta. Catarina Cruz Acalpixcan near Xochimilco (ACA) and from Tomacoco near Amecameca (TOM).

The ACA figure is engraved on a rock outcrop and partially damaged by a 'maqueta' or 'map', i.e. a figure with terraces, steps, temples and other buildings, canals, and so on. It is thought that the *maqueta* was a miniature model of the site. The *maqueta* is attributed to the Late Postclassic period (Aztec) and covers the cross-circle motif.

According to Cook de Leonard (1955: 177) the Aztec *maqueta* could refer to a sacred place where a kind of auto-sacrifice was performed. She associates the *maqueta* with calendrical and/or astronomical meaning. In the vicinity of ACA there is a well known archaeological site with the vestiges of buildings and rock carvings with calendrical and directional connotations (see Beyer 1965[1924]; Noguera 1972; Marcus 1982). Marcus (*ibid.*: 485) assumes that the site was one of the sacred places or shrines to be visited annually by the pilgrims. However, the 'map' is situated at the limits of the greater centre, not far from the cultivated fields. By representing the local landscape it could rather have served the native community. It could have been used for divinatory purposes: any liquid dispersed on the rock flows inside the canals engraved, so that the future fortune of a specific place might be determined.

My second example comes from the area of Tomacoco ex-hacienda. The cross-circle design pecked on a boulder was recently discovered by Maupomé (1988), a few metres from a well known huge boulder with a calendrical relief. This was described by Palacios (1931) as a solstice marker and by Sejourne (1981) as the place where ceremonies of the month of *II Tlacaxipehualiztli* (around Mar 20) were performed.

The site is Late Postclassic (Late Aztec) (Parsons *et al.* 1982: 162–64, figs. 14b, 15) and considered an *'isolated ceremonial precinct'*. The calendrical freeze of glyphs represents the first thirteen days, or *trecena*, of the Aztec 260-day ritual *tonalpohualli* (it starts with the day of *I Cipactli*). The site is located amongst cultivated fields and might have served for calendrically scheduled agricultural rituals. In both cases the Teotihuacan cross-circle figures presumably precede the Mexican (Aztec) artefacts, and the continuity of the use of the sites for agricultural-divinatory rituals suggests that they were considered as mantic places of some sort.

The cross-circle figure as a symbol of the quadripartite division of the world

Some of the outstanding features in the landscape such as springs, cliffs and mountains, assumed importance because special mantic properties were attributed to them. The places where the circle-cross motifs were engraved may have been chosen with some care because they demonstrated special values. The locations of cross-circle figures were considered sacred: the most outstanding example comes from the top of the Cerro Azteca, where a figure of cross-circle design is situated close to the Christian chapel. At the same time, they refer to the values that were shared by local communities: for example, they defined sacred space by establishing world directions.

The rituals in question clearly refer to the quadripartite division of the site. The officiant probably had to administer to the four rain-mountain gods at the four corners of the world or to the four world directions.

The cross-circle as mandala

The term 'mandala' has several religious connotations (Tucci 1975). Mandala is an iconic temporal-spacious representation of the world, or *imago mundi*, and in this sense symbolises the image of the (local) world. However, mandala is also a vehicle for a religious transformation of a meditating subject. At this stage of analysis we can only accept the first connotation, so we must be cautious with the use of the term.

The cross-circle as a game board

The evidence presented so far suggests that by the time of the Mexicas (Aztecs) the cross-circle figure was no longer in use. This fact may be easily explained: all the necessary information on scheduling agricultural rituals or divining the weather was inserted into codices. These iconic symbols became obsolete, incomprehensible. Since old and obsolete symbols and rituals tend to group at cultural peripheries (as games, riddles, and so on) then our figures could serve as game boards in later Postclassic times.

Conclusions

In this paper I have attempted to demonstrate that the Mesoamerican cross-circle figure was a numerical device related to the cultural image of the world in which certain astronomical or calendrical cycles were encoded in order to schedule agricultural and divinatory ceremonies in which the rain-mountain gods at the four world directions were honoured. The ceremonies were held in the fields.

Cross-circle figures played a role in precodical times. When the information was encoded in the written record the figures became obsolete and used as game boards.

References

Aveni, A.F. (1988). The Thom paradigm in the Americas: the case of the cross-circle designs. In *Records in stone*, ed. C.L.N. Ruggles, pp. 442–72. Cambridge: Cambridge University Press.

Aveni, A.F. (1989). Pecked cross petroglyphs at Xihuingo. *Archaeoastronomy* no. 14 (supplement to *Journal for the History of Astronomy* **20**) S73–115.

Aveni, A.F. and Hartung, H. (1986). Maya city planning and the calendar. *Transactions of the American Philosophical Society* **76**(1), 1–87.

Aveni, A.F. and Hartung, H. (1989). Uaxactun, Guatemala, Group E and similar assemblages: an archaeoastronomical reconsideration. In *World archaeoastronomy*, ed. A.F. Aveni, pp. 441–61. Cambridge: Cambridge University Press.

Aveni, A.F., Hartung, H. and Buckingham, B. (1978). The pecked cross symbol in ancient Mesoamerica. *Science* **202**, 267–79.

Beyer, H. (1965[1924]). Los bajorelieves de Sta. Cruz Acalpixcan. *El México Antiguo* **10**, 105–23.

Bricker, V.R. (1986). *A grammar of Mayan hieroglyphs*. New Orleans LA: Tulane University Press.

Broda, J. (1971). Las fiestas aztecas de los dioses de la lluvia. *Revista Española de Antropología Americana* **6**, 245–327.

Broda, J. (1982). El culto mexica de los cerros y del agua. *Multidisciplina* **3**(7), 45–56.

Broda, J. (1993). Astronomical knowledge, calendrics and sacred geography in ancient Mesoamerica. In *Astronomies and cultures*, eds. C.L.N. Ruggles and N.J. Saunders. Niwot CO: University Press of Colorado. In press.

Cook de Leonard, C. (1955). Una maqueta prehispánica. *El México Antiguo* **8**, 169–88.

Drucker, D.R. (1977). A solar orientation framework for Teotihuacan. In *Los procesos de cambio (en Mesoamérica y areas circunvecinas)*, vol. 2, pp. 277-84. Guanajuato: Sociedad Mexicana de Antropología e Historia (XV Mesa Redonda).

Grube, N. and Stuart, D. (1987). Observations on T110 as the syllable ko. *Research Reports on Ancient Maya Writing* **8**, 1–14.

Iwaniszewski, S. (1989). Exploring some anthropological theoretical foundations for archaeoastronomy. In *World archaeoastronomy*, ed. A.F. Aveni, pp. 27–37. Cambridge: Cambridge University Press.

Iwaniszewski, S. (1991). La arqueología y la astronomía en Teotihuacan. In *Arqueoastronomía y etnoastronomía en Mesoamérica*, eds. J. Broda, S. Iwaniszewski and L. Maupomé, pp. 269–90. Mexico City: Instituto de Investigaciones Históricas, Universidad Nacional Autónoma de México.

Iwaniszewski, S. (n.d.). The mountain cult at Mt. Tlaloc, Mexico: an archaeoastronomical approach. Unpublished manuscript.

Kelley, D.H. (1976). *Deciphering the Maya script*. Austin TX: University of Texas Press.

Marcus, J. (1982). The Aztec monuments of Acalpixcan. In *Prehistoric settlement patterns in the southern Valley of Mexico: the Chalco-Xochimilco region*, eds. J.R. Parsons, E. Brumfield, M.H. Parsons and D.J. Wilson, pp. 475–85. Ann Arbor MI: University of Michigan (Memoirs of the Museum of Anthropology, no 14).

Maupomé, L. (1988). Un petroglifo astronómico-calendárico descubierto en las cercanías de Amecameca. *Anuario del Observatorio Astronómico Nacional* **108**, 157–64.

Millon, R., Drewitt, B. and Bennyhoff, J.A. (1965). *The pyramid of the sun at Teotihuacan: 1959 investigations*. Philadelphia PA: American Philosophical Society (Transactions of the American Philosophical Society, 55).

Noguera, E. (1972). Antigüedad y significado de los relieves en Acalpixcan, D.F. (México). *Anales de Antropología* **9**, 77–94.

Palacios, E.J. (1931). El relieve solsticial de Amecameca. *Universidad de México* **2**(7), 181–97.

Parsons, J.R., Brumfield, E., Parsons, M.H. and Wilson, D.J., eds. (1982). *Prehistoric settlement patterns in the southern Valley of Mexico: the Chalco-Xochimilco region*. Ann Arbor MI: University of Michigan (Memoirs of the Museum of Anthropology, no. 14).

Pasztory, E. (1974). *The iconography of the Teotihuacan Tlaloc*. Washington DC: Dumbarton Oaks and Trustees for Harvard University (Studies in Pre-Columbian Art and Archaeology, no 15).

Ruggles, C.L.N. and Saunders, N.J. (1984). The interpretation of the pecked cross symbols at Teotihuacan: a methodological note. *Archaeoastronomy* no. 7 (supplement to *Journal for the History of Astronomy* **15**), S101–10.

Sejourne, L. (1981). *Pensamiento náhuatl cifrado por los calendarios.* Mexico City: Siglo XXI Editores.

Sharer, R.J. (1985). Archaeology and epigraphy revisited. *Expedition* **27**(3), 16–19.

Thompson, J.E.S. (1972). *A commentary on the Dresden Codex.* Philadelphia PA: American Philosophical Society.

Tichy, F. (1980). Jahresanfänge mesoamerikanischer Kalender mit 20-Tage-Perioden. *Indiana* **6**, 55–70.

Tichy, F. (1983). El patrón de asentamientos con sistema radial en la meseta central de México: 'sistemas ceque' en Mesoamérica? *Jahrbuch für Geschichte von Staat, Wirtschaft und Gesellschaft Lateinamerikas* **20**, 61–84.

Tucci, G. (1975). *Teoría y práctica del mandala.* Buenos Aires: Editorial Dédalo.

Von Winning, H. (1988). *La iconografía de Teotihuacan.* Mexico City: Universidad Nacional Autónoma de México.

27

Were the Incas able to predict lunar eclipses?

MARIUSZ S. ZIÓŁKOWSKI AND ARNOLD LEBEUF

Astronomy occupied a privileged place in the administration of the Inca state and its religious doctrine. Special officers were dedicated to the astronomical observation and interpretation of celestial phenomena. Specially important decisions connected with the reform of calendars were attributed to rulers themselves (Zuidema 1981; 1982a; 1982b; Ziółkowski 1985; 1987; 1988; 1991). But although the Inca state was called the Empire of the Sun, the moon was a far more important celestial body than is generally claimed. The lunar cycle co-ordinated many social activities and the full moon was supposed to be particularly propitious, for example for military purposes. This fact was realised by the Spaniards and used to their advantage during the siege of Cuzco in 1536, where it greatly contributed to their victory over the Manco Inca army (Ziółkowski 1985: 159–61).

It is remarkable that the Incas, considered mostly as sun worshippers, feared lunar eclipses much more than solar ones. Lunar eclipses were generally considered an illness of the moon and the first phase of the collapse of the universe.[1] According to Cobo, the Incas attributed this disease of the moon to the attack of two celestial beasts, a lion and a serpent:

---

[1] 'Al eclipse de la luna, viéndola ir negreciendo, decían que enfermaba la luna, y que si acababa de escurecerse, habia de morir y caerse del cielo, y cogerlos a todos debajo y matarlos, y que se habia de acabar el mundo; por este miedo en empezando a eclipsarse la luna, tocaban trompetas, caracoles, atambales y atambores, y cuantos instrumentos podian haber que hiciesen ruldo; ataban los perros grandes y chicos, dábanles muchos palos para que ahullasen y llamasen la luna, que por cierta fábula que ellos contaban, decian que la luna era aficionada a los perros, por cierto servicio que le habian hecho...'

'Mandaban a los muchachos y niños que llorasen y diesen grandes voces y gritos, llamándola Mama Quilla, que es madre luna, rogándola que no se muriese

They used to say than when the moon started to eclipse, a lion or a serpent attacked it to tear it into pieces. Therefore when it started eclipsing they were yelling and beating dogs to make them bark and howl. Armed men were standing and blowing horns, beating drums, yelling fiercely, shooting arrows and throwing javelins in the direction of the moon, menacing with their spears as if they wanted to hurt the lion and the serpent. They used to say that in this way they were threatening out and frightening them, making them unable to tear the moon.

Some of our priests used to predict the eclipses to liberate them from these absurdities.

That is how the Spaniards acquired an opinion of great sages. There is a great admiration for us among them because we are able to predict the eclipse with such an accuracy that we warn them not only about the night when it will take place, but even about the time of the beginning and lasting about the part of the moon that will be eclipsed.[2]

---

[2] porque no pereciesen todos. Los hombres y las mujeres hacían lo mismo. Había un ruido y una confusión tan grande que no se puede encarecer. Conforme al eclipse grande o pequeño, juzgaban que había sido la enfermedad de la luna. Pero si llegaba a ser total, ya no había que juzgar sino que estaba muerta, por momentos temían el caer la luna y el perecer de ellos. Entonces era más de veras el llorar y planir, como gente que veía al ojo la muerte de todos y acabarse el mundo.' (Garcilaso de la Vega 1960: L.II ch. XXIII 74 *et seq.*)

This is our free translation of Bernabé Cobo:

'Acerca del eclipse de la luna tenían tantas boberías como del sol: decían, cuando se eclipsaba, que un león o serpiente la embestía para despedazarla; y por esto, cuando comenzaba eclipsarse, daban grandes voces y gritos y azotaban los perrros para que ladrasen y aullasen.

'Poníanse los varones a punto de guerra, tañendo sus bocinas, tocando atambores y dando grandes alaridos, tiraban flechas y varas hacia la luna y hacían grandes ademanes con lanzas, como si hubiesen de herir al león y sierpe; porque decían que desta manera los asumitaban y ponían espanto para que no despedazasen la luna.

'Lo cual hacían, porque tenían aprehendido que si el león y sierpe hiciese su efecto, quedarían en oscuridad y tinieblas...

'Suelen algunos de sus curas, para apartallos deste su desvarío prevenirles los eclipses... con el cual también han cobrado para con ellos muy grande opinión de sabios los españoles, porque es notable la admiración que les causa ver que podemos nosotros alcanzar a saber los eclipses antes que vengan, con tanta puntualidad que antes les avisamos no sólo de la noche en que suceden, sino hasta de la hora en que han de comenzar, la cantidad de la luna que se escurecerá y el tiempo que durarán. Y la verdad, no comprendiendo ellos las causas de un efecto tan admirable, quedan como fuera de sí de ver que nosotros lo podamos saber antes que suceda.' (Cobo 1964: L.II ch. VI 158–59).

As similar legends and myths about two animals devouring the moon during eclipse are known universally as symbolical representations of the nodes of the lunar orbit, we suggest here that the Incas knew about the nodes. But were they able to predict lunar eclipses and use this knowledge for socio-technical purposes, as was the case in other civilisations with a similar level of technological achievement?

The prediction of eclipses in pre-Columbian America could not have involved a level of accuracy comparable to that in the Old World. The Incas did not possess a system of dividing time into hours, nor any adequate mathematical apparatus. Therefore in speaking about the prediction of an eclipse we are thinking of rudimentary methods for predicting the night when it would occur rather than the precise hour or the magnitude. We consider evidence concerning the Inca system of horizon observations.

It is well known that by observing the azimuth of the moon at standstill we can define the ecliptic longitude of the node and thus predict eclipses. When archaeoastronomers look for significant orientations, the only ones to which a possible lunar significance is attached tend to be the major and minor standstills. But solar solstitial azimuths are also limiting monthly azimuths of the moon; they are what we might call the 'central' ones, occurring roughly halfway between major and minor standstill. At these times, when the lines of nodes of the moon's orbit has rotated through roughly 90° or 270° since the last major standstill,[3] the effect of perturbation in the inclination of the lunar orbit disappears, and it is possible to define the moon's longitude with the greatest accuracy. Why do we suspect that apparent orientations upon solar solstitial azimuths might in fact have been used for lunar observation?

Dearborn *et al.* (1987) studied the cave of Intimachay, but were puzzled by the presence of an apparently precise orientation upon December solstitial sunset when the rainy season would make the observation almost impossible. Moreover, assuming that the climate has not radically changed during the past five centuries, it is difficult to see why sunset was chosen in preference to sunrise, given that the evening is generally more cloudy than the morning. Observations of moonset, on the other hand, do not suffer from this limitation; furthermore, the moon setting at this azimuth is full precisely in the season of best visibility.

If the Incas really were, as we suggest, keen observers of the moon and were interested in the determination of the longitude of the nodes, then they would surely have been interested in other details of the lunar cycle. But where might indications of such an interest be found? A possibility is to reconsider other orientations which have hitherto been assumed to be of solar significance.

---

[3] Actually 102° and 292° (Lebeuf 1989: 56).

Were the Incas able to predict lunar eclipses?

```
      2 IX              18 VIII            3 VIII
      ⊙                   ⊙                  ⊙
    ─────────────────────────────────────────────
     8° 08'             13° 18'            17° 40'
```

FIG. 1. The solar orientations of the four pillars of Picchu, according to Zuidema.

The four pillars on the Picchu mountain of Cuzco have been investigated by many scholars. Zuidema (1981; 1982a) and Aveni (1981) claim that they were oriented upon sunset on Aug 3, Aug 18 and Sep 2. If equinox occurs at sunset on Mar 21, then the declination of the sun at sunset on these dates is respectively 17° 40', 13° 18' and 8° 8'. According to Zuidema, the Pillars of Picchu were framing the sun at those dates (Fig. 1).

According to Zuidema (1982a: 68), this arrangement served to observe the sunset and to establish important dates such as the start of the planting season and nadir passage. However, he also remarks that the argument is weak because 'August 4th, of special importance to the Incas as the official date on which the Inca himself opened the agricultural season, is preserved today for similar purposes, but the date has equal importance all over the Andes, from Southern Ecuador to Northern Chile' (*ibid.*: 69). This is correct and has been confirmed by other authors and by fieldwork including the calculation of Ziółkowski and Sadowski (1985; 1991) at Ingapirca. Concerning the nadir passage on Aug 18 he also remarks that it is only an 'accident of the latitude of Cuzco' (*ibid.*: 68).

In the same article Zuidema also mentions a local interest for the moon setting over Cuzco Picchu in August and wonders how the full moon could be used to determine nadir passage, a phenomenon not directly observable. He also underlines the difficulty caused by the imprecision arising from the five degree inclination of the moon's orbit to the ecliptic:

> A full moon Zenith Passage can occur in Cuzco from August 3rd to September 2nd and from April 11th to May 9th. August 3rd or 5th in Cuzco at the beginning of the year might derive its practical importance also from this observation of the moon. (Zuidema 1981: 324).

Dearborn and Schreiber (1986) criticise this explanation. 'Because of the inclination of the moon's orbit to the sun, zenith passage can occur when the sun has a declination from 8·5 to 18·5, so this defines a broad period and not a particular date, like the sun passing a pillar does' (*ibid.*: 36).

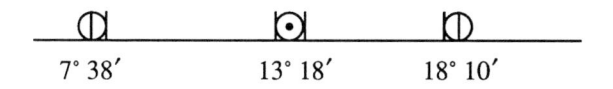

```
      ☽                   ⊙                  ☾
    ─────────────────────────────────────────────
     7° 38'             13° 18'            18° 10'
```

FIG. 2. An alternative explanation of the four pillars of Picchu.

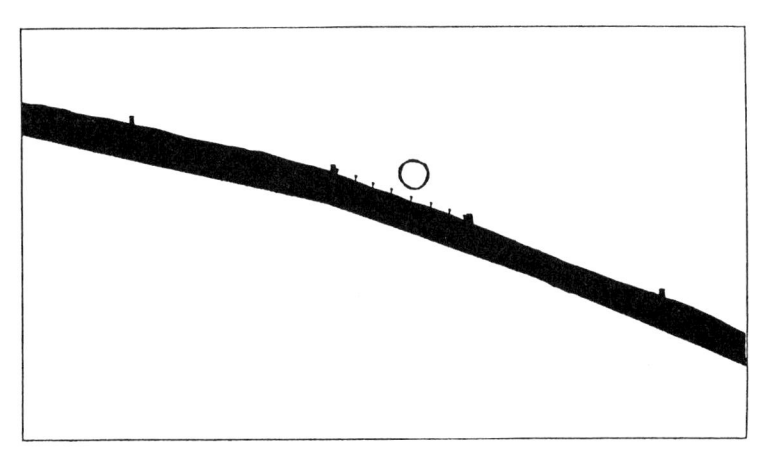

FIG. 3. Dearborn and Schreiber's (1986: fig. 4) reconstruction of the four pillars of Picchu.

Zuidema's explanation, although very ingenious, is too complicated to be convincing. On the other hand, though, with the information he offers and the dates he proposes, we can try another explanation of the pillars of Cuzco. Let us consider the difference in the declination of the sun between Aug 18 and Sep 2. The result is not bad because 5° 10′ is within half a minute of the mean inclination of the moon's orbit to the ecliptic. If now we place the moon outside the outer pillars, we obtain the declinations shown in Fig. 2.

We now notice that the difference between 18° 10′ and 7° 38′, 10° 32′, is within 3 arc minutes of the amplitude of the moon's motions about the ecliptic (twice 5° 17′). Furthermore, the lunar declination of the Aug 3 pillar is 18° 10′, half a minute from that of minor standstill with maximum inclination. Thus, by observing the oscillations of the minor standstill moon about the pillar we can deduce the variation in the inclination of the moon's orbit. In our schematic reconstruction (Figs. 1 and 2) we have worked only on the basis of calendrical information but according to the Dearborn and Schreiber's (1986) reconstruction (Fig. 3) the central pillars are some 2°·5 apart.

We wonder why the Incas would choose to have such a distance between the central pillars if the problem was, as Zuidema (1981) suggests, merely to frame the sun at the date of anti-zenith passage.

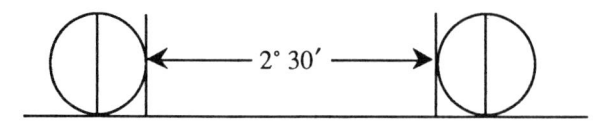

FIG. 4. A possible explanation of the central pillars at Picchu.

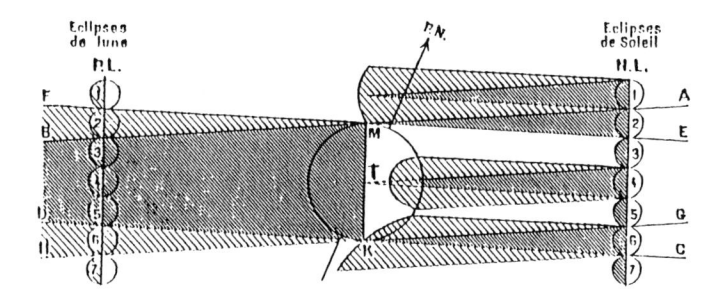

FIG. 5. The size of the Earth's shadow at a lunar eclipse (left) and of the sun's shadow at a solar eclipse (right).

Another possibility is presented in Fig. 4. The distance between the centres of the moon's disc when just touching the pillars is about three degrees, approximately equal to the change in azimuth from one moonset to another when the moon passes this part of the ecliptic. So when we observe that the moon reaches one of the pillars at setting time, the next moonset will occur close to the other pillar. If this happens in the period of opposition near to the date of solar zenith passage (Feb 13 and Oct 30), the observer will be sure that the moon will be partially eclipsed, because $2°·5$ is also the size of the shadow of the Earth (see Fig. 5).

Support for this hypothesis comes from the situation of Picchu mountain. Viewed from the Haucaypata plaza, the profile of the Picchu mountain is about 7° above the mathematical horizon (Aveni 1981). If the moon touches the western horizon at the time of sunrise at the mathematical horizon (this can be established by observing the first sunbeams on peaks of surrounding mountains), it still has 7° to move in longitude before it is in exact opposition (full moon). As the moon moves 13° in longitude in 24 hours, the observer will know that a lunar eclipse will be visible the next evening, after moonrise.

We do not, of course, assume that the Incas had a theoretical model to explain this (as we have done above); the above-mentioned relation between the positions of the moon and the sun could have been established by practice. It is certainly true that all the technical possibilities yielded by the particular situation of the pillars on the Picchu mountain do not seem to be merely the product of a simple coincidence.

While Zuidema (1982a: 68) tells us that the date of Aug 18 is an accident of the latitude of Cuzco,[4] we may ask instead if the establishment of the capital city of the Inca Empire was not deliberately chosen to fit the centre of a cosmological system related to practical astronomical use, as some myths may suggest.[5]

If the Incas were indeed able to predict eclipses, the quotation of Cobo can be understood differently. The Incas were not so much impressed by the ability of Spaniards to predict eclipses, as by the precision with which they could achieve the prediction. If it was the case that the Incas had knowledge and made use of eclipse prediction, then this must have been a very secret state affair and it is no wonder that they themselves would not inform their enemies on such matters.

Our hypothesis throws some light on a rather puzzling report of Diego Rodriguez de Figueroa (1910[1565]), a missionary sent as an ambassador of the Spanish authorities at Cuzco to negotiate with Titu Cusi Yupanqui, the ruler of the last independent Inca bastion at Vilcabamba. At that time, the political situation was at breaking point. On the one hand, the Spaniards, who controlled almost the whole of the former Inca Empire, could no longer endure the continuous razzias and raids launched from Vilcabamba. There were strong pressures on the central authorities to organise a military expedition against the Vilcabamba state. On the other hand, the ruler of Vilcabamba himself was obliged by the Inca religious and political system to legitimize his position by victories over the enemies (Ziółkowski 1983; 1988). Thus the troops of Inca Titu Cusi were pushing him into military action, although the Inca had good reasons and experience to be more careful, remembering the tragic destiny of his own father who was not able to win over the Spaniards with much greater forces.

Rodriguez de Figueroa started from Cuzco on Apr 8 1565 but was considerably delayed in his trip. Inca first stopped him at the border and then suddenly called on him at a very precise date: 'E luego otro día que fueron honce de mayo, recíbí otra carta del ynga...que yo fuesa á otro pueblo más adelante, que se llama Banbacona, para que mss presto nos

---

[4] Zuidema's greatly ingenious hypothesis was based on the results of fieldwork undertaken together with the astronomer Anthony Aveni, although there are some incoherence between the separate papers of Aveni (1981) and Zuidema (1981) arising from this fieldwork. Zuidema has since changed his mind concerning the dates from the pillars of Cuzco, pushing them back by two weeks (Zuidema 1988).

[5] For example, a golden stick was offered to the first Inca ruler by the (solar?) divinity, which sunk vertically in the ground and disappeared at the very place where the capital was to be established (Sarmiento de Gamboa 1960: 213–18; Garcilaso de la Vega 1960: L.I ch. IV 25–29; Cobo 1964: L.XII ch. III 62).

Were the Incas able to predict lunar eclipses? 305

viwsemos, y que él sería de ay s dos dhas' (Rodriguez de Figueroa 1910[1565]: 97). On May 14[6] they met at Bambacona, in the free Inca territory. The meeting was diplomatically polite but accompanied by an impressive military parade and demonstration of power. Some soldiers dared to threaten the Ambassador openly, but the Inca himself calmed them down.

Father Rodriguez was sent home overnight and locked in his estate outside the Inca settlement with a strong personal guard offered by the Inca, to protect rather than to watch him. During that night he heard the clamour of the Inca troops under arms and the sound of drums and flutes.[7] The next morning the negotiation started and after some unexpected imbroglios and a succession of friendly and aggressive attitudes from the Incas, the two parties came to establish a peace agreement.

The analysis of the detailed report of the Spanish Ambassador raises two questions:

• Why was Titu Cusi apparently so interested in receiving the Spanish emissary on a precise date?

• How did he succeed in calming down the aggression of his own troops and officers?

The eclipses table (Sadowski 1989) may offer a partial explanation:

No.	Date	Mag.	Moon–Zenith λ	ϕ	Duration of the partial eclipse
4280	1565 May 14	7·2	-70°	-21°	82 mins

---

[6] The Spanish ambassador dated his diary carefully (see the note above), the only incoherence appearing when he calls two successive days May 14. Thus 'A catorze de mayo los yndios de Bambacona tenían hecho una cassa grande en un fuerte alto ...Y así me dixeron esperase allí, que de allí vería la entrada del ynga...E luego empeçó á entrar la jente con el ynga en la plaça'. (Rodriguez de Figueroa 1910[1565]: 98) and after this 'A catorze de mayo luego por la mañana me enbió á llamar el ynga a su cassa...' (*ibid.*: 103). However, the first quotation comes directly after the description of the day May 13: 'A treze de mayo yo enbié al camino...' (*ibid.*: 98). Thus the subsequent error could simply be due to the emotion of the Spaniard during his first meeting with the Inca.

[7] The chronicler, locked inside his estate, explains this behaviour by the change of guards. But this explanation does not clarify why this agitation took place only on that specific night. Rather, we should remember a fragment of the information of Father Cobo, cited above: 'Poníanse les varones a punto de guerra, tañendo sus bocinas, tocando atambores y dando grandes alaridos, tiraban flechas y varas hacia la luna y hacían grandes ademanes con lanzas, como si hubiesen de herir al león y sierpe...'

As can be seen, the lunar eclipse took place on the very night following the arrival of the Spaniard at Bambacona. Moreover, the moon was then not far from the local Zenith. Now let us remember that Titu Cusi Yupanqui was not only an Inca ruler but the self-declared arch-priest of the state religion.[8] It is possible, then, that he found a solution for his diplomatic and military problems in the second of his functions. Knowing in advance that a lunar eclipse would take place on the night of May 14/15, he arranged for the arrival of the Spanish emissary together with the general mobilisation of his troops to take place precisely on that date.

As we know, the full moon was for the Incas a very favourable moment to start military action. When the eclipse took place, Titu Cusi had a chance to convince his army that the gods were not favourable to any warfare against the Spaniards, as they had manifested themselves so dramatically at the arrival of the Ambassador. Such an explanation allowed the Inca to negotiate without losing his prestige. Furthermore, the very impressive military parade had properly prepared the Spanish emissary to treat peace at a rather low price.

Inca Titu Cusi knew well enough that if conditions are unfavourable, the best way not to lose is not to fight.

Bibliography

Aveni, A.F. (1981). Horizon astronomy in Incaic Cuzco. In *Archaeoastronomy in the Americas*, ed. R.A. Williamson, pp. 305–18. Los Altos CA: Ballena Press.

Cobo, B. (1964). *Historia del Nuevo Mundo, vol. II*. Madrid: Biblioteca de Autores Españoles, vol. 92.

Dearborn, D.S.P. and Schreiber, K.J. (1986). Here comes the sun: the Cuzco-Machu Picchu connection. *Archaeoastronomy* (Center for Archaeoastronomy) **9**, 15–37.

Dearborn, D.S.P., Schreiber, K.J. and White, R.E. (1987). Intimachay: a December solstice observatory. *American Antiquity* **52**, 346–52.

Garcilaso de la Vega (1960). El Inca, I—Comentarios reales de los Incas, primera parte. Madrid: Biblioteca de Autores Españoles, vol. 133.

Lebeuf, A. (1989). L'observatoire astronomique de la Cathédrale Saint-Lizier de Couserans. *Publication de l'Observatoire Astronomique de Strasbourg, série 'Astronomie et Sciences Humaines'* **3**, 39–77.

---

[8] 'E que así él era sumo çacerdote en lo que llamamos nosotros acá espiritual...e así mesmo ereduana el señorío temporal...' (Rodriguez de Figueroa 1910 [1565]: 110)

Rodriguez de Figueroa, Diego (1910[1565]). Relación del camino e viaje desde la ciudad del Cuzco a la tierra de guerra de Mango Inca. In *Bericht des Diego Rodrigues de Figueroa uber seine Verhandlungen usW*, ed. R. Pietschmann. Göttingen: Akademie des Wissenschaften.

Sadowski, R.M. (1989). The sky above the Incas: an abridged astronomical calendar for the 16th century. In *Time and calendars in the Inca empire*, ed. M.S. Ziółkowski and R.M. Sadowski, pp. 75–106. Oxford: British Archaeological Reports (BAR International Series 479).

Sarmiento de Gamboa, P. (1965). Historia Indica. In *Obras del Inca Garcilaso de la Vega, vol. IV*, pp. 159–279. Madrid: Biblioteca de Autores Españoles, vol. 135.

Ziółkowski, M.S. (1983). La piedra del cielo: algunos aspectos de la educación e iniciación religiosa de los príncipes incas. *Ethnologia Polona* **9**, 219–34. Reprinted (1984) in: *Anthropologica PUCP* [Pontífica Universidad Católica de Peru] **2**, 45–65.

Ziółkowski, M.S. (1985). Hanan pachap unanchan: las 'señales del cielo' y su papel en la historia andina. *Revista Española de Antropología Americana* **15**, 147–82.

Ziółkowski, M.S. (1987). Las fiestas del calendario metropolitano inca: Primera parte. *Ethnologia Polona* **13**, 183–217.

Ziółkowski, M.S. (1988). Los cometas de Atawallpa: acerca del papel de las profecías en la política del Estado Inka. *Anthropologica PUCP* [Pontífica Universidad Católica de Peru] **6**, 85–109.

Ziółkowski, M.S. (1989). El calendario metropolitano del Estado Inka. In *Time and calendars in the Inca Empire*, eds. M.S. Ziółkowski and R.M. Sadowski, pp. 129–66. Oxford: British Archaeological Reports (BAR International Series 479).

Ziółkowski, M.S. and Sadowski, R.M. (1985). Informe de la segunda temporada de investigaciones arqueoastrónomicas en Ingapirca, Ecuador. In *Memorias del Primer Simposio Europeo sobre Antropología del Ecuador*, ed. S.E. Moreno Yanez, pp. 91–116. Quito: Ediciones Abya-Yala (Instituto de Antropología Cultural de la Universidad de Bonn).

Ziółkowski, M.S. and Sadowski, R.M. (1991). Investigaciones arqueoastronómicas en el sitio de Ingapirca, prov. de Cañar, Ecuador. In *Colloquio Internazionale Archeologia e Astronomia*, ed. G. Romano and G. Traversari, pp. 151–62. Rome: Giorgio Bretschneider Editore (Supplementi alla RdA, 9).

Zuidema, R.T. (1981). Inca observations of the solar and lunar passages through the zenith and anti-zenith at Cuzco. In *Archaeoastronomy in the Americas*, ed. R.A. Williamson, pp. 319–42. Los Altos CA: Ballena Press.

Zuidema, R.T. (1982a). The sidereal lunar calendar of the Incas. In *Archaeoastronomy in the New World*, ed. A.F. Aveni, pp. 59–107. Cambridge: Cambridge University Press.

Zuidema, R.T. (1982b). Catachillay: the role of the Pleiades and the Southern Cross and Alpha and Beta Centauri in the calendar of the Incas. In *Ethnoastronomy and archaeoastronomy in the American tropics*, eds. A.F. Aveni and G. Urton, pp. 203–29. New York NY: Annals of the New York Academy of Sciences, 385.

Zuidema, R.T. (1988). The pillars of Cuzco: which two dates of sunset did they define? In *New directions in American archaeoastronomy*, ed. A.F. Aveni, pp. 143–69. Oxford: British Archaeological Reports (BAR International Series 454).

28

Callanish: maximising the symbolic and dramatic potential of the landscape at the southern extreme moon

MARGARET CURTIS AND RONALD CURTIS

Introduction

Many megalithic sites—standing stones in rings or rows, single standing stones, burial cairns, and rock carvings—are concentrated at Callanish, Isle of Lewis, Scotland. At this latitude (58°N) the extreme lunar rising and setting positions in the north and south, which for zero altitude correspond to azimuths of approximately 23°/337° and 157°/203°, lie considerably outside the solar extremes of 40°/320° and 140°/220°. Our research over many years in the vicinity of Callanish has shown that the four extreme lunar rising and setting positions were marked in megalithic times at individual sites and by pairs of sites, although there are few records of the solar extremes (Ponting and Ponting 1981; Ponting 1988).

Continuation of this work has revealed that the precise locations and sizes of the six known ring sites (Curtis 1988) were determined not only by the visibility of the south extreme moon's path—rise, set and partial reappearance ('re-gleam')—in relation to two conspicuous and dramatic ranges of hills, but also by the need for ceremonial activities performed at the time of the southern extreme. Our suggestion is that these involved seeing the moon rise out of an 'earth mother' while framed by the stones of the ring; and seeing the moon's brief last (re-)appearance, also through the ring, with a human silhouette (that of the astronomer priest/ priestess?) set against the moon's orb (Pls. 1, 2, 3).

Astronomy

Knowledge of the moon's movements was essential in order to synchronise any such ceremonies. At the Standing Stones of Callanish this could have been achieved by observation of moonrise and set in relation to the spacing of the stones of the east side of the avenue (Ponting and Ponting 1981: 77–83). This would allow the prediction, on a daily, fortnightly, and monthly basis for several years on either side of the extreme, of the height of the moon's path, the rise, set and reappearance positions, and the proximity to the real extreme.

The prehistoric inhabitants of Callanish are unlikely to have isolated the 173-day lunar perturbation from the other irregularities. They may, however, have been interested in the extreme positions of the moon and, consequently, puzzled by the apparent inconsistencies that we know to be caused by the perturbation, rapid change of declination, refraction, temperature and precession of the ecliptic. Repeated observations at any of the ring sites could have brought these minor variations to light. Simultaneous observations at several ring sites would have demonstrated minor variations from site to site due to the slightly differing azimuths and altitudes.

Radiocarbon dates suggest use of some of the Callanish sites as early as 3000 BC and as late as 800 BC, a prolonged period of continuous or intermittent use. Over the centuries, precession would have caused the positions of lunar rise, set and reappearance as observed from a given spot to alter appreciably.

Significant ranges of hills

The range of hills from which the moon, when near to its southern extreme, rises in the vicinity of Callanish is known locally in English as 'The Sleeping Beauty' and in Gaelic as 'Cailleach na Mointeach'— literally 'The Old Woman of the Moors' (Fig. 1). Across Europe there are ancient and widespread legends about a Mother Goddess or an Earth Mother. The full extent of the 'sacred' horizon from which the moon rises includes her 'pillows', Beinn Mhor and Roineaval.

The range into which the southern moon first sets is the Clisham Hills (Fig. 2). These contain a deep V-shaped valley in which, from suitable vantages, the moon reappears briefly and dramatically.

Opposite page: FIG. 1 (top left): 'The Sleeping Beauty' range. FIG. 2 (top right): The Clisham Hills range. FIG. 3 (lower centre left): Plan of Cnoc Fillibhir Bheag ('Callanish III') and vicinity. FIG. 4 (bottom): Plan of Cnoc Ceann a'Gharaidh ('Callanish II') and vicinity. PL. 1 (upper centre left): The Sleeping Beauty range framed by the stones of Cnoc Fillibhir Bheag as viewed from the stone setting at *D*. PL. 2 (upper centre right): The Clisham Hills and the deep valley appearing behind the stones of Cnoc Fillibhir Bheag as viewed from backsight *A*. PL. 3 (lower centre right): A human silhouette set against the moon's orb, observed at the standing stones of Callanish.

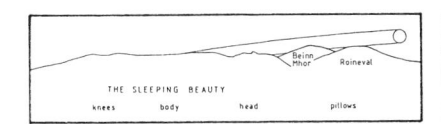

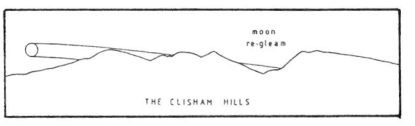

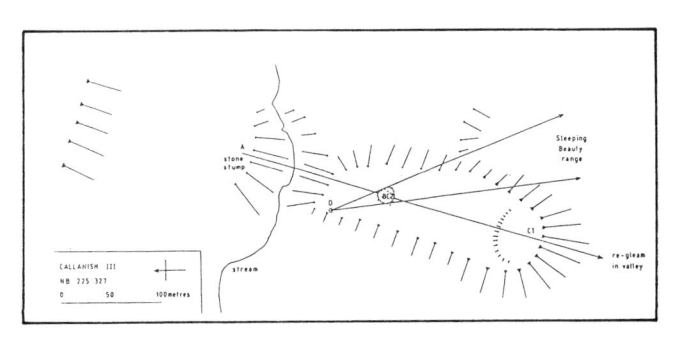

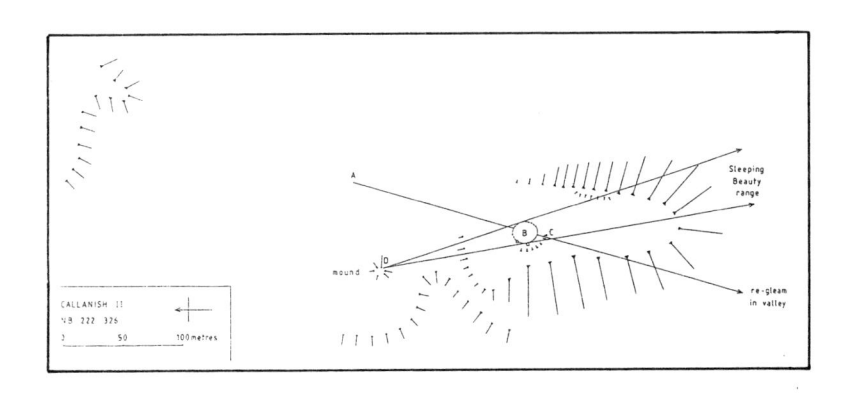

Callanish: symbolic and dramatic potential of the landscape

311

Site location

The six known stone rings in the vicinity of Callanish are that in the centre of the Standing Stones of Callanish ('Callanish I'), Cnoc Ceann a'Gharaidh ('Callanish II'), Cnoc Fillibhir Bheag ('Callanish III'), Ceann Hulavig ('Callanish IV'), Na Dromannan ('Callanish X') and Achmore ('Callanish XXII') (Curtis 1988). When these sites were examined it was found that each of them was located where the local topography permitted two very specific viewing lines towards lunar rise and set, and that it was virtually impossible to have moved the rings along or across the ridges and still achieve these lines.

Cnoc Fillibhir Bheag ('Callanish III')

This double ring of standing stones is set part way along a ridge which has a distinct summit at its South end, and a small stream valley at its North end (Fig. 3).

The rise of the moon from the Sleeping Beauty range is visible from the ring. The viewing line runs up from a sub-circular stone setting at D, 57m from the centre of the ring on the extreme west edge of the ridge, and crosses the ridge diagonally. The ring itself (B) is at the highest point on this alignment. The position of the setting clearly defines the width of horizon to be viewed through the ring (Pl. 1). A viewer at D would have seen the Sleeping Beauty range framed within the height of the standing stones and within the width of the ring, and through this would have seen the gradual rise of the moon near standstill from first appearance until finally clearing the horizon. The fact that Roineaval appears within this width suggests that the site may have been in use about 3000 BC.

The set and reappearance of the near-standstill moon in the Clisham hills and the deep valley are visible from the ring, over the southern summit of the ridge. The backsight position A is marked by a stone stump, 148m from the centre of the ring. From here, the standing stones straddle the deep valley where the moon reappears (Pl. 2). A person on the summit at position $C1$ would appear as a small figure within the moon. A person at position $C2$ in the ring would have their head silhouetted against the moon.

Opposite page: FIG. 5 (top): Plan of Na Dromannan ('Callanish X') and vicinity. FIG. 6 (centre): Plan of Ceann Hulavig ('Callanish IV') and vicinity. FIG. 7 (bottom): Plan of Achmore ('Callanish XXII') and vicinity.

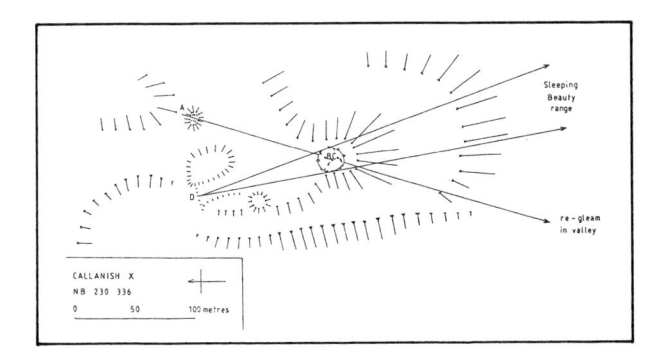

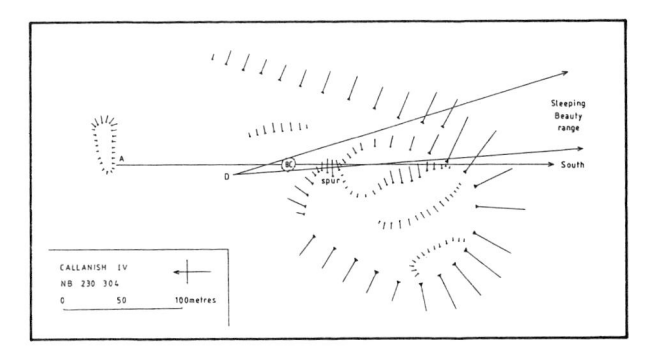

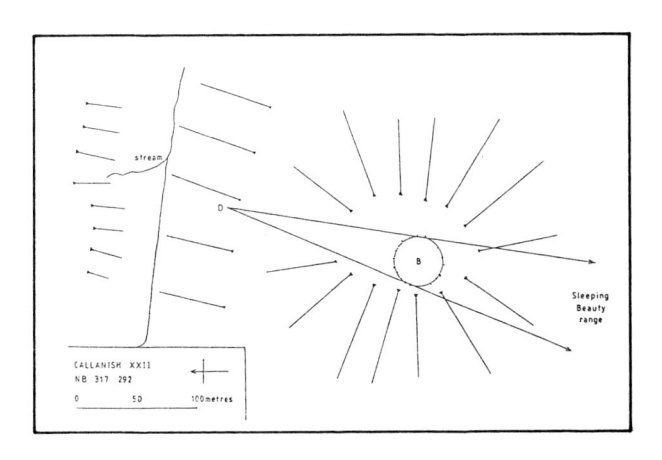

Callanish: symbolic and dramatic potential of the landscape 313

Cnoc Ceann a'Gharaidh ('Callanish II')

This elliptical ring of standing stones is set near the southern end of a ridge a little way from the summit (Fig. 4). The viewing lines for the rise and reappearance are similar to those at Cnoc Fillibhir Bheag (Callanish III), the rise being observed from a single low rock mound.

Na Dromannan ('Callanish X')

This double ring of standing stones, now fallen, is set at the southern end of a rocky ridge whose crest falls gently to the north (Fig. 5). The viewing lines for the rise and reappearance are similar to those for Cnoc Fillibhir Bheag, the rise being observed from D through a shallow gully, and the reappearance being observed from the twin humps of a natural rocky outcrop at A, 133m from the centre of the ring.

Ceann Hulavig ('Callanish IV')

This elliptical ring of standing stones is set on the eastern shoulder of a sloping ridge a little below the summit at its southern end. There is a small easterly rock spur between the ring and the summit. The ridge runs down to the north and broadens into a gently sloping peaty hillside (Fig. 6). The viewing line for the rise is similar to that for Cnoc Fillibhir Bheag, except that parts of the Sleeping Beauty range are masked by the spur and by a hill in the middle distance. It is impossible to see Clisham and the valley from the ring, but from the backsight position D the near-standstill moon set very close to due south, where a person within the ring at C would have been silhouetted in front of it.

Achmore ('Callanish XXII')

A variation was possibly achieved at Achmore, 11 km east of Callanish, where a large circular ring of stones, most of them fallen, stands on a flat peaty hill summit (Fig. 7). From this ring the Sleeping Beauty range of hills has a different appearance from that when seen from Callanish. The Beinn Mhor hill no longer forms her 'pillow' but her 'belly' giving the Sleeping Beauty the appearance of a pregnant woman, literally an earth mother. A viewer at the backsight position A, 169 metres from the centre of the ring, position B, would have seen the standing stones of the ring framing the Sleeping Beauty and would therefore have seen the moon set into her 'belly' within the ring. A reappearance might have occurred at her neck or brow, depending on the date of observation.

Opposite page: FIG. 8 (top): Plan of the Standing Stones of Callanish and vicinity. FIG. 9 (centre): Profile of the Callanish standing stones, viewed from the west. PL. 4 (bottom): The reappearance of the near-standstill moon at the foot of the central menhir at Callanish, as viewed from the northern end of the avenue.

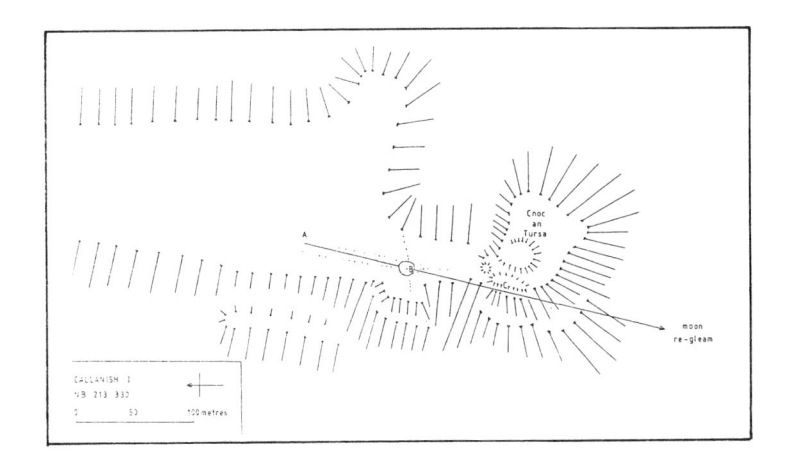

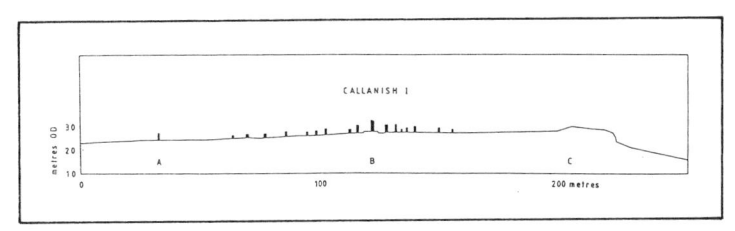

Callanish: symbolic and dramatic potential of the landscape 315

Drama and astronomy at the Standing Stones of Callanish

This complex site straddles the crest of a glacially sculpted north-south ridge with a rocky hillock, Cnoc an Tursa, at its southern end. A grand avenue, 83m long, leads to a ring with radiating rows and a massive central monolith (Fig. 8). An intervening hill in the middle distance hides all but the top of the deep valley. However, it is possible that the prehistoric builders stage-managed an artificial reappearance drama.

Seen from north end of the avenue (A), the moon rises gently from the Sleeping Beauty range, passes low at due south, skims some stones in the east row, dips into Cnoc an Tursa, and disappears. However, it then reappears briefly within the circle, at the foot of the largest stone of all and at the head of the burial cairn (Pl. 4).

This dramatic view of the moon's reappearance at ground level within the circle is achieved because the ground slopes up from the end of the avenue (A) to the circle (B) at roughly the same angle as the altitude of the moon, somewhat less than $2°$ (Fig. 9). Beyond the circle, the ground dips into a slight hollow, passes the west edge of Cnoc an Tursa, then rises to regain the same viewing line at a high point (C). The ground then drops away steeply. To a viewer at A, a person standing at C at the moment of the moon's reappearance would just fit within its orb (Pl. 3). This is because at this distance (175m) a person subtends about the same angle as the moon.

Conclusion

Considering the large number of criteria to be satisfied, it is truly remarkable that the prehistoric inhabitants of this area found so many sites at which they could locate rings for maximising the symbolic and dramatic potential of the landscape in relation to a lunar event which occurred monthly for only a few years in the 18·6-year cycle.

References

Curtis, G.R. (1988). The geometry of some megalithic rings. In *Records in stone*, ed. C.L.N. Ruggles, pp. 351–77. Cambridge: Cambridge University Press.

Ponting, M.R. and Ponting, G. (1981). Decoding the Callanish complex —some initial results. In *Astronomy and society in Britain during the period 4000–1500 BC*, ed. C.L.N. Ruggles and A.W.R. Whittle, pp. 63–110. Oxford: British Archaeological Reports (BAR British Series 88).

Ponting, M.R. (1988). Megalithic Callanish. In *Records in stone*, ed. C.L.N. Ruggles, pp. 423–41. Cambridge: Cambridge University Press.

29

The Bush Barrow gold lozenge: a solar and lunar calendar for Stonehenge?

ARCHIBALD S. THOM

Introduction

A lozenge, of 0·5 mm-thick beaten gold, was excavated in 1808 from The Bush Barrow, 1 km from Stonehenge. Its carefully inscribed markings are believed to be identifiable as a calendar fashioned for use at Stonehenge (Thom et al. 1988).[1] Found over the breast of a skeleton of a tall man lying from north to south, fixed to a thin piece of wood over the edges of which it had been lapped, its symmetrical shape and correct corner angles make it appear probable that it had something to do with the four cardinal points and solstitial sunrises and sunsets.

By fixing the flat lozenge on a table at eye level and orienting it with its shorter diagonal on the meridian, an observer could use an alidade while watching sunrise or sunset throughout the year. Were the bronze rivets, intermixed with wood and some thin bits of bronze, found nearby, the remains of the alidade? Markings exist on the plaque which indicate that the sixteen-month calendar was in use. Guide lines exist for inserting the intercalary leap day. Eight additional lines can be identified as indicating moonrise and moonset at standstills. Using actual horizon altitudes at Stonehenge and azimuths shown by the lozenge, calculation shows that the average discrepancy of the solar calendar lines is 0·36 days and that it was made in about 1600 BC.

---

[1] This paper is complementary to the earlier report and contains additional data. In Fig. 3 and Table 1 of the earlier report, two errors exist: solar azimuths 49°·66 and 310°·59 should be 49°·68 and 310°·44 respectively.

	Historical name	Epoch no.	Days in ½ or one month	Nominal day n	Fractional day $n \times 365\frac{1}{4}/365$
Vernal equinox Mar 21		0		0	0
			11		
		½	11½*	11·5	11·51*
			12		
		1		23	23·02
			23		
May Day	Beltane	2		46	46·03
			23		
		3		69	69·05
			23		
Midsummer June 21		4		92	92·06
			23		
		5		115	115·08
			23		
Lammas	Lug Lughnas	6		138	138·09
			22		
		7		160	160·11
			11		
		7½		171	171·12
			11		
Autumnal equinox Sep 21		8		182	182·12
			11		
		8½		193	193·13
			11		
		9		204	204·14
			23		
Martinmas Halloe'en	Samain or Sowain	10		227	227·16
			23		
		11		250	250·17
			23		
Midwinter Dec 21		12		273	273·19
			23		
		13		296	296·20
			23		
Candlemas Feb 1	Imbolc Bride	14		319	319·22
			23		
		15		342	342·23
			12		
		15½	11½*	353·5	353·74*
			11		
Vernal equinox		16		365	365·25

TABLE 1. The calendarer's division of the year with normal integral and fractional days. An asterisk indicates that half-months of 11½ days have been used for comparison.

Variation of the sun's declination

Fig. 1 shows the variation of the sun's declination throughout a year. Thom and Thom (1982) have given the equation for this variation in 1800 BC. Here the same equation is used, but with $\varepsilon = 23°\cdot9168$, $P = 216°\cdot363$ and $e = 0\cdot01816$, all three evaluated for $T = 38$ centuries before AD 1900, namely 1900 BC.

Alexander Thom's megalithic calendar (1967: 110; 1982: 76) indicates the division of the year into sixteen months of 22 or 23 days, with 182 and 183 days in the summer and winter seasons respectively (see Table 1). This particular arrangement he called 'Calendar Scheme A'. He stated that there is slight but inconclusive evidence for the further division of the months into half-periods of eleven or twelve days.

Numbers 1 to 16(=0) refer to epochs, zero epoch being arbitrarily chosen at sunrise on the day of the Vernal Equinox, sometimes referred to as the Neolithic equinox. The term 'epoch' here refers to an instant in time, i.e. to a particular date or day in the year. 'n' or 's' is attached to the epoch number to show whether the sun is moving to the north or south. On the lozenge, four lines exist at the half-epochs $\frac{1}{2}$, $7\frac{1}{2}$, $8\frac{1}{2}$ and $15\frac{1}{2}$; these are included in this analysis.

A radiating line on the lozenge is deemed to be *compatible* if it is within the annual variation of the rising or setting sun on the horizon. Sixteen compatible lines are discernible. A radiating line with a compatible solar declination shows two dates. Autumn (a) is here defined as the period from midsummer to midwinter; spring (s) as that from midwinter to midsummer.

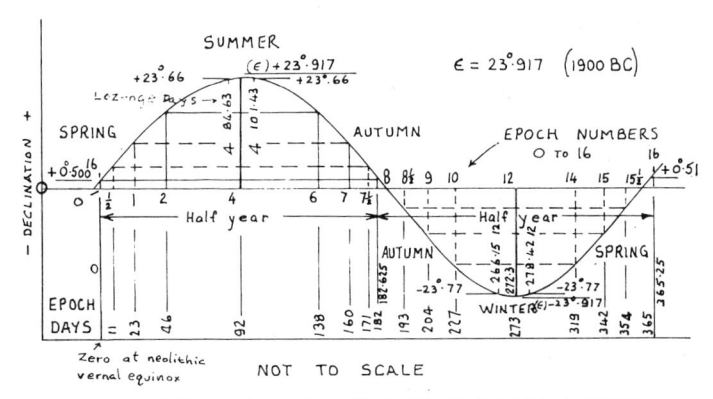

FIG. 1. Declination of the sun throughout a year.

Procedure

1. An azimuth from the lozenge is used to evaluate the 'lozenge declination', i.e. the indicated declination of the sun's centre at the latitude of Stonehenge using the observed horizon altitude corrected for refraction.

2. The lozenge declination is then used in the 1900 BC equation for the sun's declination throughout the year and the relevant lozenge days d_a or d_s calculated.

3. The lozenge days are compared with the nominal calendar day n in scheme A, with 23 and 22 day 'months' (see Table 1).

Fourteen radiating lines

Taking compatible line 2W as an example, its azimuth is seen from Table 2 to be 296°·55, showing a declination of +16°·17 (see also Thom *et al.* 1988: fig. 3). This line indicates epoch 2, when the sun set in the late spring in the fourth quadrant; it also indicates sunset at epoch 6 in early autumn (in Table 2, this epoch is shown in brackets to indicate a pairing). The dates in lozenge days for epoch 2 are, respectively, d_s = 44·28 days and d_a = 140·68 days. These last two day numbers correspond to the nominal days n = 46 and n = 138 on the calendar. We consider these two days to be paired. The differences between nominal day n and lozenge day d are given as t/a and t/s, referring respectively to the autumn and spring halves of the year. The algebraic mean of t/a and t/s for the line 2W is –0·48 days.

The two 'incompatible' lines 4E and 12W yield declinations that are greater than ε = 23°·917, the obliquity in 1900 BC, and so no lozenge day can be calculated. These lines are not included in the averaging

---

TABLE 2 (opposite page). Comparisons between the lozenge indications and Alexander Thom's 'Calendar Scheme A'.

The lozenge azimuth and lozenge declination data are repeated from Thom *et al.* (1988: table 2). The two incompatible and two unusable lines are italicised, leaving fourteen compatible lines in the last three columns.

The nominal day n is the integral number of days according to the epoch arrangement shown here and in Thom *et al.* (1988: table 2). Asterisks in the 'nominal day' column indicate where half-months of $11\frac{1}{2}$ days have been used. Nominal calendar day numbering n stops at 365. (See Table 1 for fractional calendar days.)

t/a and t/s are the algebraic difference between the nominal day and the lozenge day in the spring and autumn half of the year, respectively.

The Bush Barrow gold lozenge 321

Loz. line no.		Line no.	Epoch no. or (pair)		Loz. azim. A °	Loz. dec. δ °	Loz. day (aut) d_a	Nominal day n	t/a days (aut) $n-d_a$	Loz. azim. A °	Epoch no. or (pair)		Loz. day (spr) d_s	Nominal day n	t/s days (spr) $n-d_s$	$t/a+t/s$ 1900 BC	2045 BC	x_1^2 1600 BC
36	32	4W	4	s		+23·66	101·43	(92)	−9·43	310·09	4	n	84·63	92	+7·37	−1·03	+1·11	−0·91
36	31	2W	(6)	s		+16·17	140·68	(138)	−2·68	296·55	2	n	44·28	46	+1·72	−0·48	−0·53	−0·39
36	30	1W	(7)	s		+9·46	160·16	(160)	−0·16	285·41	1	n	23·80	23	−0·80	−0·48	−0·50	−0·39
36	29	½W	(7½)	s		+4·31	173·29	(171)	−2·29	276·83	½	n	9·90	11½*	+1·60	−0·34	−0·36	−0·32
36	28	0/16W	(8)	s		+0·07	183·66	(182)	−1·66	269·99	16/0	n	364·15	365	+0·85	−0·41	−0·41	−0·32
28	32	4E			49·58	+23·9664												−0·41
28	33	6E	6	s	63·06	+16·37	140·02	138	−2·02		(2)	n	44·99	(46)	+1·01	−0·50	−0·53	−0·46
28	34	7E	7	s	74·08	+10·01	158·68	160	+1·32		(1)	n	25·36	(23)	−2·36	−0·52	−0·54	−0·32
28	35	7½E	7½	s	82·92	+4·38	173·13	171	−2·13		(½)	n	10·07	(11½)*	+1·43	−0·35	−0·36	−0·39
28	36	8E	8	s	89·99	−0·10	184·07	182	−2·07		(16/0)	n	363·72	(365)	+1·28	−0·40	−0·40	−0·35
28	37	8½E	8½	s	96·99	−4·52	194·87	193	−1·87		(15½)	n	352·38	(353½)*	+1·22	−0·32	−0·31	−0·23
28	38	9E	9	s	105·96	−10·11	209·04	204	−5·04		(15)	n	337·32	(342)	+4·68	−0·18	−0·16	−0·18
28	39	10E	10	s	116·95	−16·67	228·00	227	−1·00		(14)	n	317·47	(319)	+1·53	+0·27	−0·30	+0·59
28	40	12E	12	s	130·47	−23·77	266·15	273	+6·85		(12)	n	278·42	273	−5·42	+0·72	+0·77	−0·36
36	43	15½W	(8½)	s		−4·27	194·24	(193)	−1·24	262·99	15½	n	352·94	353½*	+0·56	−0·34	−0·33	−0·26
36	42	15W	(9)	s		−9·67	207·89	(204)	−3·89	254·05	15	n	338·53	342	+3·47	−0·21	−0·19	+0·17
36	41	14W	(10)	s		−16·35	226·98	(227)	+0·02	243·13	14	n	318·54	319	+0·46	+0·24	+0·28	
36	40	12W				−23·93				229·84								
												Sum of 14		+5·20	+5·04	+4·64	
												Mean of 14		+0·37	+0·36	+0·33	
											Algeb. mean of 14		−0·29	−0·29	−0·26	
												r.m.s		0·11	0·11	0·09	

process,[2] and assessment of the artefact as a calendar must rely on the evidence of the remaining fourteen lines.

For the fourteen compatible lines the mean difference between n and d for the year 1900 BC is 0·36±0·11 days, irrespective of sign (see the foot of Table 2). The algebraic mean is -0·29 days, which could be related to an erroneous orientation. Similar calculations were undertaken for 2045 and 1600 BC. Inspection of Table 2 indicates 1600 BC to be the best date; the mean of the fourteen lines is slightly smaller, being 0·33±0·09 days.

The lines at half-months greatly strengthen the calendrical hypothesis. They are at the easily discernible part of the sun's passage at equinox; they also strengthen the case for fixing calendar zero at an equinox.

The calendarer would have a running count of days which would show him, while using the lozenge, when to insert the intercalary day; any difference would show up better at equinoxes than at solstices, because the rate of movement of the sun on the horizon at equinoxes is about 1·2 diameters per day. The half-month lines would assist, along with the small angles 15, 04, 16 and 17, 04, 16 (see Thom *et al.* 1988: figs. 1 and 2), each of which represents the sun's daily change in azimuth.

Calculations were also made for 1900 BC using fractional days (see Table 1) in place of nominal (integral) days (*ibid.*: 498). The resulting average discrepancy worked out to be 0·28 days as opposed to 0·36 days (see Table 2). This strengthens the hypothesis that the lozenge designer was using the intercalary day.

With regard to the possibility of the observer having used the alidade in the reverse direction (*ibid.*: 498), calculations for 1900 BC gave the same mean of 0·36 days, but an algebraic mean of –0·31 as opposed to –0·29. Table 3 shows the necessary horizon altitudes in the reverse direction.

---

[2] The details are as follows:

	Line 12E		Line 4W
Lozenge azimuth at sunrise	130° 28' 12"	at sunset	310° 5' 24"
Sun's azimuth at sunrise	130° 45' 12"	at sunset	310° 26' 24"
Azimuth difference	17' 00"		21' 0"
Lozenge declination	–23°·7692		+23°·6631
Sun's declination at solstice	–23°·9168		–23°·9168
Declination difference	0°·1476		0°·2537

The differences in azimuth are, for line 12E, 17' and, for line 4W, 21'. The sun's semi-diameter is about 15'·5. While examining Table 2, because the two lines both miss the sun's orb, it would appear better, while averaging discrepancies, not to use 12E and 4W. The lozenge days involved, namely 84·63 and 101·43 for line 4W and 266·15 and 278·42 for line 12E, are indicated in Fig. 1.

Lozenge line number	Lozenge azimuth	Observed altitude	
28-36	89°·99	22'·6	e
29-36	96°·83	19'·9	e
30-36	105°·41	17'·5	
31-36	116°·55	19'·2	e
32-36	130°·09	15'·0	
32-38	229°·58	24'·4	e
33-28	243°·06	35'·5	
34-28	254°·08	45'·0	
35-28	262°·92	37'·0	
36-28	269°·99	34'·5	
37-28	276°·99	31'·5	
38-28	285°·96	21'·0	
39-28	296°·95	22'·4	e
40-28	310°·47	20'·2	e
40-36	49°·84	28'·3	e
41-36	63°·13	21'·3	e
42-36	74°·05	22'·0	
43-36	82°·99	25'·6	e

TABLE 3. Observed horizon altitudes on reverse azimuths of 18 lines. *e* indicates that the altitude in question was estimated because of trees on the horizon. The latitude of Stonehenge is 51° 10' 42".

Was this ceremonially buried gold artefact a copy of a more robust working calendar or was it the original master copy? Was the lozenge a means whereby observed angular measurements could be recorded and subsequently retrieved years later without recourse to writing? Was it a 'text book' for making the calendar, a 'reference encyclopaedia'?

References

Thom, A. (1967). *Megalithic sites in Britain*. Oxford: Oxford University Press.

Thom, A. and Thom, A.S. (1982). Statistical and philosophical arguments for the astronomical significance of standing stones with a section on the solar calendar. In *Archaeoastronomy in the Old World*, ed. D.C. Heggie, pp. 53–82. Cambridge: Cambridge University Press.

Thom, A.S., Ker, J.M.D. and Burrows, T.R. (1988). The Bush Barrow gold lozenge: is it a solar and lunar calendar for Stonehenge? *Antiquity* **62**, 492–502.

30

New evidence concerning possible astronomical orientations of 'Tombe di Giganti'

EDOARDO PROVERBIO

1 Introduction

The existence is well known in Sardinia of numerous megalithic funerary structures, from pre-Nuraghic dolmens dating back to the Neolithic and the early Bronze Age (c. 2500–1500 BC) to the small funerary grottoes ('domus de janas') which, originating in the Copper Age, continue on to the early Nuraghic and Iron Ages (c. 1600–800 BC). More highly developed forms of megalithic tomb deriving from the dolmen are represented by the covered corridors of the allée couverte (gallery) type; these have an orthostatic structure with a flat roof and an entrance that is usually cut into the lower part of the front stele. The origins of these go back to the early pre-Nuraghic period (the Monte Claro and Bonnannaro cultures c. 2200–1700 BC). The development of these tombs, with the addition around the front stele of a characteristic exhedral structure forming an arc of a circle composed of massive stone, leads to the so-called 'tombe di giganti'[1] which date from the end of the Bronze Age to the middle Nuraghic (c. 1500–800 BC).

The 'tombe di giganti', with their rectangular corridors and rectangular or trapezoidal cross sections, have, as has been said, front stelae of varying sizes with openings out into them at the bottom. These stelae

---

[1] The popular name for these collective tombs, characteristic of agro-pastoral communities, appears to derive from the belief that these large burial monuments (the Goronna tomb at Paulilatino could contain up to two hundred bodies) were built for the burial of giants.

are situated at the centre of the exhedra. While the semi-circular form of the exhedra may be associated with the crescent of the new moon, some have seen in the overall shape of the tools the suggestion of a stylised bull's head;[2] such a combination in a burial area of a bull symbol with the subterranean element of the tomb is widely diffused throughout burial cults of agricultural peoples of the Late Neolithic. It is also interesting to note that the symbols of either the crescent or the bull's head is present in Sardinia in the oldest burials in artificial grottoes attributed to the Neolithic, as for example, in some tombs in the Anghelu Ruju necropolis. This combination of bull and moon symbolism has been interpreted not only in the visual sense (bull's horn – crescent moon) but also as reflecting the presence of an astral element within the framework of burial cults associated with the belief in the recovery of life and 'rebirth' after death. Christian Zervos, an expert in Sardinian cultures, arrives at this conclusion when, on associating the principle of generation with that of fecundity, he declares that: 'la tête taurine ou ses cornes n'étaient que les signes de la déesse-lune' (Zervos 1954: 267).

The still meagre knowledge we have at present concerning both the allées couvertes and the tombe di giganti themselves[3] often makes it difficult for the investigator to associate these monuments chronologically with other megalithic structures, whether funerary or not, in the same area. In fact, only a limited number of tombe di giganti situated in central Sardinia (Barbagia) (Lilliu 1981: 88–93, 117–28; Atzeni 1988) and in the Sassari and Cagliari areas (Atzeni 1985; Castaldi 1968; 1969) have been excavated and accurately investigated by modern archaeologists. In other cases we are compelled to resort to morphological and structural criteria in an attempt to date the tombs. Lilliu (1967: 172) suggests a possible chronology of megalithic tombs on the basis of changes in the form of the central stele. By associating the large cambered stelae divided across the middle into two parts, which are present in certain tombs, with analogous decorative elements carved onto the front of hypogean grottoes (domus de janas), Lilliu concluded that these tombs are to be attributed to the early Nuraghic period (c. 1500 BC). The large cambered stelae with no dividing line presumably belong to a later period. The stelae that are made up of an architrave supporting the stones forming the exhedra, and which have a trapezoidal form with dentils (dentilled stelae), are associated with still more evolved forms (c. 1200–1000 BC).

---

[2] The stylised plan of the 'tombe di giganti' in Sardinia was compared by Giovanni Lilliu, the leading authority on the pre-Nuraghic and Nuraghic societies in Sardinia, to a bull's head 'col muso arrotondato costituito dal muro di fondo in curva e con le corna disegnate nell'ampia esedra a mezzaluna' (Lilliu 1967: 309).

[3] Lilliu (1967: 216), on referring to the tombe di giganti, speaks of 'scarsa conoscenza di materiali di questa forma di sepoltura collettiva'.

The suggestion that pre- and protohistoric burial monuments may have been oriented in the direction of points on the horizon where the brighter celestial bodies rose and set finds support in the existence of ancient luni-solar or stellar burial rites associated with astral religious cults. Furthermore, the existence of well-defined astronomical orientations has recently been demonstrated in Neolithic (Cantacuzino 1967) and Bronze Age (Barlai 1980) tombs and cemeteries, and even in those dating from historical periods (Parisot and Petrequin 1982).

The determination of the orientations of five tombe di giganti near Madau (near Fonni) in the Barbagia area of central Sardinia, carried out in 1985 (Proverbio *et al.* 1987), supplied interesting indications that justified a second campaign of measurements, which was carried out in May 1986 (Proverbio *et al.* 1991a). To obtain additional data from another region on the island, in October 1989 the author, in collaboration with Michael Hoskin, carried out a third campaign of measurements in an area of central Sardinia farther to the west, including Paulilatino, Sedilo and Birori.

2 The orientations of the tombe di giganti

The tomb orientations taken into consideration (Table 1) were those of the longer axis in the direction of the ideal centre of the exhedra, that is to say in the direction leading from the inside to the outside of the tomb through the entrance.

Accurate measurement of the direction of the axes was made difficult by the deterioration of many of the tombs. Whenever possible (specifically in tombs III and IV) we determined the direction of the centre-line and also of the left-hand and right-hand walls of the same tombs, these being marked by means of stakes. Elsewhere, the best preserved of these three directions was measured. In general, independent measurements of the orientations were taken using a military compass and a theodolite. However, because of the high degree of deterioration of tombs II and VII, their orientations were determined by means of the compass alone. The orientation of tomb XV was also determined by compass alone because cloud cover obscured the sun.

The measurement of the astronomical azimuth with the theodolite was in general carried out by observing three successive positions of the sun, which allowed us to calculate the standard deviation of each single astronomical alignment observed, this proving to be on average below

TABLE 2 (opposite page). Values of the observed azimuths (A) and calculated azimuths (A_c) corresponding to the occurrence of the celestial phenomena indicated and to the dating interval given in table 1.

NTG	Town	A	h	Type of stele	Dating interval
I	Sedilo	173°·4	+2°·5	Dentilled	1200–1000
II	Aidomaggiore	173°·8	+2°·7	Dentilled?	1200?–1000?
III	Aidomaggiore	157°·6	+1°·1	Cambered	1500–1200
IV	Aidomaggiore	110°·1	0°·0	Cambered	1500–1200
V	Aidomaggiore	69°·4	+1°·0	Cambered?	1500?–1200?
VI	Aidomaggiore	77°·2	0·°0	Cambered	1500–1200
VII	Borore	111°·3	0°·0	Cambered	1500–1200
VIII	Borore	81°·7	+0°·5	Architrave	1000–800
IX	Birori	121°·3	0°·0	Architrave	1000–800
X	Sindia	185°·1	+2°·2	Architrave	1000–800
XI	Abbasanta	123°·9	+0°·4	Cambered	1500–1200
XII	Abbasanta	113°·4	+0°·5	Cambered	1500–1200
XIII	Paulilatino	127°·9	+0°·5	Dentilled	1200–1000
XIV	Paulilatino	87°·3	–0°·5	Cambered	1300–1000
XV	Paulilatino	102°·3	–0°·5	Cambered	1300–1000

TABLE 1. Azimuths A of the directions of the tombe di giganti, heights h of the natural horizon in these directions, and dating intervals.

NTG	A	Celestial phenomenon	A_c		Declination	
I	173°·4	Rise of α Cru	168°·5	171°·5	–45°·9	–46°·9
II	173°·8	Rise of α Cru	167°·1	169°·5	–45°·9	–46°·9
III	157°·6	Rise of α Cru	157°·7	162°·1	–44°·4	–45°·9
IV	110°·1	Rise of β Ori	113°·8	112°·6	–18°·4	–17°·1
V	69°·4	Moonrise	66°·8	66°·9	+(ε–i)+s	
VI	77°·2	Rise Pleiades	79°·1	77°·7		
VII	111°·3	Rise of β Ori	113°·8	112°·6	–18°·4	–17°·1
VIII	81°·7	Rise of α Tau	82°·4	81°·1	+5°·7	+6°·7
IX	121°·3	Sunrise	121°·5	121°·5	–ε+s	
X	185°·1	Setting α Cru	189°·6	185°·6	–46°·9	–48°·4
XI	123°·9	Sunrise	122°·2	122°·1	–ε+s	
XII	113°·4	Rise of β Ori	114°·3	112°·5	–18°·4	–17°·1
		Rise of α CMa	114°·0	113°·0	–18°·2	–17°·6
XIII	127°·9	Moonrise	130°·0	130°·6	–(ε+i)+s	
XIV	87°·3	Rise of α Ori	88°·7	87°·2	+0°·3	+1°·4
XV	102°·3	Rise of ε Ori	101°·2	99°·7	–9°·3	–8°·1

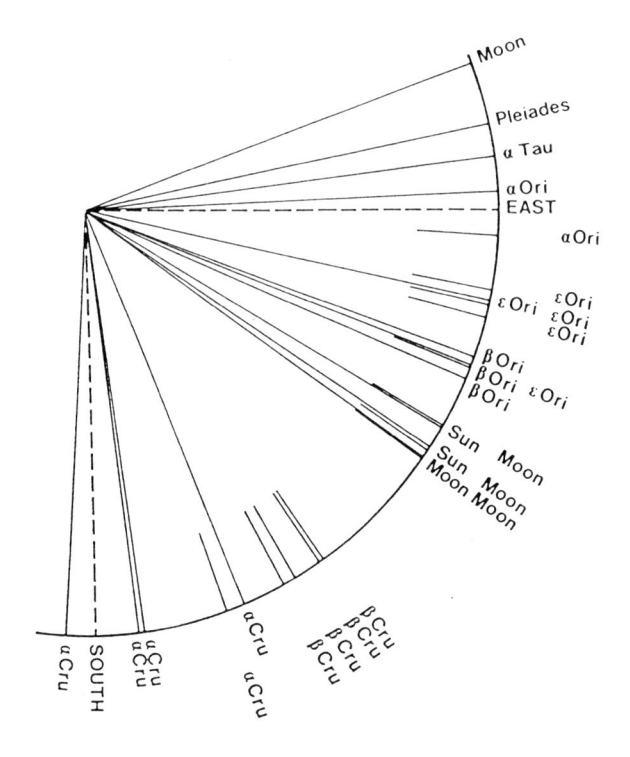

FIG. 1. Orientations of tombe di giganti in Sardinia. Short lines represent thirteen tombs measured by Proverbio, Romano and Aveni in 1986 (Proverbio *et al.* 1991a); long lines represent fifteen tombs measured by Proverbio and Hoskin in 1989.

$\pm 0°\cdot 1$. The astronomical azimuth of the centre-line of tomb I was taken twice, with the aligning stakes in different positions. This led to two independent values for the observed azimuth: $173°\cdot 8$ and $172°\cdot 9$.

3 Analysis of the results

The fifteen tombe di giganti situated in central-western Sardinia appear all to be facing in a direction corresponding roughly to the south-east quadrant. In Table 2 we compare the azimuths observed (A) with those calculated (A_c) for the following astronomical events at the two extremes of the dating interval: solstitial sunrise (upper limb); major and minor standstill moonrise (upper limb); Pleiades rise; the rise of Aldebaran (α Tau); and the rise of the brightest stars of Orion (α Ori, β Ori, ε Ori) and the Southern Cross (α Cru, β Cru). Fig. 1 combines the orientation data from the second and third campaigns of measurements.

LCT	α Cru		α Ori
	Rise	Set	Rise
20:00 hrs	Mar 15	May 30	Oct 5
22:00 hrs	Feb 15	Apr 30	Sep 5
24:00 hrs	Jan 15	Mar 30	Aug 5
02:00 hrs	Dec 15	Feb 28	Jul 5
04:00 hrs	Nov 15	Jan 30	Jun 5

TABLE 3. Approximate time of year when α Ori rose and α Cru rose and set at given times of night, expressed in local civil time (LCT), around the year 1200 BC.

It can be seen that the orientations of the tombs are roughly clustered around the following four sectors:

- azimuths between about 190° and 140°, corresponding to α and β Cru;
- azimuths between about 130° and 120°, corresponding to the southern standstill moon and/or winter solstitial sun;
- azimuths between about 115° and 85°, corresponding to α, β and ε Ori or α CMa; and
- azimuths between about 85° and 65°, corresponding to the northern minor standstill moon and/or α Tau and/or the Pleiades.

The first and third cluster, and possibly also the fourth, suggest that the stellar reference points in the burial rites of these ancient proto-Nuraghic civilisations were not so much single stars as groups of stars, such as Orion's belt, a compact asterism composed of three easily distinguished stars, and the entire Southern Cross.

It is interesting to note that around 1200 BC the rising of the brightest stars of Orion was observable during the short, summer nights whereas the rising (and setting) of the Southern Cross took place during the winter months (see Table 3). The brightest and southernmost star of the Southern Cross, α Cru, culminated at about 4 degrees of altitude; its diurnal semi-arc was thus quite brief (about 2·5 hours) and its points of rising and setting were very close together and well to the south. This suggests that in the course of the year the prehistoric communities of central and western-central Sardinia presumably had to refer to different asterisms during ceremonies associated with astral burial rites.

4 Discussion

It is clearly of interest to investigate whether there is corroborating evidence of an interest in the asterisms apparently evidenced from the orientations of the tombe di giganti.

Unfortunately, there are very few data on possible astronomical orientations of Sardinian funerary monuments older than the tombe di giganti, such as dolmens and domus de janas. In recent work (Proverbio et al. 1991b), the author was able to show that the 'sanctuary' of Monte d'Accoddi—one of the oldest and most interesting prehistoric monuments in Sardinia, presumably of a religious-funerary character, dating back to the Late Neolithic (4500–4000 BC) and rebuilt during the Neolithic (about 1800 BC)—seems to be oriented in the direction of the rising of southern minor standstill moonrise and the rise of ε Ori. This result appears to confirm the persistence in Sardinia of ceremonial cults and burials in some way associated with the rising of celestial objects, presumably as manifestations of a belief in the 'resurrection' and 'regeneration' of the dead.

Important evidence in favour of the existence of astronomical orientations of ceremonial megalithic monuments in the western Mediterranean area is Hoskin's (1989) demonstration of the existence of a presumed astral cult or religion within the framework of the talayotic culture (c. 1300–800 BC) characteristic of the island of Menorca. The temples (or 'taulas') in the south of the island are invariably sited on elevated ground and have an uninterrupted view of the horizon. Except for one unusual monument, all are oriented within 37° of due south, that is, towards the sector of the sky where in talayotic times the constellation Centaurus (including, notably, the Southern Cross and α and β Cen) rose and set. Furthermore, amongst the statues found by the excavator of one of the sites is one of the Egyptian god Imhotep, whose Greek counterpart Asclepius was tutor to the Centaur.

These results, when and if they receive further confirmation in the course of future measurement campaigns, may lead to new interpretations of the culture of prehistoric communities in this area of the Mediterranean and may contribute to supplying more accurate answers concerning the nature of these native cultures and the possible influences exerted on them by the great astral religions of the eastern Mediterranean.

Acknowledgements

The author wishes to thank Dr Lello Fadda for his invaluable assistance in locating and measuring the orientations of the tombe di giganti, and Mr Peppino Calledda for his help during the measurement campaign and the reduction of the data.

Special and heartfelt thanks are offered to Dr Michael Hoskin for his close collaboration in the course of the campaign, during which he determined the orientations of all the tombs with a high-precision compass, and for the invaluable suggestions and advice offered when the project was being drawn up.

References

Atzeni, E. (1985). *Tombe eneolitiche nel cagliaritano: Studi in onare di Giovanni Lilliu.* Cagliari: Dipartimiento di Archeologia.

Atzeni, E. (1988). Tombe megalitiche di Laconi (Nuoro). *Rassegna di Archaeologia* **7**, 526–27.

Barlai, K. (1980). On the orientation of graves in prehistoric cemeteries. *Archaeoastronomy* (Center for Archaeoastronomy) **3**(4), 29–32.

Cantacuzino, Gh. (1967). Necropoli preistorica de la Cernica si locul ei in neoliticul roma nesc si europeau. *Studii și cerecetări de Istorie veçhe* **3**, 379.

Castaldi, E. (1968). Nuove osservazioni sulle tombe di giganti. *Bullettino di Paletnologia italiana* **77** (year XIX), 7–91.

Castaldi, E. (1969). Tombe di giganti nel sassarese. *Origini* **3**, 119–274.

Hoskin, M.A. (1989). The orientations of the taulas of Menorca (1): the southern taulas. *Archaeoastronomy* no. 14 (supplement to *Journal for the History of Astronomy* **20**), S117–36.

Lilliu, G. (1967). *La civiltà dei Sardi.* Turin: ERI.

Lilliu, G. (1981). *Monumenti antichi barbaricini.* Gaggari: Dessi (Quaderni Ministero Beni Culturali e Ambientali, no. 10).

Parisot, J.P. and Petrequin, P. (1982). Orientations astronomiques des tombes mérovingiennes du cimitière de Soyria (Jura). *Archaeoastronomy* no. 4 (supplement to *Journal for the History of astronomy* **13**), S41–48.

Proverbio, E., Romano, G. and Aveni, A.F. (1987). Astronomical orientations of five megalithic tombs at Madau, near Fonni, in Sardinia. *Archaeoastronomy* no. 11 (supplement to *Journal for the History of astronomy* **18**), S55–66.

Proverbio, E., Romano, G. and Aveni, A.F. (1991a). Astronomical orientations of 'tombe dei giganti' in Barbagia (Sardinia). In *Colloquio Internazionale Archeologia e Astronomia,* ed. G. Romano and G. Traversari, pp. 52–59. Rome: Giorgio Bretschneider Editore (Supplementi alla RdA, 9).

Proverbio, E., Romano, G. and Aveni, A.F. (1991b). Astronomical orientations of Monte d'Accoddi (Sassari in Sardinia). In *Colloquio Internazionale Archeologia e Astronomia,* ed. G. Romano and G. Traversari, pp. 38–41. Rome: Giorgio Bretschneider Editore (Supplementi alla RdA, 9).

Zervos, C. (1954). *La civilisation de la Sardaigne.* Paris: Cahiers d'Art.

V

EDUCATION AND DISSEMINATION

31

An image database for learning archaeoastronomy

CLIVE RUGGLES

> It is...the pursuit of the whole past, not just the comfortably preferred elements of it, that should be the preoccupation of all who profess an interest in antiquity. No less than architecture and artefacts, astronomy and its alignments are a legitimate part of that pursuit.
>
> Burl (1988: 200)

Introduction

Leicester is the only University in Britain which has an undergraduate course in archaeoastronomy. This is run as a final-year option in the single-subject Archaeology degree, and is based around a module of ten two-hour lectures. The course does not concentrate upon technical details, but focuses instead upon questions of archaeoastronomy's aims and objectives, scope, and methodology. It draws initially from British archaeoastronomy, and then proceeds to use a variety of examples from world archaeoastronomy and ethnoastronomy.

In order to be able to develop a suitable critical perspective, students need to be able to examine a broad range of visual evidence. This is particularly clear in the case of alignment data. A critical issue with regard to putative astronomical alignments is how objectively they have been selected. In order to appreciate the scale of the problem and to assess the possible degree of data selectivity in any particular case—say, a particular megalithic alignment upon a particular horizon notch—it is necessary to examine the whole site and horizon from as many vantages as possible. The best, and until recently the only, way to do this is actually to visit the site: several authors such as Gingerich (1981) have referred to their disillusionment when, upon actually visiting a 'classic' site rather than just seeing the illustrations of selected views and horizon features presented in the early works on archaeoastronomy, they were confronted with the data disregarded in those works—scores of alignments and horizon notches that do not readily fit any astronomical explanation.

This short paper describes the use at Leicester of a database of images, accessed using interactive video, as a way of supplying sufficiently large quantities of visual information that they can effectively substitute for first-hand data, and hence help provide the necessary critical perspectives in the context of a short course. General issues in the storage and access of image data are considered first, then we proceed to describe the particular resource set up at Leicester. Finally, we consider how this resource will be used, expanded, and could be disseminated in the future.

Storing and accessing image data

There are essentially two problems to consider in the storage and access of image data. The first is the physical (low-level) problem of how to manage a computer-based resource consisting of a large indexed collection of images. The second is the abstract (high-level) problem of how to manage conceptual structures that link them together in a meaningful way and hence allow a user to browse the collection—in other words, how to create an 'image database'.

Managing the image resource

Advances in optical storage technology during recent years—advances which continue apace—allow the possibility of holding large indexed collections of still video images on computer-based media. These images might be maps, aerial photographs, satellite images, or photographs taken from a known location.

During the late 1980s, interactive videodiscs became established as an effective way of holding a comprehensive collection of visual information (see, e.g., Martlew 1988). Images are held in analogue form. The advantages of interactive videodiscs include large capacity (a single videodisc can hold over 50,000 high-quality images on each side) and fast access (each image on a side is directly accessible within at most a few seconds). Furthermore, moving video sequences with or without sound may be stored in place of stills. Disadvantages include the very large cost of producing a videodisc—this includes assembling the entire collection of visual material on videotape and the subsequent production of a master videodisc, and can run into hundreds of thousands of dollars (copies, however, are cheap)—and the additional costs to the end-user of extra equipment such as videodisc players and video merge cards (cards that enable analogue output from the videodisc to be merged with the standard digital output from a computer), which can themselves cost as much as the PC workstation they are fitted to.

Digital storage media hold the key to the future, and these are rapidly developing, in the form of CD-ROMs, WORM (write once, read many times) drives, rewritable optical discs and many more. The storage and retrieval of raster images, as well as that of other 'multimedia' resources, on computer-based media is fast becoming commonplace. At the same time, network bandwidths are rapidly increasing and the transmission of multimedia resources around networks is becoming feasible. While there are many problems to be overcome—standards for compression, storage and transmission being a key issue—it is clear that the problem of physically managing large-scale image resources will soon be carried out of the domain of the individual user and into that of the manufacturer and system manager. In the surprisingly near future, we will be able to handle multimedia resources on computer as effortlessly and transparently as we handle text and standard graphics at present.

Managing the image database

Effective management and retrieval of a collection of images requires more than a simple indexing system; the system needs to know the 'abstract' relationships of the images (i.e. relationships that will be perceived naturally and interpreted easily by the user) both to each other and to other information held within the system so that appropriate images can be retrieved and the user can move from image to image in a logical and transparent manner.

As an example, consider a structure in which each image represents the view seen by an observer at a given position (say, at an archaeological site) facing in a given direction. The structure is organised so that neighbouring images represent the views obtained by turning or moving (i.e. side-stepping) to the left or right, or by moving forwards, from the current position. This sort of structure, which we might term a 'walkabout' or 'surrogate walk', was featured in the BBC 'Domesday' system for locating and retrieving geographical information within the British Isles (Rhind and Mounsey 1986) and subsequently in a number of specific multimedia products.

The important point, however, is that such a structure should be specifiable in a generic way, so that a *user* (rather than just the producer of a fancy but specific product) can configure existing collections of images into it. This parallels the way in which a conventional database is constructed by first identifying entities and relationships in a generic way, and then loading specific sets of data within this structure.[1]

The present author (Ruggles 1991; 1992) has identified several such generic structures in the context of spatially related images. These include grids (used, e.g., for navigating around contiguous map segments at the same scale), hierarchical grids (for navigating around map segments at different scales), maps with views (which allow a user to see the view from places identified on a map), walkabouts, and walkabouts with finder maps.

The 'Archaeology Disc' as a resource for learning archaeoastronomy

Contents of the 'Archaeology Disc'

A videodisc containing a large collection of images relating to British prehistory, and including much of specific interest to the Leicester archaeoastronomy course, was produced in 1988 (Martlew 1991). Photographs were kindly contributed by a number of institutions and personally by several leading authorities in British prehistory and in archaeoastronomy. In addition, special trips were made to a few sites in order to produce collections of pictures from which walkabouts could be configured. These were Stonehenge; Avebury, including The Sanctuary and the Avenue; Stanton Moor, Derbyshire; the Kilmartin Valley, including Temple Wood and the Nether Largie stones; Balnuaran of Clava; and the Callanish area.

Extensive pictorial archives were included on the videodisc of certain sites of special archaeoastronomical interest, such as Ballochroy, Kintraw, and the Ring of Brodgar (for which a complete horizon pan at approximately 12° intervals was included). An attempt was also made to include at least one picture of every known British stone circle—over 2,500 pictures in all. This material was complemented by aerial photographs and plans wherever possible. Finally, a few clips of motion video were donated by Yorkshire Television, including a helicopter fly-past of Brodgar and a sequence showing Michael O'Kelly viewing midwinter sunrise at Newgrange.

---

[1] Nonetheless, and rather surprisingly, this approach is fairly revolutionary in the multimedia context. It is certainly at variance with the hypertext approach currently popular in multimedia authoring software, where the user is encouraged to develop specific links between particular pieces of data (e.g. 'buttons' relating one display field, such as an image, to another) rather than to identify generic structures and relationships.

Software to access the Archaeology Disc was developed in the context of the LIVE (Leicester Interactive Video in Education) project (Ruggles et al. 1990; 1991). This included the implementation of grids and walkabouts. In addition, there were facilities for authoring structured presentations and interactive courseware modules.

Use of the Archaeology Disc in the Leicester course

For two years, students on the archaeoastronomy course have had access to the Archaeology Disc on an individual basis, but only on a single machine shared with other users.

The main benefit was gained from authored presentations which emphasised data selection issues. For example, a particularly successful presentation on Brodgar presented preliminary materials (site photographs, plan and an aerial film-clip), then presented the four lunar horizon foresights drawn by Thom and Thom (1978: 124–25) alongside a picture of the horizon around which the student could pan and attempt to identify the proposed foresights.

The site walkabouts in general were popular and found helpful.

On the other hand, the general image resource has been under-used by the archaeoastronomy students. The main reason seems to be that, because of the size of the database, there was no easy way to gain fast and efficient access to pictures relevant to a student's interests at a particular time.

Following a needs analysis of the course and a review of its aims, a number of further targeted presentations have now been set up, featuring sites such as Newgrange, Ballochroy, and Nether Largie/ Temple Wood (Nikolic 1992).

A further benefit was suggested by a use of the videodisc resources that was popular and beneficial on other courses. This was to give students the task of preparing their own thematic presentations on a particular topic, in other words of preparing their own 'multimedia essays'.

Future and wider use as an open learning resource

In the near future, the image resources on the Archaeology Disc will be converted into digital format and transferred to central facilities at Leicester, under the auspices of the STILE (Students' and Teachers' Integrated Learning Environment) Project. STILE is a large project, funded under the UK Government's Teaching and Learning Technology Programme (TLTP), which is setting up campus-wide open-learning facilities spanning four Universities in the English Midlands. One benefit of this will be to make the image resource currently available on the Archaeology Disc generally available to student workstations around the campus and, eventually, to other sites, potentially world-wide.[2] Another is that it will dispense with the one-off mastering process, enabling further images to be scanned individually and added to the system as required. A third is that STILE will implement the remaining image structures not tackled by LIVE.

A further, and more fundamental, benefit of the transfer will be the opportunity to use STILE's ability to pinpoint rapidly those resources most likely to be relevant to a student's interests at a particular time. The mechanism for doing this is based upon a novel metadata system design (Ruggles and Newman 1991; Walker *et al.* 1992). The educational benefit of such a system is that it will encourage a more open style of learning amongst undergraduate students in general. In the context of the archaeo-astronomy course it will increase usage of the basic resource of site pictures and plans. (This resource will also, of course, be useful to others in a research context.)

At the same time, it seems likely that thematic presentations based on the image resource, such as explorations of key archaeoastronomical sites, could well be of value in communicating some of the basic principles of archaeoastronomy—as well as its more spectacular results—to other groups, such as secondary school students or museum visitors.

Discussion

The basic principles of archaeoastronomy involve developing a critical perspective on the selection of data, especially in the case of alignments. In order to examine 'classic' sites in this context it is necessary to have a broad range of visual materials—multiple views; plans and finder maps; the ability to explore sites, say by 'walking around' them; pans of horizons; and so on—that cannot be provided in books. Computer technology already has the capability of providing this, and this capability will become increasingly accessible in the future.

In addition, recent technological developments are starting to provide tools of more specialised interest to archaeoastronomy, as well as to landscape archaeology as a whole. For example, with a combination of Geographical Information System (GIS) and three-dimensional modelling functionality we may be able to do better than visiting the site itself: we could view the local topography without obstruction by vegetation, afforestation etc., and examine horizon profiles without being reliant upon favourable weather conditions. These suggest a new research tool at least as much as a learning tool, and such avenues are already being explored (e.g. Ruggles *et al.* 1993).

---

[2] This clearly raises a number of tricky issues. Prominent amongst these is the problem of copyright, which is currently being investigated by a number of TLTP and other projects and agencies.

References

Burl, H.A.W. (1988). 'Without sharp north...': Alexander Thom and the great stone circles of Cumbria. In *Records in stone*, ed. C.L.N. Ruggles, pp. 175–205. Cambridge: Cambridge University Press.

Gingerich, O. (1981). Comment on 'Stone age science in Britain?', by A. Ellegård. *Current anthropology* **22**, 121–22.

Martlew, R.D. (1988). Optical disc storage: another can of worms? In *Computer and quantitative applications in archaeology 1987*, ed. C.L.N. Ruggles and S.P.Q. Rahtz, pp. 265–68. Oxford: British Archaeological Reports (BAR International Series 393).

Martlew, R.D. (1991). Every picture tells a story: the 'Archaeology Disc' and its implications. In *Computer and quantitative applications in archaeology 1990*, ed. K. Lockyear and S.P.Q. Rahtz, pp. 15–19. Oxford: Tempus Reparatum (BAR International Series 565).

Nikolic, L.P. (1992). *Undergraduate teaching materials in archaeology: a pilot study on producing courseware for the STILE programme*. MSc dissertation, School of Archaeological Studies, University of Leicester.

Rhind, D. and Mounsey, H. (1986). The landscape and people of Britain: a Domesday record. *Transactions of the Institute of British Geographers* New Series **11**, 315–25.

Ruggles, C.L.N. (1991). *Structuring spatially-related image data within a multi-media information system*. Research Report no. 29, Midlands Regional Research Laboratory, University of Leicester and Loughborough University.

Ruggles, C.L.N. (1992). Structuring image data within a multimedia information system. *International Journal of Geographical Information Systems* **6**(3), 205–22.

Ruggles, C.L.N., Lauder, I., Pringle, J.H., Hayles, S. and Huggett, J. (1990). Future-proofing for interactive multi-media courseware: the Leicester structure model, metafile and courseware development system. *The CTISS File* **10**, 21–22.

Ruggles, C.L.N., Huggett, J, Hayles, S, Pringle, J.H. and Lauder, I. (1991). LIVE update: archaeological courseware using interactive video. In *Computer and quantitative applications in archaeology 1990*, ed. K. Lockyear and S.P.Q. Rahtz, pp. 23–28. Oxford: Tempus Reparatum (BAR International Series 565).

Ruggles, C.L.N., Medyckyj-Scott, D.J. and Gruffydd, A. (1993). Multiple viewshed analysis using GIS and its archaeological application: a case study in northern Mull. In *Computer Applications and Quantitative Methods in Archaeology 1992*, ed. J. Andresen, T. Madsen and I. Scollar. Århus: University of Århus, in press.

Ruggles, C.L.N. and Newman, I.A. (1991). The MRRL's meta-information retrieval and access system. In *Metadata in the Geosciences*, ed. D.J. Medyckyj-Scott, I.A. Newman, C.L.N. Ruggles and D.R.F. Walker, pp. 187–210. Loughborough: Group D publications.

Thom, A. and Thom, A.S. (1978). *Megalithic remains in Britain and Brittany*. Oxford: Oxford University Press.

Walker, D.R.F., Newman, I.A., Medyckyj-Scott, D.J. and Ruggles, C.L.N. (1992). A system for identifying relevant datasets for GIS users. *International Journal of Geographical Information Systems* **6**(6), 511–27.

Appendix

Abstracts of papers published in the companion Oxford 3 volume 'Astronomies and cultures'

1 The study of cultural astronomy
Clive L.N. Ruggles and Nicholas J. Saunders

Anthropologists and archaeologists are concerned with retrieving data on cultural systems, identifying patterns in these data and attempting to elicit meanings from these patterns. An important part of this is how people perceive their environment. The sky is the only resource within the environment that is not subject to physical alteration by human beings. The study of repetitive astronomical phenomena has a major advantage over studies of other aspects of people's interaction with their environment: the raw resource is directly accessible to us. Within certain limits of uncertainty, we can construct directly a part of the everyday environment of a set of people in any remote culture.

The celestial resource is of almost universal concern, from the simplest groups of hunter-gatherers to the most complex of state societies. This universality permits wide-ranging structuralist studies of topics such as the correlation between astronomy and culture, cross-cultural parallels, and the ways in which different cultures interpret and manipulate the same immutable resource for different political reasons and ends. This in turn gives a special significance to enquiries about cultural astronomy within archaeology and anthropology as a whole.

Serving as an introduction to the contents of the volume as a whole, this paper charts the rise of archaeo- and ethnoastronomy during the past two decades, addressing a number of key issues such as their role within archaeology and anthropology as a whole, their scope and goals, and the various conflicts of approach that have led to recent attempts to build secure methodological foundations. The paper concludes with a vision of the role of cultural astronomy studies in the 1990s and beyond.

2 The *Yáo Diǎn* 尧典 and the origins of astronomy in China
Chen Cheng-Yih 程貞— and Xi Zezong 嘴澤示

The *Yao Dian* ('The Canon of Yao'), which survived through its incorporation in the *Shang Shi* ('The Book of Documents'), is an ancient document containing written descriptions of early calendar-making and the determination of seasons based on the meridian passage of four star-groups (equatorial compartments). J.B. Biot first proposed in 1862 that the celestial phenomena described in the *Yao Dian* could be used to determine its date based on the precession of the equinoxes. Taking the hour of observation to be 6 p.m., Biot obtained a date of c. 2400 BC in agreement with the traditional dating of the Yao period. Hashimoto later pointed out in 1928 that such calculations depend critically on the hour of observation. By taking the hour of observation to be 7 p.m. he was able to reduce the date to 800 BC. This led to a long period of debate on the origin of astronomy in China.

The discovery of pictogram inscriptions on ceremonial pottery unearthed from the Dawenkou cultural stratum (C^{14}-dated to c. 2500 BC) in Shandong reveals that the account of official activities of the Yao astronomical officers given in the *Yao Dian* was probably based on historical facts.

This independent discovery supports Biot's dating. Recent archaeological discoveries of the symbols for star-group regions, *long* (dragon) and *fu* (tiger) images, made of shells along the sides of a human corpus (C^{14}-dated to c. 6000 BC) in Henan and of the same two images together with the Big Dipper in a fifth century BC astronomical diagram of the 28 *xiu* (equatorial compartments) in Hubei suggest a long history in the development of an equatorial system. This paper investigates the significance of these new discoveries on the origin of astronomy. It examines, in particular, the onset of calendrical science and the development of positional astronomy in Chinese civilisation.

3 The riddle of red Sirius: an anthropological perspective
Roger Ceragioli

Astronomers have long been concerned about the colour of Sirius in antiquity. Ptolemy seems to record this as red, and he appears to be confirmed by Seneca and Horace. Yet modern astrophysicists have been unable to explain how the Sirius system could have evolved from red to white in only two thousand years. Moreover, researchers have pointed to other early sources that call Sirius white.

This paper outlines some results of a thorough study of the Greco-Roman sources on Sirius, showing how the figure of Sirius is involved in a complex pattern of myth, ritual and folklore, of which variations between the colours white and red, equated with purity/health and impurity/disease respectively, are essential manifestations. Ptolemy, Seneca and Horace are alluding to the dangerous and hence more notable pole of this colour variation.

This explanation not only resolves the paradox of a red, yet white star, but points to the importance of anthropological analysis for understanding the astronomy of traditional societies.

4 Astronomies and rituals at the dawn of the Middle Ages
Stephen C. McCluskey

An increasingly central concern of archaeo- and ethnoastronomy has been how astronomy defines the sacred time and space of religious ritual. Historians of Medieval religion have arrived at the same concern in investigating the process by which Europe became Christian. James Van Engen recently wrote:

> The real measure of Christian religious culture on a broad scale must be the degree to which time, space, and ritual observances came to be defined and grasped essentially in terms of the Christian liturgical year.

There are four elements defining the intersection of Christian ritual with astronomical concepts of time and space. The first of these relates elements of the Christian liturgical year to pre-Christian calendrical practices at the solstices, the equinoxes, and the mid-quarter days. Detailed examples can be seen in the Fifth Century Gallo-Roman culture of Southern France and the Celtic culture of Seventh Century Ireland. The next element of Christian astronomical practice concerns the problem of determining the date of Easter. Irish scholars from the region associated with the Christianised calendar feast of St. Brigit of Kildare first transformed this problem from a question of ritual symbolism to one of astronomical computation, based on the rudimentary data of late Roman texts.

Related to the astronomical content of the ritual calendar is the (largely unexplored) practice of orienting churches to sunrise (or sunset) on the feast of the church's dedication. Late antique and early Medieval writers provide insights into motives for the orientation of religious structures. The final element is the tradition of monastic timekeeping by observation of the stars, a practice documented in a series of monastic texts from the earliest western monasteries of the fifth century through to the monastic reformers of the eleventh.

Medieval Europe thus displays many of the relationships between astronomy and ritual that we have seen in other cultures. Yet here we can trace the historical transformation as Celtic feasts blend into saints days; quantitative data from Greek astronomy are employed to compute the date of Easter; traditions of orienting pagan temples undergo subtle changes in Christian churches, and pagan constellations are used to determine the times for monastic prayer. This paper illustrates the relationships between the adoption of Christian beliefs and practices under the influence of pre-existing pagan rituals and the adoption of Mediterranean mathematical astronomy under the influence of traditional folk techniques.

5 Folk astronomy in the service of religion: the case of Islam
David A. King

Modern accounts of Islamic astronomy, with their emphasis upon technical achievements, tend to ignore the scientific folklore which flourished alongside mathematical astronomy in the Islamic world throughout the medieval period. This folk tradition was far more widely practised than the mathematical tradition, yet the sources available for studying it have received far less attention in modern times. These sources are mainly written ones, but the application of folk astronomy in the Islamic world sometimes led also to the use of astronomical alignments in Islamic religious architecture. Furthermore, such alignments are actually advocated in texts that are still extant. Thus the study of Islamic folk astronomy provides an important case study in which ethnohistoric and archaeoastronomical evidence are intricately combined.

This paper presents a brief overview of Islamic folk astronomy, identifying the kind of sources in which it is documented and showing how it was used in practice. No attempt is made to survey Islamic mathematical astronomy or to describe the way in which the astronomers dealt with these same three problems using mathematical procedures, tables and instruments.

6 Cosmos and kings at Vijayanagara
John McKim Malville and John Fritz

The capital of the greatest Hindu empire of medieval India, Vijayanagara, was founded in the fourteenth century and sacked in 1565. At its zenith the city had a population of between one and five hundred thousand. Its kings were apparently influenced by models for royal behaviour which were directed towards increasing power and influence and maintaining social and moral order.

A major north-south axis divides Vijayanagara's royal centre into areas of performance and residence, affirming the dualities of kingship. Matanga Hill, the most significant of the four sacred hills of the area, lies on the north-south axis. On its summit the Virabhadra temple is cardinally aligned. Ramachandra temple, the major temple of the royal enclosure, lies to the east of the north-south line and is rotated away from cardinality towards Matanga Hill.

This paper describes this and a range of other evidence indicating that the layout of Vijayanagara was influenced by the origin myths of the city, the evolving sectarian preferences of the kings, and the local sacred topography.

7 Medicine Wheel astronomy
David Vogt

Medicine Wheels are enigmatic boulder configurations created by nomadic bison hunters in the Great Plains of North America during the time period since the last Ice Age. Current theories ascribe to them a variety of purposes, including use for burial, commemorative and other ceremonial functions or even as astronomical observatories, and also speculate on the use of geometrical methods in their layout. This paper presents results from the first systematic analysis of Medicine Wheels and possible Medicine Wheel astronomy. The conclusions are that Medicine Wheels display an orientational patterning that is almost certainly astronomical in origin, and that a central gnomonic device, either a standing *tipi* or pole, was the most probable means by which orientations were achieved.

8 Venus and astrologically-timed ritual warfare in Mesoamerica
John B. Carlson

Since their discovery in 1975, the spectacular pre-Columbian murals of Cacaxtla have been found to rival the world-famous Maya 'Bonampak' murals both in state of preservation and importance. Located in Tlaxcala in the Mexican Highlands not far from Puebla and Cholula and dating to the Epiclassic Period after the fall of Teotihuacan, they exhibit a unique mixture of motifs and styles reflecting influences from Teotihuacan, Oaxaca, Xochicalco, the Gulf Coast, and, most significantly, the Maya area. The previously-discovered murals depict sacrificial scenes associated with the culmination of a bloody battle between triumphant Jaguar/Rain God warriors and Blue Bird-costumed opponents who would appear to be Maya. The central themes involve the association of warfare and blood sacrifice with rain and fertility.

During 1987 and 1988, brilliant new murals were discovered at Cacaxtla. Two outer door-jamb panels show a life-sized male and female pair of winged, blue-painted rain/fertility deities who dance within Teotihuacan star-band borders and hold stars in their jaguar-paw gloved hands. The large Venus glyph buckles they wear identify them as Venus manifestations which can be shown to link them to the pan-Mesoamerican cult of Venus-regulated ritual warfare. These new data are presented in a review of current research on Maya Venus almanacs and their use in 'Star Wars' events, both in the Maya Lowlands and Mexican Highlands in the context of Highland-Lowland trade, particularly in the period of ferment during and after the fall of the Classic civilisations of Mesoamerica.

9 Astronomical knowledge, calendrics and sacred geography in ancient Mesoamerica
Johanna Broda

This paper explores the relationship between astronomical knowledge, the ancient Mesoamerican calendar system and the concept of sacred geography that is postulated to have existed in that culture area. It refers to the advances that archaeoastronomical studies on Mesoamerica have made in the past two decades, further pointing out the valuable contribution that the geography of cultural landscapes has made in this context, and relates these important new results to the structure of the Mesoamerican solar calendar that was intimately linked to social, economic and ritual activities. While this interrelationship is explored in detail in the case of Aztec society, it is assumed that this example has wider implications and is deeply rooted in a common Mesoamerican tradition that derived from pre-Classic roots.

The paper also describes the author's ongoing investigation of the ceremonial landscape of the Valley of Mexico in which the Aztecs, from their capital of Tenochtitlan—an island in the middle of a lake surrounded by high mountain ranges—related sacred mountains, sanctuaries to the rain and earth deities, and ancient historic settlements of the Valley (belonging to other ethnic groups) in a grand scheme of sacred geography based on the observation of solar horizon astronomy that was tied to calendrical periods and rituals that were performed at these sacred places. The hypothesis is proposed that the essential concepts of this solar and geometric scheme derived from ancient, probably pre-Classic, roots within the Basin of Mexico.

10 The Pleiades in comparative perspective: the Waiwai *Shirkoimo* and the Shipibo *Huishmabo*

Peter G. Roe

This paper examines the key role of the Pleiades in lowland South Amerindian ethnoastronomy and cosmology on the basis of myths and astronomical observations recorded during ethnographic fieldwork in two sites sixteen hundred kilometres apart on the north-eastern and south-western peripheries of the Amazon Basin. The two culturally and ecologically contrasting study groups—the interfluvial and upper tributary Cariban Waiwai of the upper Essequibo, Guyana on the latitude 1°N, and the riverine Panoan Shipibo of the central Ucayali, Peru on latitude 8°S—retain remarkably similar versions of the Pleiades myth.

Together with the associated constellations of the Hyades and Orion, the Pleiades set the timing of the slash-and-burn horticultural cycle on both sides of the equator. In astral myth, the peregrinations of these constellations act as metaphors for, and harbingers of, the seasonal oscillations which bracket temporal subsistence patterning for both riverine and interfluvial jungle Amerindians. The key ritual components are the seasonal maxima and nadirs of protein capture (fishing and hunting) and the carbohydrate production cycle (gardening and gathering).

INDEX

A

Aboriginal
 art, 137
 bark paintings, 139, 142, 147
 engraved art, 136
 message-sticks, 137, 140, 145
 paintings, 136
 sky mapping, 136–49
academic body for archaeoastronomy, 7
achronal risings, 159, 164
agricultural calendar, 70, 110,124, 126
 ancient Chinese, 124
 Roman, 126
agricultural ceremonies
 and cross-circles, 292–94
 see also seasonal festivals
ahu (raised platform on Polynesian temple), 128, 129, 130
air mass, 157
Aitutaki, 129, 134
Aldebaran, 118, 328
alignments, 1, 24, 25, 163–64, 185
 assessing the credibility of, 24
 lunar, 201, 202, 207, 208, 212–13
 Maya, 54
 of standing stones, 134, 163–64, 185–95
 selection of, 199–200, 335–36
 solar, 201–2, 207, 208, 212–213
 stellar, 164
 theoretical limitations on the accuracy of, 163–64, 178–83
 upon cardinal directions, 165, 244, 246, 247, 249
 upon prominent peaks, 187–90, 274, 278
 upon solstices, 283, 284
 upon Venus extremes, 270, 271–74
 see also orientations, astronomical alignments
allées couvertes, 324, 325
altitude, 157
 geocentric, 179
 maximum, of sun on equinoxes, 198
 maximum, of sun on solstices, 198, 283
 observed, determination of, 179
 of first visibility, *see* extinction angle
Amazonia, 26
Anasazi, 20, 219–26, 235, 236, 237, 238, 242–49
 rock art, 235
ancient Egypt, 107, 165
angular unit, ancient Mesoamerican, 282, 283, 284, 285–86
'antap (Chumash elite religious cult), 253
Antares, 124–26, 148
 heliacal rise of, 124
archaeoastronomy, 278

Index

and mainstream anthropology, 3–4, 8, 20, 343
and mainstream archaeology, 3–4, 8, 27, 28, 343
and the history of art, 4
and the history of religions, 4, 22
and the history of science, 3, 6, 7, 8, 34, 95
Andean, 6, 23–26
applications of mainstream astronomy in, 117, 121, 155–71
approaches to, 17, 19–20
as an interdiscipline, 2
British, 335
data selection criteria in, 199–201, 208, 212
definition of, 1
education and training in, 8–9, 335–40
future of, 2, 7–10, 343
green and brown, 5, 10, 16
Icelandic, 69
image database in, 336–40
in Bulgaria, 107–15
in Europe, 2
in Mesoamerica, 20–23
in the Americas, 2, 15–32
in the disciplinary literature, 18–19
megalithic, 16, 88
methodological foundations of, 1, 4–5, 9, 343
methodological issues, 121, 195–96, 199–201
multi-disciplinary approaches, 21–22, 149
primary developments since 1986, 15
publications in, 3
relationship to ethnoastronomy, 1, 5–6, 16, 26
renamed 'cultural astronomy', 6–10, 343
academic body for, 7
scope and goals of, 3–4, 5–7
'Archaeology Disc' (interactive videodisc), 338–39
Ardnacross, 185, 186–87, 189–90, 191, 192, 193, 195
Argo, 145
Ari the learned, 71
art
  Aboriginal, 137
  see also bark paintings, engraved art, paintings, rock art, sand paintings
Asclepius, 330
astral myths, 26, 145, 147
astral symbolism, 101–2
  in rock art, 20, 147, 227–38
  Teutonic, 101–2
astronomical alignments
  in Islamic religious architecture, 346
  Irish passage grave cemeteries, 201, 208, 212
  of Basque stone octagons, 80
  Thracian, 114
  western Mediterranean, 330
  see also orientations, alignments
astronomical knowledge
  Aboriginal 138, 146
  as part of cognitive mapping, 136, 143–47, 149
  Egyptian, 107
  expressed in art, 149
  prehistoric north-eastern Mexico, 266
  transmission of, 35
  see also astronomy, folk astronomy
astronomical symbols, sacred, 251
astronomy
  amateur, 170
  amongst hunter-gatherers, 251
  Anasazi, 243
  ancient Chinese, 123–26, 162, 168, 169, 344
  and navigation, 70, 134, 146
  and ritual in Medieval Europe, 345–46
  Arabic, 168
  as definer of sacred space, 345
  Babylonian, 6, 107, 160
  Chumash, 251–61
  Classic Maya, 6
  correlation with cultural variables, 4, 45, 249, 343
  Hindu, 168

in Bulgarian lands, 107–15
Inca, 298–306
Indian, 168
Islamic, 160, 167–68, 346
Mayan, 170
Puebloan, 20
traditional cultures of America, 33–35
Viking, 70
see also astronomical alignments, astronomical knowledge, folk astronomy, mainstream astronomy
atmospheric extinction, 157, 159, 163, 180, 182
atmospheric refraction, 157, 162, 163, 179–80, 183
Aurora Borealis, *see* Northern Lights
Avebury, 338
Aveni, Anthony, 278
ayantu (Borana sky experts) 118, 120
azimuth, 179
   derivation from declination, 163
   uncertainty due to finite size of body, 182
   variation with declination, 180
   variation with latitude, 180
Aztecs, 17, 23, 37

B

Babylonians, 107
Baffin Island, 59
Bajlovo, 110–12
Balliscate, 185, 189–90, 191, 192–93, 195
Ballochroy, 338, 339
Bambacona, 305–6
bark paintings
   Aboriginal, 139, 142, 147
Basque Country, 6, 77–89
Bayesian statistics, 5
Ben More (Mull) 187–89, 191, 192, 194–95
Bithynians, 113
Boca de Potrerillos, 264, 265, 267–68
Book of Odes, 126
Borana, 117–21
Boyne Valley, 198–99, 201–8, 210, 213, 214, 215

geometrical characteristics, 204–7, 214–15
brightness
   of moon, 156
   of planets, 156
   of shadows, 158
   of sky, 156
   of stars, 156
Broda, Johanna, 278
Brodgar, Ring of, 338, 339
bull, as lunar symbol, 325
burial monuments, possible astronomical orientation of, 326
Burro Flats, 255–56
Bush Barrow lozenge, 317–23
Byzantium, 167

C

Cabeza de Vaca, 268
Cacaxtla murals, 347–48
Caesar, Julius, 99
calendar sticks, 170, 265
calendar
   Bush Barrow lozenge as, 317–23
   evolution from lunar to solar, 110
   Holly ruins petroglyph panel as, 222–25, 226
calendars
   agricultural, 70, 110, 124, 126
   Anasazi, 20
   ancient Egyptian, 96, 109, 110
   and the length of the year, 71–72
   Babylonian, 109
   Borana, 117–21
   Celtic, 345
   Christian, 94, 345
   Chumash, 252
   Cushitic, 119
   Gallo-Roman, 94, 345
   Greek, 94
   Inca, 25
   Indo-European, 92–96
   Islamic, 167–68, 170
   lunar, 109–10, 111, 118, 119–20, 145, 160, 167–68, 252, 266
   luni-solar, 92–96, 109, 112
   Mesoamerican, 278, 348
   old Icelandic, 70, 72, 74
   old Teutonic, 99–100

Index

Pueblo, 20
sidereal, 125, 165
sixteen-month, 317, 319
solar, 317
traditional cultures of America, 33–34
Californian Indians, 251–61
Callanish, 309–316, 338
Canis Major, 145
Canon of Yao, 125, 344
Canyon de Chelly, 227, 234, 237
Caracol (Chichen Itza), 271
cardinal directions, 35, 37, 114, 131, 134, 147, 170, 182, 317
   alignments upon, 165, 244, 246, 247, 249, 347
   correlation of nomenclature for with social variables, 47–49, 53
   markers of, 268
   skewed, 38, 275
Carpathian Basin, 98–105
Carrasco, Davíd, 17
Carrowkeel, 199, 208–10, 213, 215
   astronomical characteristics, 215
   geometrical characteristics, 210, 215
cave sanctuaries, 110–12
caves, as sacred places, 244–46
Cehtzuc, 272
celestial myths, 251
celestial north pole, *see* north celestial pole
'celestial visibility', as a discipline, 171
Centaurus, 330
ceque system, 25–26, 284
ceremonies
   Aboriginal, 137–38, 149
   Chumash, 252
   female menstruation, 140
   male initiation, 140, 148
   Navajo, 235
   see also seasonal ceremonies, seasonal festivals
Cerro La Bola, 265
Chac, 272, 292–93
Chaco Canyon, 170, 182, 183, 242, 246, 249
chants, Navajo, 235, 236
Cheops, Great Pyramid of, 165

Chichen Itza, 21, 24, 271
   serpent hierophany at, 23
Chimney Rock, 249
China, ancient, 123–26
Cholula, 284
Chou period, 123–26
Christmas, 94
Chumash, 251–61
church orientations, 345
   Valley of Mexico, 280
Cillchriosd, 194–95
Codex Bodley, 283
Codex Dresden, 292
Codex Mendoza, 23, 285
Codex Selden, 283
cognitive mapping, astronomical knowledge as part of 136, 143–47, 149
Coligny calendar, 94
colour
   of planets, 156
   of Sirius, in antiquity, 344–45
   of stars, 156
comparative studies, 198, 349
conjunctions of moon and Venus, 166
Constantinople, fall of 166–67
Cook Islands, 128, 129, 134
Copan, 21, 53, 270
   Venus cult at, 21
coral trilithon, 128, 129, 132–34
Corvus, 145
cosmology
   ancient Mesoamerican, 279
   Andean, 37
   Arctic, 59–68
   Basque, 81
   Chumash, 252
   correlation with structure of society, 36
   Inuit, 59–68
   Mesoamerican, 36
   Navajo, 237
   Pueblo, 237, 244
   South Amerindian, 349
   traditional societies of America, 35
cosmovisión, 22, 26, 54, 274
crescent moon, *see* lunar crescent
Crete, 114
cromlechs, 86

cross-circles, 22, 288–95
  and agricultural ceremonies, 292–94
  as architectural benchmarks, 22, 290
  as calendrical markers, 290–91
  as counting devices, 288
  as game boards, 294
  as symbols of the quadripartite division of space, 294
  as Tlaloc-like images, 291
  axial orientations of, 290
cross-circles, *see also* petroglyphs, rock art
Crucifixion, date of, 164–65
Crux, *see* Southern Cross
'cultural astronomy', 6–10, 343
cup-and-groove engravings, 141, 143, 144, 149
Cuzco, 24–26, 284
  latitude of, 304
  pillars of, 25, 301–3
  siege of, 298
cyclic time, 95

D

dances, Chumash, 252
dark sun, 95
data selection criteria, 199–201
  Irish passage grave cemeteries, 201, 208, 212
day-names, 120
declinations, 162, 164, 183, 190, 199
  conversion to azimuth, 163
  determination of, 179
  of prominent peaks, 191, 194–95
definition of archaeoastronomy, 1
Dervaig, 185, 189, 193
Dionysiac religion, 113
directional symbolism, structuralist analysis of, 52–53
directions
  of sunrise and sunset, 47, 73
  of sunrise and sunset at solstices, 37, 38–39, 52, 54
Doggett, LeRoy, 156
dolmens, 86, 113, 324, 330
domus de janas, 324, 330
dot configurations, 265–66
  see also dot-and-tally petroglyphs, dot-and-line symbols, tally counts
dot-and-line symbols, 136–37, 138, 141–43, 145, 149
  see also dot-and-tally petroglyphs, dot configurations, tally counts
dot-and-tally petroglyphs, 264–67
  see also dot-and-line symbols, dot configurations, tally counts
Dowth, 201–8
Dreamtime Ancestors, 137, 142, 148
Dzibilchaltun, 21

E

earthen mounds, 129
Easter Island, 128, 129
Easter, date of, 345
eclipses
  lunar, 99, 103, 105, 161, 164, 165, 166, 298, 303, 306
  total solar, 161
education and training in archaeoastronomy, 8–9, 335–40
Egypt, ancient, 107, 165
Egyptian calendar, 110
Elvina Track, 144
engraved art, Aboriginal, 136
equinoxes, 23, 112, 319, 322, 345
  alignments upon, 114, 130–32, 268
  daily change in sun's azimuth at, 182
  directions of sunrise and sunset at, 178
  gnomon angles at local noon, 202–4, 207, 208, 210, 213
  light and shadow displays at, 23, 222
  maximum altitude of sun on, 198
  observations of, 72, 74
  sunrise and sunset at, 322
Ereñotzu, 85, 88
Ethiopia, 117
ethnoastronomy, 117, 120–21, 335, 345, 349
  relationship to archaeoastronomy, 1, 5–6, 16, 26
ethnographic accounts, problems with, 118

Index

Evening Star, 270, 271, 272, 273, 275
   ceremonies 139
   see also Morning Star, Venus
excavation, 185, 193
   at Ardnacross, 186–87
   at Glengorm, 186
extinction angle, 159, 163, 164
extinction, atmospheric, 157, 159, 163, 180, 182

F

Fajada Butte, 24
female menstruation ceremony, lunar associations 140
Fenrir, 103
Fiji, 128
'Fire Star' (Antares), 124–26
folk astronomy
   Islamic, 346
   Medieval European, 346
Fomalhaut, 170
four-directionality of space, *see* quadripartite division of space
'fourness of things', 36–37
Frazer, J.G., 94
full moon, 301, 306
   and siege of Cuzco, 298
   nearest to summer solstice, 192, 193–94
   see also lunar phase cycle
future of archaeoastronomy, 2, 7–10, 343

G

gender specialisation, correlation with nomenclature for cardinal directions 48
geocentric altitude, 179
Geographical Information Systems, 194, 195, 340
'geometric foot', 78
geometrical perception, Aboriginal, 148
geometry
   data selection criteria, 199–201, 208, 212
   in ancient Mesoamerica, 278–86
   relationship to orientation, 198
Gervase of Canterbury, 169
glare, 158

Glengorm, 185, 186, 187, 189, 193
glyphs, Venus 272
gnomon angles, 200, 204, 207, 215
   Irish passage grave cemeteries, 202–4, 207, 208, 210, 213
gnomons 155, 165, 347
Gobernador Phase star ceilings, 229
gods
   Egyptian, 96, 330
   Greek, 330
   Indian, 96
   Indo-European, 95–96
   Mayan, 272
   mountain, 291
   solar, 51, 113
   Teutonic, 99
   Thracian, 113
   Venus, 52
Goodman Point, 243, 247–49
Gotland, 101
Governor, House of the (Uxmal), 272–73, 275
grave orientations, Inuit 62
Great North Road (Chaco Canyon), 170
Great Pyramid of Cheops, 165
Great Star Chant (Navajo), 235–36
Greeks, ancient, 81, 167
   accounts of Thracians 107
green flash, 170

H

Harrington, John Peabody, 252
Hawaii, 128, 130
Heggie, Douglas, 17
heiau (Polynesian temples), 128, 130, 132
heliacal rising
   of Antares, 124
   of Sirius, 165
heliacal risings, 155, 170, 267–68
   ancient Chinese observations of, 169
   representation in rock art, 112
   theoretical calculation of, 159
Herodotus, 107, 113
Holly ruins, 219–26
Homo Erectus, 143
Hopi, 33, 34, 35–36, 37–43, 225, 233

initiation ceremony, 39
horizon
  distance, variation with azimuth, 190, 191
  notches, 42
  visibility, variation with azimuth, 191
Hoskin, Michael, 326
Hovenweep, 219, 225
Hudson Bay, 59
Hudson, Travis, 33, 252
Huexotla, 273–74
human eye, capabilities of, 155
Hyades, 349
hymns, Vedic, 95

I

image data
  on interactive videodisc, 336–37, 338–39
  digital, 337
  storage and access, 336–38
image databases, 336, 337–38
image structures, 337–38
Imhotep, 330
Inca calendar, 25
Indo-Europeans, 88, 92–96, 100
  conceptualisation of time, 95
  see also pre-Indo-Europeans
Indonesia, 93
Ingapirca, 301
initiation ceremony, Hopi 39
interactive videodiscs, 336–37, 338–39
intercalary periods, 95, 96
  day, 317, 322
  month, 94, 100, 120
Intimachay, 300
Inuit
  cosmology, 59–68
  grave orientations, 62
  myths, 60, 66–68
  seasonal activities, 60
  seasonal festivals, 63
  shamans, 63, 66
Iran, 167
Ireland, 198–215
Irish passage grave cemeteries
  astronomical characteristics of, 214–15
  geometrical characteristics of, 214–15
Islam, 33, 165–68
  astronomical determination of prayer times 168
  folk astronomy, 346

J

jade discs, Chinese 169
Jordanes, 107
Judge, James, 19–20, 28
Jupiter, 139, 169
  see also planets

K

Kenya, 117, 119, 120
kerb cairns, 186, 187
Killichronan, 194
Kilmartin Valley, 338
King of Tonga, 128, 132–33
king-priest, Thracian, 113
Kintraw, 338
kivas, 244
  function of, 249
Knowth, 201–8
Kuhn, Thomas, 34

L

La Mula, 266–7
Lakota, 36
landscape archaeology, 340
language
  Aboriginal, 137, 146
  Basque, 88
  Inuit, 59
  Mayan, 53
  terms for east and west in, 47, 53
Large Magellanic Cloud, 134
Legesse, Asmarom, 117–20
Leicester University, 8, 335–36, 339
Leo, 145
light-and-shadow displays
  association with ritual containers, 254, 255–56, 257–60
  at equinox, 23
  at equinoxes, 222
  at summer solstice, 220–26, 254–61

Index 357

at winter solstice, 255
Lines of Nazca, 25–6
locational analysis, 185, 194
Loughcrew, 199, 212, 214, 215
   astronomical characteristics of, 215
   geometrical characteristics of, 213–14, 215
Lounsbury, Floyd, 17
lozenge, Bush Barrow, 317–23
lunar alignments, 201, 202, 207, 208, 212–13
lunar calendars, 109–10, 111, 118, 119–20, 145, 160, 167–68, 252, 266
   Aboriginal, 145
   prehistoric north-eastern Mexico, 266
lunar crescent
   length of, 161
   visibility of, 156, 160–61, 164, 166, 170
lunar cycles, 99
lunar eclipse, possible, at Crucifixion 165
lunar eclipses, 99, 103, 105, 161, 164, 165, 166, 298, 303, 306
   as omens, 104–5
   visibility of, at low altitude, 161
lunar lodges, 168, 171
lunar nutation, 163
lunar observatories, megalithic, 163
lunar occultations, 156
   ancient Chinese records of, 169
   visibility of, 170
lunar phase cycle, 92–94, 99–100, 107, 109, 118, 143, 145, 147, 148, 149, 252, 298
   and Palaeolithic art, 143
   representation of in rock art, 110–11
   see also lunar synodic month, see also full moon
lunar standstills, 164, 300
   alignments upon, 328
   declinations of, 199
   directions of moonrise and moonset at, 178, 182, 317
   see also major standstill, minor standstill

lunar symbolism
   in Aboriginal art, 139, 148
   in Scottish stone rows, 195
   Teutonic, 100–2
lunar synodic month, 118, 120, 140, 264, 266
   see also lunar phase cycle
lunar year, 94, 96
lunar-associated numbers, on Aboriginal artefacts, 140–41, 148, 149
luni-solar calendars, 92–96, 109, 112

M

Macedonia, ancient, 113
Magura, 112
mainstream astronomy
   applications in archaeoastronomy, 117, 121, 155–71
major standstill, 192, 193, 309
   azimuth of moonrise and moonset at, 309
   see also minor standstill, lunar standstills
male initiation ceremony, lunar associations 140, 148
mandalas, 294
Maol Mor, 185, 189, 193
Maori sites, 130
marae (Polynesian temples), 128–34
markers, solstitial, 40–41
Mars, 139
   see also planets
Maya, 37, 170
   Late Preclassic, 49–54
   orientations, 22, 51
   present-day, 34
   codices, 20
   hieroglyphs, 53
Medicine Wheels, 170, 347
'megalithic astronomy', 16, 88
'megalithic lunar observatories', 163
megalithic tombs
   Sardinia, 324–30
   Scandinavian, 199
Menorca, 330
menstrual sticks, 140
meridian determination, 165
Mesa Verde, 242–43, 249

Mesoamerican cosmologies, 36
message-sticks
    Aboriginal, 137, 140, 145
    and lunar-associated numbers, 140
meteor showers, 126
methodological foundations of archaeoastronomy, 1, 4–5, 9, 343
methodology of data selection, 199–201
metrological traditions in western Europe, 79–80
mid-quarter days, 282, 345
Milky Way, 149
minor standstill, 195, 302
    see also major standstill, lunar standstills
Misminay, 37
Monte d'Accoddi, 330
Montezuma Basin, 242–49
month
    intercalary, 94, 120
    sidereal, 120, 267
    synodic, 120
moon
    brightness of, 156
    bull as symbol for, 325
    conjunctions with Venus, 166
    cult, Teutonic, 98–105
    declination limits, 119
    in bark paintings, 147
    observations of in relation to stars, 118
    visibility of, during total solar eclipse, 161
    zenith passage of, 301
'Moon Dreaming', 140
Moon Man (Inuit), *see* Moon Spirit (Inuit)
Moon Spirit (Inuit), 59, 60, 62, 64, 66, 67
moonrise and moonset, 178, 180, 185
    alignments upon, 300
    at the standstills 178, 182, 309, 317
    azimuth of, 182, 300
    in relation to prominent peaks, 192–94
    observations of, 310–16
    orientation upon, 192–94
Moorea, 130–31

Morning Star, 141, 148, 271, 272
    ceremony, 139, 148
    see also Evening Star, Venus
mountain gods, 291
Mull, 185–96
Mulloy, William, 128
Mummy Lake, 242
Mursi, 121
Musca, 145
Mutau Flat, 257–61
myths
    Aboriginal, 136, 137, 139, 144, 145, 146, 147, 149
    ancestral sky-heroes, 144, 147
    ancient Chinese, 123
    astral, 26, 145, 147
    celestial, 251
    Chumash, 252
    Greco-Roman, 345
    Hindu, 347
    Inca, 304
    Indian, 96
    Inuit, 60, 66–68
    Mid-Eurasian, 103
    native American, 36
    Navajo, 235, 236–37
    old Teutonic, 100
    Scandinavian, 102
    South Amerindian, 349
    symbolic representation of lunar nodes in, 300

N

nadir passage, solar, 301, 302
Namoratung'a II, 119
Navajo, 227–38
    myths, 235
    Pueblo contact with, 233–35, 237
navigation, 70
Nazca lines, 6, 25–26
Nether Largie, 338, 339
New Guinea, 128
New Zealand, 128, 130
Newgrange, 199, 201–8, 338, 339
Night of Power, the, 166, 168
Nocuchich, 273
Nohpat, 272
north celestial pole, 179
    location of, 169

Index

Northern Lights, 63
numerical perception, Aboriginal, 147–148
nutation, lunar, 163

O

O'Kelly, Michael, 338
observations
    anticipatory, of solstices, 41–42
    of solstitial sunrise and sunset, 40–42, 252
observed altitude, theoretical determination of 179
Occam's razor, 72, 214
Oddi, *see Star-Oddi*
Oddi's tale, 72–74
Odin, 103
olin (Olmec sign), 279–82, 285–86
orientation
    calendars 22, 119, 267–68, 278, 291, 348
    relationship to geometrical design, 198
orientations
    ancient Egyptian, 165
    churches, 280, 345
    in ancient Mesoamerica, 270–75
    Mayan, 22, 51
    of burial monuments, 326
    of churches, 280
    of Irish passage graves, 198
    of Medicine Wheels, 347
    of Medieval religious structures, 345
    of Mesoamerican ceremonial centres, 283–84
    of Scottish stone rows, 186, 192–94, 195
    of tombe di giganti, 326–29
    Polynesian, 129–35
    upon Venus, 270–75
    see also alignments, astronomical alignments
Orion, 118, 123–24, 132, 142, 144–45, 147, 328–29, 349
Orpheus, 113
Ovcharovo, 110, 112
'Oxford 1', 2, 16, 264
'Oxford 2', 2, 16, 25

proceedings, reviews of, 16–17
'Oxford 3', 2, 7

P

paintings
    Aboriginal, 136
    lunar symbolism in, 139, 148
    solar symbolism in, 139
Paleocastro, 112–13
parallax, 163, 179–80
passage graves, 198–215
    'centres' of 200
    see also Irish passage grave cemeteries
pecked cross circles, *see* cross-circles
Penistaja star ceiling, 229–31, 233–34
petroglyph panel (Holly ruins), 219–26
    as calendar, 222–25, 226
petroglyphic counting, relation to terrestrial navigation, 268
petroglyphs, 20
    dot-and-tally, 264–67
    in Valley of Mexico, 265
petroglyphs, *see also* cross-circles, rock art
Pico Tres Padres, 274
Pillars of Cuzco, 25, 301–3
Plains Indians, 193
planetaria, 8
planets
    brightness of, 156
    colour of, 156
    motions of, 107
    rising and setting positions of, 178, 179, 180
    visibility of, near moon, 161
    see also Venus, Mars, Jupiter
Pleiades, 118, 142, 144, 147, 180, 182, 328, 349
Plutarch, 126
Poljanitsa, 112
Polynesia, 128–35
Ponce de León, A., 273
Popol Vuh, 37
prayer sticks, 38, 40
    Hopi, 40
pre-Indo-Europeans, 88–89
    see also Indo-Europeans
Presa de La Mula, 264

Prokopius, 98
prominent peaks, 215
  alignments upon, 187–90, 274
  as solstitial markers, 193–94
  astronomical potential of, 195
  astronomical significance of, 274
  declinations of, 191, 194–95
  in relation to moonrise and moonset, 192–94
  in relation to sunrise and sunset, 193–94
  in ritual landscape, 195, 274
  location of stone rows in relation to, 192–94
  visibility of, 190, 195
publications in archaeoastronomy, 3
Pueblo, 20, 233–35, 237
  cosmology, 244
  contact with Navajos, 233–35, 237
Pueblo Bonito, 170
Pythagorean triangles, 202, 204, 207, 210

Q

quadripartite division of space, 23, 35, 36–43, 52–53, 54, 279, 294
  correlation with cultural variables, 45–49
  cross-circles as symbols of, 294
Quauhtepetl, 274
Quetzalcóatl, 274
  relationship to Venus, 274
Quinish, 185, 189, 193

R

recumbent stone circles, 190, 191
refraction, atmospheric, 157, 162, 163, 179–80, 183
refraction, terrestrial, 156
Regulus, 164
religion, Medieval, 345
right angles, 204, 210, 213
Rindi, 93
Ring of Brodgar, 338, 339
Río Azul, 53
ritual containers, association with light-and-shadow displays 254, 255–56, 257–60
ritual landscapes, 195, 348

  see also sacred geography
rock art
  Aboriginal, 136, 138, 141–43, 149
  Anasazi, 235
  and solstice observation, 252
  astral symbolism in, 20, 147, 227–38
  Chumash, 252
  lunar symbolism in, 110–12, 138, 147
  Navajo, 227–38
  solar symbolism in, 112, 113, 225, 254–55
  water symbolism in, 225
rock art, *see also* cross-circles, petroglyphs
rock sanctuaries, Thracian, 112–14
rock-cut tombs, Thracian, 113
Rodriguez de Figueroa, Diego, 304–5
Romania, 167
Roscher, W.H., 92
'rune calendar', 100

S

sacred geography, 274
  Hindu, 346–47
  Valley of Mexico, 348
  see also ritual landscapes
sacred geometry, 284
sacred hills, 347
sacred knowledge, Aboriginal, 138
sacred space, defined by astronomy, 345
Sagittarius, 145
Samoa, 128, 129
San Cristobal star ceiling, 231–34, 237
Sand Canyon Pueblo, 243, 249
sand paintings, 20, 237
Santa Catarina Cruz Acalpixcan, 293
Santa Rosa Xtampak, 273
sarobe, *see* stone octagons
Satrae, 113
Saturnalia, 94
Saunders, Nicholas, 17
scope and goals of archaeoastronomy, 3–4, 5–7
Scorpio, 123, 124, 144–45, 147
Scotland, 156, 164, 185–96
Sea Woman (Inuit), 59, 60, 64, 67

Index

seasonal activities
    Inuit, 60
    US south-west, 225
seasonal ceremonies, 251
    Aboriginal, 145
    Chumash, 253
    see also seasonal festivals
seasonal festivals
    Anasazi, 246
    ancient Mesoamerica, 291
    Christian, 94
    Chumash, 252
    Indo-European, 94
    Inuit, 60, 63
    Niman (Hopi), 39
    present-day Mesoamerica, 275
    see also agricultural ceremonies, seasonal ceremonies
selection criteria for archaeoastronomical data, 199–201, 208, 212
septarian number system, 79
settlement pattern, correlation with nomenclature for cardinal directions, 47–48
shadows, 155
    brightness of, 158
    edge of, 158, 165
    see also light-and-shadow displays
shamanism, and solstice observation, 252
shamans, 251
    Chumash, 253, 255
    Inuit, 63, 66
Shipibo, 349
sidereal calendars, 125
    ancient Egyptian, 165
sidereal month, 120, 267
signs of the Zodiac, 107
Sirius, 118, 344–45
    colour of, in antiquity, 344–45
    heliacal rising of, 165
sky
    as a cultural resource, 343
    brightness of, 156
sky mapping, Aboriginal, 136–49
sky-heroes, ancestral, 144, 147
Slatino model furnace, 109–110, 112
Society Islands, 128, 132
solar calendars, 317

solar cult in south-eastern Europe, 112–13
solar eclipses, total, 161
solar gods, 51
solar solstices, *see* solstices
solar symbolism
    in Aboriginal art, 139
    in US south-west, 225
    Teutonic, 100–2
solstice observation
    and rock art, 252
    and shamanism, 252
solstices, 35, 51, 282, 322, 345
    alignments upon, 24, 51, 114, 130–32, 133, 198, 244–46, 247–49, 283, 284, 300, 328
    anticipatory observations of, 41–42
    ceremonies at, 51
    declinations of, 199
    directions of sunrise and sunset at, 37, 38–39, 52, 54, 73, 178, 182, 280
    gnomon angles at local noon, 202–4, 207, 208, 210, 213
    marked by prominent peaks, 193–94
    markers of, 40–41, 244–46, 293
    maximum altitude of sun on, 198, 283
    observations of, 72–74, 252
    observations of sunrise and sunset at, 40–42, 182, 252, 317
    orientation upon, 128
    see also summer solstice, winter solstice
songs
    Aboriginal, 137, 139
    Chumash, 252
Southern Cross, 145, 148, 328–29, 330
Southern Triangle, 145
space
    east/west organisation of, 244
    four-directionality of, 35, 36–43, 52–53, 54, 279
    quadripartite division of, 23, 35, 36–43, 52–53, 54, 279, 294
    sacred and profane, 46
    vertical division of, 35–36, 52

space-time
    four-fold categorisation of, 23
    in relation to number, 148
    ritual organisation of, 23
Spain, 33
Spica, 132
standing stones, 129, 147, 309
    alignments of, 129, 195
    centroids of, 200
    orientation of, 147
standstills, *see* lunar standstills, major standstill, minor standstill
star and crescent symbol, 165–67
star ceilings
    Navajo, 227–38
    Pueblo, 234
star-cross motif
    Navajo, 227–38
    Pueblo, 234–35
Star-Oddi, 69, 72, 74
stars
    as protective symbols, 235–37, 238
    association with birds, 235
    association with war, 235
    brightness of, 156
    colour of, 156
    visibility at rise and set, 155, 178, 180, 182–83
    visibility of, near moon, 155, 161
statistical investigations
    of Irish passage grave cemeteries, 198–215
    of megalithic alignments, 185
    of Scottish stone rows, 190–92
statistics, Bayesian, 5
Stephen, Alexander, 33, 38
STILE Project, 339–40
stone alignments, 134
stone circles, 80, 338
    recumbent, 190, 191
    see also stone rings
stone octagons (Basque), 6, 77–78, 80–89
    design of, 82–84
    extent of, 84
stone rings 309, 310, 312–16
    see also stone circles
stone rows, 185–95, 309
    horizon visibility distribution, 191
    location in relation to prominent peaks, 192–94
    orientation of, 192–94
stone statues, 113, 128
stone tombs, 129
Stonehenge, 24, 164, 317, 320, 338
Strabon, 107
structuralist analysis
    of directional symbolism, 52–53
    of Inuit cosmology, 59–68
structuralist studies, 4, 9, 343
subsistence economy, correlation with nomenclature for cardinal directions, 47–48
summer solstice, 204, 253–54
    full moon nearest to, 192, 193–94
    light and shadow displays at, 220–26, 254–61
    ritual containers for offerings at, 253, 255, 257, 260
    see also winter solstice, solstices
sun
    annual motion of, 107, 114, 319
    at zenith, 54
    daily change in azmiuth at equinoxes, 182
    houses of, 40–41, 42
    in bark paintings, 147
    nadir passage of, 301, 302
    personification of, 113
    reverse movement of, 95
    temples of the, 112–13
    zenith passage of, 282, 303
Sun Dance, 193
'sun-serpents', 220–24
sun-stick, Chumash, 253
sunrise and sunset, 178, 180
    alignments upon, 119, 291
    annual motion of, 72, 282, 319
    at the equinoxes, 178, 182, 268, 280, 322
    at the mid-quarter days, 280, 284
    at the solstices, 35, 178, 182, 252, 280, 317, 338
    azimuth of, 182, 278
    church orientations upon, 345
    direction of, 280–82
    in relation to prominent peaks, 193–94

Index

observations of, 74, 278, 317
symbolic representation of, 112
time of, 155
sunspots, 155–56, 168
visibility of, 162
Sydney (Australia), 136
symbolism
astral, 20, 101–2, 147, 227–38
dot-and-line, 136–37, 138, 141–143, 145, 149
lunar, 100–2, 139, 148, 325
sacred astronomical, 251
solar, 100–2, 139
star and crescent, 165–67
synodic month, *see* lunar synodic month

T

Tacitus, 99
Tahiti, 130–31
tally counts, 264, 266
association with hunting motifs, 266
see also dot-and-tally petroglyphs, dot-and-line symbols, dot configurations
Taputapu-atea, 132
taulas, 330
telescope optics, 157
Temple Wood, 338, 339
Tenga, 194
Tenochtitlan, 22
Teotihuacan, 22, 280, 284, 285, 288, 290–92
Tlaloc-like god at, 291
terrestrial navigation, relation to petroglyphic counting, 268
terrestrial refraction, 156
Teutons, moon cult of, 98–105
Tewa Indians, 233–34
Tezcatlipoca, 17
Thom, Alexander, 33, 156, 319
Thorsteinn the black, 71, 72
Thracian rock sanctuaries, 112–14
solar observations at, 114
Thracians, 107–8, 112–15
Thule culture, 59
Tikal, 21, 49, 285
time

cyclic, 95
measurement of, by Thracians, 107
conceptualisation of, by Indo-Europeans, 95
time-reckoning
Borana, 117
Vikings, 70
timekeeping, monastic, 345
Titu Cusi Yupanqui, 304–6
Tlalancaleca, 288
Tlaloc, 291
Tlaloc-like god at Teotihuacan, 291
Todosio Canyon star ceiling, 229
Tomacoco, 293
'tombe di giganti', 330
orientations of, 326–29
tombs of giants, *see* 'tombe di giganti'
tonalpohualli, 294
Tonga, 128, 129, 132–34
Torreón (Machu Picchu), 24
Tostarie, 194–95
traditional astronomies, 34–35
and religion, 35
Triangulum, 118
Triangulum Australe, *see* Southern Triangle
trilithin, coral, 128, 132–34

U

Uaxactun, 49–54
Uluvalt, 195
Underhay, Ernest, 252
units of measurement, 78–80
septarian, 78, 79–80, 81
Uxmal, 272–73, 275

V

Vedic hymns, 95
Venus, 21, 139
and rain and maize, 275
and ritual warfare in Mesoamerica, 347–348
apparent motion of, 270
conjunctions with moon, 166
glyphs, 272
gods, 52
orientations upon, 270–75
relationship to Quetzalcóatl, 274
visibility of, 170

Venus extremes, 270, 274
   alignments upon, 270, 271–74
   correlation with the seasons, 271, 272, 275
   see also Evening Star, Morning Star, planets
videodiscs, interactive, 336–37, 338–39
Vijayanagara, 346–47
Vikings, 69–70
Vilcabamba, 304
visibility
   human factors, 158
   of lunar crescent, 160–61, 164, 166, 170
   of lunar eclipses at low altitude, 161
   of lunar occultations, 170
   of lunar surface, during total solar eclipse, 161
   of planets, near moon, 161
   of prominent peaks, 190, 195
   of stars, near moon, 161
   of sunspots, 162
   of Venus, 170

W

Waiwai, 349
Walpi pueblo, 41
Weather Spirit (Inuit), 59, 64, 66
Wheat, J.B., 244
winter solstice, 204, 249, 253
   light and shadow displays at, 255
   sunrise at, 338
   see also summer solstice, solstices
women's knowledge
   Aboriginal, 138, 140
world directions, 294

X

Xihuingo, 265, 266
Xochicalco, 285

Y

Yao, Canon of, 125, 344
year
   sidereal, 125
Yellow Jacket, 243, 247, 249
Yemen, 167

Yucca House, 249

Z

zenith, 137, 179
zenith passage
   lunar, 301
   solar, 303
zenith passage, solar, 282
zenith sun, 54
zenith-antizenith sun
   alignments upon, 24
Zodiac
   signs of the, 107